Two Manuals Combined into One Book

Ranger Survival

And

Medical Evacuation

Department of Defense

Pentagon
Publishing

RANGER CREED

Recognizing that I volunteered as a Ranger, fully knowing the hazards of my chosen profession, I will always endeavor to uphold the prestige, honor, and high esprit de corps of the Rangers.

Acknowledging the fact that a Ranger is a more elite Soldier who arrives at the cutting edge of battle by land, sea, or air, I accept the fact that as a Ranger my country expects me to move further, faster, and fight harder than any other soldier.

Never shall I fail my comrades I will always keep myself mentally alert, physically strong, and morally straight and I will shoulder more than my share of the task whatever it may be, one hundred percent and then some.

Gallantly will I show the world that I am a specially selected and well trained Soldier. My courtesy to superior officers, neatness of dress, and care of equipment shall set the example for others to follow.

Energetically will I meet the enemies of my country. I shall defeat them on the field of battle for I am better trained and will fight with all my might. Surrender is not a Ranger word. I will never leave a fallen comrade to fall into the hands of the enemy and under no circumstances will I ever embarrass my country.

Readily will I display the intestinal fortitude required to fight on to the Ranger objective and complete the mission, though I be the lone survivor.

STANDING ORDERS, ROGERS' RANGERS

1. Don't forget nothing.
2. Have your musket clean as a whistle, hatchet scoured, sixty rounds powder and ball, and be ready to march at a minute's warning.
3. When you're on the march, act the way you would if you was sneaking up on a deer. See the enemy first.
4. Tell the truth about what you see and what you do. There is an army depending on us for correct information. You can lie all you please when you tell other folks about the Rangers, but don't never lie to a Ranger or officer.
5. Don't never take a chance you don't have to.
6. When we're on the march we march single file, far enough apart so one shot can't go through two men.
7. If we strike swamps, or soft ground, we spread out abreast, so it's hard to track us.
8. When we march, we keep moving till dark, so as to give the enemy the least possible chance at us.
9. When we camp, half the party stays awake while the other half sleeps.
10. If we take prisoners, we keep' em separate till we have had time to examine them, so they can't cook up a story between' em.
11. Don't ever march home the same way. Take a different route so you won't be ambushed.
12. No matter whether we travel in big parties or little ones, each party has to keep a scout 20 yards ahead, 20 yards on each flank, and 20 yards in the rear so the main body can't be surprised and wiped out.
13. Every night you'll be told where to meet if surrounded by a superior force.
14. Don't sit down to eat without posting sentries.
15. Don't sleep beyond dawn. Dawn's when the French and Indians attack.
16. Don't cross a river by a regular ford.
17. If somebody's trailing you, make a circle, come back onto your own tracks, and ambush the folks that aim to ambush you.
18. Don't stand up when the enemy's coming against you. Kneel down, lie down, hide behind a tree.
19. Let the enemy come till he's almost close enough to touch, then let him have it and jump out and finish him up with your hatchet.

--MAJOR ROBERT ROGERS, 1759

RANGER HISTORY

The history of the American Ranger is a long and colorful saga of courage, daring and outstanding leadership. It is a story of men whose skills in the art of fighting have seldom been surpassed. Only the highlights of their numerous exploits are told here.

Rangers primarily performed defensive missions until Benjamin Church's Company of Independent Rangers from Plymouth Colony proved successful in raiding hostile Indians during King Phillip's War in 1675. In 1756, Major Robert Rogers, a native of New Hampshire, recruited nine companies of American colonists to fight for the British during the French and Indian War. Ranger techniques and methods of operation were an inherent characteristic of the American frontiersmen; however, Major Rogers was the first to capitalize on them and incorporate them into the fighting doctrine of a permanently organized fighting force.

The method of fighting used by the first Rangers was further developed during the Revolutionary War by Colonel Daniel Morgan, who organized a unit known as "Morgan's Riflemen". According to General Burgoyne, Morgan's men were "....the most famous corps of the Continental Army, all of them crack shots."

Francis Marion, the "Swamp Fox," organized another famous Revolutionary War Ranger element known as "Marion's Partisans." Marion's Partisans, numbering anywhere from a handful to several hundred, operated both with and independent of other elements of General Washington's Army. Operating out of the Carolina swamps, they disrupted British communications and prevented the organization of loyalists to support the British cause, substantially contributing to the American victory.

The American Civil War was again the occasion for the creation of special units such as Rangers. John S. Mosby, a master of the prompt and skillful use of cavalry, was one of the most outstanding Confederate Rangers. He believed that by resorting to aggressive action he could compel his enemies to guard a hundred points. He would then attack one of the weakest points and be assured numerical superiority.

With America's entry into the Second World War, Rangers came forth to add to the pages of history. Major William O. Darby organized and activated the 1st Ranger Battalion on June19, 1942 at Carrickfergus, North Ireland. The members were all hand-picked volunteers; 50 participated in the gallant Dieppe Raid on the northern coast of France with British and Canadian commandos. The 1st, 3rd, and 4th Ranger Battalions participated with distinction in the North African, Sicilian and Italian campaigns. Darby's Ranger Battalions spearheaded the Seventh Army landing at Gela and Licata during the Sicilian invasion and played a key role in the subsequent campaign which culminated in the capture of Messina. They infiltrated German lines and mounted an attack against Cisterna, where they virtually annihilated an entire German parachute regiment during close in, night, bayonet, and hand-to-hand fighting.

The 2nd and 5th Ranger Battalions participated in the D-Day landings at Omaha Beach, Normandy. It was during the bitter fighting along the beach that the Rangers gained their official motto. As the situation became critical on Omaha Beach, the division commander of the 29th Infantry Division stated that the entire force must clear the beach and advance inland. He then turned to Lieutenant Colonel Max Schneider, Commander of the 5th Ranger Battalion, and said, "Rangers, lead the way." The 5th Ranger Battalion spearheaded the breakthrough and thus enabled the allies to drive inland away from the invasion beaches.

The 6th Ranger Battalion, operating in the Pacific, conducted Ranger type missions behind enemy lines which involved reconnaissance and hard-hitting, long-range raids. They were the first American contingent to return to the Philippines, destroying key coastal installations prior to the invasion. A reinforced company from the 6th Ranger Battalion formed the rescue force which liberated American and allied prisoners of war from the Japanese prison camp at Cabanatuan.

Another Ranger-type unit was the 5307th Composite Unit (Provisional), organized and trained as a long-range penetration unit for employment behind enemy lines in Japanese occupied Burma. The unit commander was Brigadier General (later Major General) Frank D. Merrill. Its 2,997 officers and men became popularly known as "Merrill's Marauders."

The men composing Merrill's Marauders were volunteers from the 5th, 154th, and 33rd Infantry Regiments and from other Infantry regiments engaged in combat in the southwest and South Pacific. These men responded to a call from then Chief of Staff, General George C. Marshall, for volunteers for a hazardous mission. These volunteers were to have a high state of physical ruggedness and stamina and were to come from jungle-trained and jungle-tested units.

Prior to their entry into the Northern Burma Campaign, Merrill's Marauders trained in India under the overall supervision of Major General Orde C. Wingate, British Army. There, they were trained from February to June 1943 in long-range penetration tactics and techniques of the type developed and first employed by General Wingate. The operations of the Marauders were closely coordinated with those of the Chinese 22nd and 38th Divisions in a drive to recover northern Burma and clear the way for the construction of Ledo Road, which was to link the Indian railhead at Ledo with the old Burma Road to China. The Marauders marched and fought through jungle and over mountains from Hukwang Valley in northwest Burma to Myitkyina and the Irrawaddy River. In 5 major and 30 minor engagements, they met and defeated the veteran soldiers of the Japanese 18th Division. Operating in the rear of the main force of the Japanese, they prepared the way for the southward advances of the Chinese by disorganizing supply lines and communications. The climax of the Marauder's operations was the capture of Myitkyina Airfield, the only all-weather strip in northern Burma. This was the final victory of "Merrill's Marauders," which disbanded in August 1944. Remaining personnel were consolidated into the 475th Infantry Regiment, which fought its last battle on February 3, and 4,1945, at Loi-Kang Ridge, China. This Infantry Regiment would serve as the forefather of today's 75th Ranger Regiment.

Shortly after the outbreak of the Korean War in June 1950, the 8th Army Ranger Company was formed of volunteers from American units in Japan. The Company was trained in Korea and distinguished itself in combat during the drive to the Yalu River, performing task force and spearhead operations. In November 1950 during the massive Chinese intervention, this small unit, though vastly outnumbered, withstood five enemy assaults on its position.

In September 1950, a Department of the Army message called for volunteers to be trained as Airborne Rangers. In the 82nd Airborne Division, five thousand regular Army paratroopers volunteered, and from that number nine hundred men were selected to form the initial eight Airborne Ranger Companies. An additional nine companies were formed from volunteers of regular Army and National Guard Infantry Divisions. These seventeen Airborne Ranger companies were activated and trained at Fort Benning, Georgia, with most receiving additional training in the mountains of Colorado.

In 1950 and 1951, some 700 men of the 1st, 2nd, 3rd, 4th, 5th, and 8th Airborne Ranger companies fought to the front of every American Infantry Division in Korea. Attacking by land, water, and air, these six Ranger companies conducted raids, deep penetrations, and ambush operations against North Korean and Chinese forces. They were the first Rangers in history to make a combat jump. After the Chinese intervention, these Rangers were the first Americans to re-cross the 38th parallel. The 2nd Airborne Ranger Company was the only African American Ranger unit in the history of the American Army. The men of the six Ranger companies who fought in Korea paid the bloody price of freedom. One in nine of this gallant brotherhood died on the battlefields of Korea.

Other Airborne Ranger companies led the way while serving with infantry divisions in the United States, Germany, and Japan. Men of these companies volunteered and fought as members of line infantry units in Korea. One Ranger, Donn Porter, would be posthumously awarded the Medal of Honor. Fourteen Korean War Rangers became general officers and dozens became colonels, senior noncommissioned officers, and leaders in civilian life. They volunteered for the Army, the Airborne, the Rangers, and for combat. The first men to earn and wear the coveted Ranger Tab, these men are the original Airborne Rangers.

In October 1951, the Army Chief of Staff, General J. Lawton Collins directed, "Ranger training be extended to all combat units in the Army." The Commandant of the Infantry School was directed to establish a Ranger Department. This new department would develop and conduct a Ranger course of instruction. The objective was to raise the standard of training in all combat units. This program was built upon what had been learned from the Ranger Battalions of World War II and the Airborne Ranger companies of the Korean conflict.

During the Vietnam Conflict, fourteen Ranger companies consisting of highly motivated volunteers served with distinction from the Mekong Delta to the DMZ. Assigned to separate brigade, division, and field force units, they conducted long-range reconnaissance and exploitation operations into enemy-held areas providing valuable combat intelligence. Initially designated at LRRP, then LRP companies, these units were later designated as C, D,E,F,G,H,I,K,L,M,N,O and P (Ranger) 75th Infantry.

Following Vietnam, recognizing the need for a highly trained and highly mobile reaction force, the Army Chief of Staff, General Abrams directed the activation of the first battalion-sized Ranger units since World War II, the 1st and 2nd Battalions (Ranger), 75th Infantry. The 1st Battalion was trained at Fort Benning, Georgia and was activated February 8, 1974 at Fort Stewart, Georgia with the 2nd Battalion being activated on October 3, 1974. The 1st Battalion is now located at Hunter Army Airfield, Georgia and the 2nd Battalion at Fort Lewis, Washington.

The farsightedness of General Abrams' decision, as well as the combat effectiveness of the Ranger battalions, was proven during the United States' invasion of the island of Grenada in October 1983 to protect American citizens there, and to restore democracy. As expected, Rangers led the way! During this operation, code named "Urgent Fury," the Ranger battalions conducted a daring, low level airborne assault (from 500 feet) to seize the airfield at Point Salines, and then continued operations for several days to eliminate pockets of resistance, and rescue American medical students.

As a result of the demonstrated effectiveness of the Ranger battalions, the Department of the Army announced in 1984 that it was increasing the strength of Ranger units to its highest level in 40 years. To do this, it activated another Ranger battalion as well as a Ranger Regimental Headquarters. These new units, the 3rd Battalion (Ranger), 75th Infantry, and Headquarters Company (Ranger), 75th Infantry, have increased the Ranger strength of the Army to over 2,000 soldiers actually assigned to Ranger units. On February 3, 1986, the 75th Infantry was renamed the 75th Ranger Regiment.

On December 20, 1989, the 75th Ranger Regiment was once again called upon to demonstrate its effectiveness in combat. For the first time since its reorganization in 1984, the Regimental Headquarters and all three Ranger battalions were deployed on Operation "Just Cause" in Panama. During this operation, the 75th Ranger Regiment spearheaded the assault into Panama by conducting airborne assaults onto Torrijos/Tocumen Airport and Rio Hato Airfield to facilitate the restoration of democracy in Panama, and protect the lives of American citizens. Between December 20, 1989 and January 7, 1990, numerous follow-on missions were performed in Panama by the Regiment.

Early in 1991, elements of the 75th Ranger Regiment deployed to Saudi Arabia in support of Operation Desert Storm.

In August 1993, elements of the 75th Ranger Regiment deployed to Somalia in support of Operation Restore Hope, and returned November 1993.

In 1994, elements of the 75th Ranger Regiment deployed to Haiti in support of Operation Uphold Democracy.

In 2000 – 2001, elements of the 75th Ranger Regiment deployed to Kosovo in support of Operation Joint Guardian.

Since September 11, 2001, the 75th Ranger Regiment has led the way in the Global War on Terrorism. In October 2001, elements of the 75th Ranger Regiment deployed to Afghanistan in support of Operation Enduring Freedom. In March 2003, elements of the Regiment deployed in support of Operation Iraqi Freedom.

The performance of these Rangers significantly contributed to the overall success of these operations and upheld the Ranger tradition of the past. As in the past, the Regiment stands ready to execute its mission to conduct special operations in support of the United States' policies and objectives.

RANGER MEDAL OF HONOR WINNERS

Millett, Lewis L. Sr	Captain	Feb 7 1951	Co. E 2/27th Infantry
Porter, Donn F.*	Sergeant	Sept 7 1952	Co. G 2/14th Infantry
Mize, Ola L.	Sergeant	June 10-11 1953	Co. K 3/15th Infantry
Dolby, David C.	Staff Sergeant	May 21 1966	Co. B 1/8th (ABN) Calvary
Foley, Robert F.	Captain	Nov 5 1966	Co. A 2/27th Infantry
Zabitosky, Fred M.	Staff Sergeant	Feb 19 1968	5th Special Forces
Bucha, Paul W.	Captain	May 16-19 1968	Co. D 3/187 Infantry
Rabel, Laszlo*	Staff Sergeant	Nov 13 1968	74th Infantry (LRRP)
Howard, Robert L.	Sergeant First Class	Dec 30 1968	5th Special Forces
Law, Robert D. *	Specialist 4	Feb 22 1969	Co. I 75th Infantry (Ranger)
Kerrey, J. Robert	Lieutenant	Mar 14 1969	Seal Team 1
Doane, Stephen H.*	1st Lieutenant	Mar 25 1969	Co. B 1/5th Infantry
Pruden, Robert J.*	Staff Sergeant	Nov 22 1969	Co. G 75th Infantry (Ranger)
Littrell, Gary L.	Sergeant First Class	April 4-8 1970	Advisory Team 21 (Ranger)
Lucas, Andre C.*	Lt Colonel	Jul 1-23 1970	HHC 2/506 Infantry
Gordon, Gary I. *	Master Sergeant	Oct 3 1993	Task Force Ranger
Shughart, Randall D. *	Sergeant First Class	Oct 3 1993	Task Force Ranger

*Awarded posthumously

SUMMARY OF CHANGES

THROUGHOUT	UPDATED doctrine, technology, terms; clarified verbiage
CHAPTER 1	UPDATED duties and responsibilities ADDED FO
CHAPTER 2	UPDATED WARNO ADDED sample WARNO UPDATED OPORD ADDED terrain model
CHAPTER 3	ADDED PLOT-CR ADDED risk estimate distances
CHAPTER 4	IDENTIFIED TL positions during movement
CHAPTER 5	REORGANIZED combat operations
CHAPTER 6	UPDATED battle drills IAW FM 3-21.8 CONSOLIDATED building clearing techniques under Battle Drill 5 (previously in Chapter 14)
CHAPTER 7	ADDED section on antennas
CHAPTER 8	UPDATED aircraft
CHAPTER 9	UPDATED rope bridge equipment UPDATED sequence and responsibilities
CHAPTER 10	UPDATED equipment and descriptions UPDATED climbing commands
CHAPTER 11	NA
CHAPTER 12	ADDED first aid techniques
CHAPTER 13	ADDED figure 'Lighting a time fuse with a match' ADDED expedient explosives ADDED table of US demolitions ADDED breaching and timber cutting charges
CHAPTER 14	MOVED building clearing techniques to Battle Drill 5 in Chapter 6 ADDED marking SOP
CHAPTER 15	NEW chapter covering vehicle convoy operations
APPENDIX A	NEW - Resources
APPENDIX B	NEW - Quick Reference Cards

TABLE OF CONTENTS

Chapter 1
LEADERSHIP

Leadership, the most essential element of combat power, gives purpose, direction, and motivation in combat. The leader balances and maximizes maneuver, firepower, and protection against the enemy. This chapter discusses how he does this by exploring the principles of leadership (BE, KNOW, DO); the duties, responsibilities, and actions of an effective leader; and the leader's assumption of command.

1-1. **PRINCIPLES**. The principles of leadership are BE, KNOW, and DO.

BE

- Technically and tactically proficient
- Able to accomplish to standard all tasks required for the wartime mission.
- Courageous, committed, and candid.
- A leader with integrity.

KNOW

- The four major factors of leadership and how they affect each other are--
  - ---Led
  - ---Leader
  - ---Situation
  - ---Communications
- Yourself, and the strengths and weaknesses in your character, knowledge, and skills. Seek continual self-improvement, that is, develop your strengths and work to overcome your weaknesses.
- Your Rangers, and look out for their well-being by training them for the rigors of combat, taking care of their physical and safety needs, and disciplining and rewarding them.

DO--

- Seek responsibility and take responsibility for your actions; exercise initiative; demonstrate resourcefulness; and take advantage of opportunities on the battlefield that will lead to you to victory; accept fair criticism, and take corrective actions for your mistakes.
- Assess situations rapidly, make sound and timely decisions, gather essential information, announce decisions in time for Rangers to react, and consider the short- and long-term effects of your decision.
- Set the example by serving as a role model for your Rangers. Set high but attainable standards; be willing do what you require of your Rangers; and share dangers and hardships with them.
- Keep your subordinates informed to help them make decisions and execute plans within your intent, encourage initiative, improve teamwork, and enhance morale.
- Develop a sense of responsibility in subordinates by teaching, challenging, and developing them. Delegate to show you trust them. This makes them want more responsibility.
- Ensure the Rangers understand the task; supervise them, and ensure they accomplish it. Rangers need to know what you expect: when and what you want them to do, and to what standard.
- Build the team by training and cross-training your Rangers until they are confident in their technical and tactical abilities. Develop a team spirit that motivates them to go willingly and confidently into combat.
- Know your unit's capabilities and limitations, and employ them accordingly.

1-2. **DUTIES, RESPONSIBILITIES, AND ACTIONS**. To complete all assigned tasks, every Ranger in the patrol must do his job. Each must accomplish his specific duties and responsibilities and be a part of the team.

PLATOON LEADER (PL)

- Is responsible for what the patrol does or fails to do. This includes tactical employment, training, administration, personnel management, and logistics. He does this by planning, making timely decisions, issuing orders, assigning tasks, and supervising patrol activities. He must know his Rangers and how to employ the patrol's weapons. He is responsible for positioning and employing all assigned or attached crew-served weapons and employment of supporting weapons.
- Establishes time schedule using backwards planning. Considers time for execution, movement to the objective, and the planning and preparation phase of the operation.
- Takes the initiative to accomplish the mission in the absence of orders. Keeps higher informed by using periodic situation reports (SITREP).
- Plans with the help of the platoon sergeant (PSG), squad leaders, and other key personnel (team leaders, FO, attachment leaders).
- Stays abreast of the situation through coordination with adjacent patrols and higher HQ; supervises, issues FRAGOs, and accomplishes the mission.
- If needed to perform the mission, requests more support for his patrol from higher headquarters.
- Directs and assists the platoon sergeant in planning and coordinating the patrol's sustainment effort and casualty evacuation (CASEVAC) plan.
- During planning, receives on-hand status reports from the platoon sergeant and squad leaders.
- Reviews patrol requirements based on the tactical plan.
- Ensures that all-round security is maintained at all times.
- Supervises and spot-checks all assigned tasks, and corrects unsatisfactory actions.
- During execution, positions himself where he can influence the most critical task for mission accomplishment; usually with the main effort, to ensure that his platoon achieves its decisive point
- Is responsible for positioning and employing all assigned and attached crew-served weapons.
- Commands through his squad leaders IAW the intent of the two levels higher commanders.
- Conducts rehearsals.

PLATOON SERGEANT (PSG)

The PSG is the senior NCO in the patrol and second in succession of command. He helps and advises the patrol leader, and leads the patrol in the leader's absence. He supervises the patrol's administration, logistics, and maintenance, and he prepares and issues paragraph 4 of the patrol OPORD.

DUTIES

- Organizes and controls the patrol CP IAW the unit SOP, patrol leader's guidance, and METT-TC factors.
- Receives squad leader's requests for rations, water, and ammunition. Work with the company first sergeant or XO to request resupply. Directs the routing of supplies and mail.
- Supervises and directs the patrol medic and patrol aid-litter teams in moving casualties to the rear.
- Maintains patrol status of personnel, weapons, and equipment; consolidates and forwards the patrol's casualty reports (DA Forms 1155 and 1156); and receives and orients replacements.
- Monitors the morale, discipline, and health of patrol members.
- Supervises task-organized elements of patrol:
  - — Quartering parties.
  - — Security forces during withdrawals.
  - — Support elements during raids or attacks.
  - — Security patrols during night attacks.
- Coordinates and supervises patrol resupply operations.
- Ensures that supplies are distributed IAW the patrol leader's guidance and direction.
- Ensures that ammunition, supplies, and loads are properly and evenly distributed (a critical task during consolidation and reorganization).
- Ensures the casualty evacuation plan is complete and executed properly.

- Ensures that the patrol adheres to the platoon leader's time schedule.
- Assists the platoon leader in supervising and spot-checking all assigned tasks, and corrects unsatisfactory actions.

ACTIONS DURING MOVEMENT AND HALTS

- Takes actions necessary to facilitate movement.
- Supervises rear security during movement.
- Establishes, supervises, and maintains security during halts.
- Knows unit location.
- Performs additional tasks as required by the patrol leader and assists in every way possible. Focuses on security and control of patrol.

ACTIONS AT DANGER AREAS

- Directs positioning of near-side security (usually conducted by the trail squad or team).
- Maintains accountability of personnel.

ACTIONS ON THE OBJECTIVE AREA

- Assists with ORP occupation.
- Supervises, establishes, and maintains security at the ORP.
- Supervises the final preparation of men, weapons, and equipment in the ORP IAW the patrol leader's guidance.
- Assists the patrol leader in control and security.
- Supervises the consolidation and reorganization of ammunition and equipment.
- Establishes, marks, and supervises the planned CCP, and ensures that the personnel status (to include WIA/KIA) is accurately reported to higher.
- Performs additional tasks assigned by the patrol leader and reports status to platoon leader.

ACTIONS IN THE PATROL BASE

- Assists in patrol base occupation.
- Assist in establishing and adjusting perimeter.
- Enforces security in the patrol base.
- Keeps movement and noise to a minimum.
- Supervises and enforces camouflage.
- Assigns sectors of fire.
- Ensures designated personnel remain alert and equipment is maintained to a high state of readiness.
- Requisitions supplies, water, and ammunition, and supervises their distribution.
- Supervises the priority of work and ensures its accomplishment.
  - Security plan.
    - Ensures crew-served weapons have interlocking sectors of fire.
    - Ensures Claymores are emplaced to cover dead space.
    - Ensures range cards and sector sketch are complete.
  - Alert plan.
  - Evacuation plan.
  - Withdrawal plan.
  - Alternate patrol base.
  - Maintenance plan.
  - Hygiene plan.
  - Messing plan.
  - Water plan.
  - Rest plan.
- Performs additional tasks assigned by the patrol leader and assists him in every way possible.

SQUAD LEADER (SL)

Is responsible for what the squad does or fails to do. He is a tactical leader who leads by example.

DUTIES

- Completes casualty feeder reports and reviews the casualty reports completed by squad members.
- Directs the maintenance of the squad's weapons and equipment.

- Inspects the condition of Rangers' weapons, clothing, and equipment.
- Keeps the PL and PSG informed on status of squad.
- Submits ACE report to PSG.

ACTIONS THROUGHOUT THE MISSION

- Obtains status report from team leaders and submits reports to the PL and PSG.
- Makes a recommendation to the PL/PSG when problems are observed.
- Delegates priority task to team leaders, and supervises their accomplishment IAW squad leader's guidance.
- Uses initiative in the absence of orders.
- Follows the PL's plan and makes recommendations.

ACTIONS DURING MOVEMENT AND HALTS

- Ensures heavy equipment is rotated among members and difficult duties are shared.
- Notifies PL of the status of the squad.
- Maintains proper movement techniques while monitoring route, pace, and azimuth.
- Ensures the squad maintains security throughout the movement and at halts.
- Prevents breaks in contact.
- Ensures subordinate leaders are disseminating information, assigning sectors of fire, and checking personnel.

ACTION IN THE OBJECTIVE AREA

- Ensures special equipment has been prepared for actions at the objective.
- Maintains positive control of squad during the execution of the mission.
- Positions key weapons systems during and after assault on the objective.
- Obtains status reports from team leaders and ensures ammunition is redistributed and reports status to PL.

ACTIONS IN THE PATROL BASE

- Ensures patrol base is occupied according to the plan.
- Ensures that his sector of the patrol base is covered by interlocking fires; makes final adjustments, if necessary.
- Sends out LP or OPs in front of assigned sector. (METT-TC dependent).
- Ensures priorities of work are being accomplished, and reports accomplished priorities to the PL and PSG.
- Adheres to time schedule.
- Ensures personnel know the alert and evacuation plans, and the locations of key leaders, OPs, and the alternate patrol base.

WEAPONS SQUAD LEADER

Is responsible for all that the weapons squad does or fails to do. His duties are the same as those of the squad leader. Also, he controls the machine guns in support of the patrol's mission. He advises the PL on employment of his squad.

DUTIES

- Supervises machine gun teams to ensure they follow priorities of work.
- Inspects machine gun teams for correct range cards, fighting positions, and understanding of fire plan.
- Supervises maintenance of machine guns, that is, ensures that maintenance is performed correctly, that deficiencies are corrected and reported, and that the performance of maintenance does not violate the security plan.
- Assists PL in planning.
- Positions at halts and danger areas and according to the patrol SOP any machine guns not attached to squads.
- Rotates loads. Machine gunners normally get tired first.
- Submits ACE report to PSG.
- Designates sectors of fire, principal direction of fire (PDF), and secondary sectors of fire for all guns.
- Gives fire commands to achieve maximum effectiveness of firepower:
  - — Shifts fires.
  - — Corrects windage or elevation to increase accuracy.
  - — Alternates firing guns.
  - — Controls rates of fire and fire distribution.
- Knows locations of assault and security elements, and prevents fratricide.
- Reports to the PL.

TEAM LEADER (TL)

Controls the movement of his fire team and the rate and placement of fire. To do this, leads from the front and uses the proper commands and signals. Maintains accountability of his Rangers, weapons, and equipment. Ensures his Rangers maintain unit standards in all areas, and are knowledgeable of their tasks and the operation. The following checklist outlines specific duties and responsibilities of team leaders during mission planning and execution. The TL leads by example:

ACTIONS DURING PLANNING AND PREPARATION

- Warning Order.
  - — Assists in control of the squad.
  - — Monitors squad during issuance of the order.
- OPORD Preparation.
  - — Posts changes to schedule.
  - — Posts and updates team duties on warning order board.
  - — Submits ammunition and supply requests.
  - — Picks up ammunition and supplies.
  - — Distributes ammunition and special equipment.
  - — Performs all tasks given in the SL's special instructions paragraph.
- OPORD Issuance and Rehearsal.
  - — Monitors squad during issuance of the order.
  - — Assists SL during rehearsals.
- Takes actions necessary to facilitate movement.
- Enforces rear security.
- Establishes, supervises, and maintains security at all times.
- Performs additional tasks as required by the SL, and assists him in every way possible, particularly in control and security.

ACTION IN THE ORP

- Assists in the occupation of the ORP.
- Helps supervise, establish, and maintain security.
- Supervises the final preparation of Rangers, weapons, and equipment in the ORP IAW the SL's guidance.
- Assists in control of personnel departing and entering the ORP.
- Reorganizes perimeter after the reconnaissance party departs.
- Maintains communication with higher headquarters.
- Upon return of reconnaissance party, helps reorganize personnel and redistribute ammunition and equipment; ensures accountability of all personnel and equipment is maintained.
- Disseminates PIR to his team.
- Performs additional tasks assigned by the SL.

ACTIONS IN THE PATROL BASE

- Inspects the perimeter to ensure team has interlocking sectors of fire; prepares team sector sketch.
- Enforces the priority of work and ensures it is properly accomplished.
- Performs additional tasks assigned by the SL and assists him in every way possible.

MEDIC

Assists the PSG in directing aid and litter teams; monitors the health and hygiene of the platoon.

DUTIES

- Treats casualties, conducts triage, and assists in CASEVAC under the control of the PSG.
- Aids the PL or PSG in field hygiene matters. Personally checks the health and physical condition of platoon members.
- Requests Class VIII (medical) supplies through the PSG.
- Provides technical expertise to and supervision of combat lifesavers.
- Ensures casualty feeder reports are correct and attached to each evacuated casualty.
- Carries out other tasks assigned by the PL or PSG.

RADIO OPERATOR Is responsible for establishing and maintaining communications with higher headquarters and within the patrol. *DUTIES DURING PLANNING* • Enters the net at the specified time. • Ensures that all frequencies, COMSEC fills, and net IDs, are preset in squad/platoon radios. • Informs SL and PL of changes to call signs, frequencies, challenge and password, and number combination based on the appropriate time in the ANCD. • Ensures the proper function of all radios and troubleshoots and reports deficiencies to higher. • Weatherproofs all communications equipment. *DUTIES DURING EXECUTION* • Serves as en route recorder during all phases of the mission. • Records all enemy contact and reports it to higher in a SALUTE format. • Reports all OPSKEDs to higher. • Consolidates and records all PIR.
FORWARD OBSERVER (FO) Works for the PL. Serves as the eyes of the FA and mortars. Is mainly responsible for locating targets, and for calling for and adjusting indirect fire support. Knows the terrain where the platoon is operating; knows the tactical situation. Knows the mission, the concept, and the unit's scheme of maneuver and priority of fires. *DUTIES DURING PLANNING* • Selects targets to support the platoon's mission based on the company OPORD, platoon leader's guidance, and analysis of METT-TC factors. • Prepares and uses situation maps, overlays, and terrain sketches. *DUTIES DURING EXECUTION* • Informs the FIST headquarters of platoon activities and of the fire support situation • Selects new targets to support the platoon's mission based on the company OPORD, the platoon leader's guidance, and an analysis of METT-TC factors. • Calls for and adjusts fire support. • Operates as a team with the radio operator. • Selects OPs. • Maintains communications as prescribed by the FSO. • Maintains the current 8-digit coordinate of his location at all times.

1-3. **ASSUMPTION OF COMMAND**. Any platoon/squad member might have to take command of his element in an emergency, so every Ranger must be prepared to do so. During an assumption of command, situation permitting, the Ranger assuming command accomplishes the following tasks (not necessarily in this order) based on METT-TC:

- *INFORMS* the unit's subordinate leaders of the command and notifies higher.
- *CHECKS* security.
- *CHECKS* crew-served weapons.
- *PINPOINTS* location.
- *COORDINATES* and *CHECKS* equipment.
- *CHECKS* personnel status.
- *ISSUES* FRAGO (if required).
- *REORGANIZES* as needed, maintaining unit integrity when possible.
- *MAINTAINS* noise and light discipline.
- *CONTINUES* patrol base activities, especially security, if assuming command in a patrol base.
- *RECONNOITERS* or, at the least, conducts a map reconnaissance.
- *FINALIZES* plan.
- *EXECUTES* the mission.

NOTES

Chapter 2
OPERATIONS

This chapter provides techniques and procedures used by Infantry platoons and squads throughout the planning and execution phases of tactical operations. Specifically, it discusses the troop-leading procedures, combat intelligence, combat orders, and planning techniques and tools needed to prepare a platoon to fight. These topics are time sensitive and apply to all combat operations. When they have time, leaders can plan and prepare in depth. If they have less time, they must rely on previously rehearsed actions, battle drills, and standing operating procedures (SOPs).

2-1. TROOP-LEADING PROCEDURES

Troop leading procedures comprise the following steps. They are what a leader does to prepare his unit to accomplish a tactical mission. The TLP starts when the leader is alerted for a mission or receives a change or new mission. He can perform Steps 3 through 8 in any order, or at the same time. (He can also use the tools of the tactician shown in Figure 2-1):

1. Receive the mission.	5. Reconnoiter.
2. Issue a warning order.	6. Complete the plan.
3. Make a tentative plan.	7. Issue the complete order.
4. Initiate movement.	8. Supervise.

Figure 2-1. TOOLS OF THE TACTICIAN RELATIONSHIP

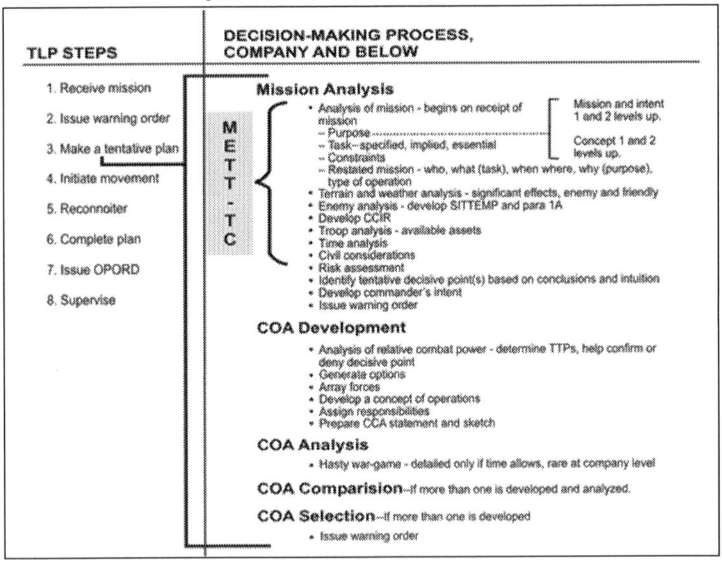

a. **STEP 1--RECEIVE THE MISSION**. The leader may receive the mission in a warning order, an operation order (OPORD), or a fragmentary order (FRAGO). He should use no more than one third of the available time for his own planning and for issuing his OPORD. The remaining two thirds is for subordinates to plan and prepare for the operation. Leaders should also consider other factors such as available daylight and travel time to and from orders and rehearsals.

b. **STEP 2--ISSUE A WARNING ORDER**. The leader provides initial instructions in a warning order. The warning order contains enough information to begin preparation as soon as possible. The warning order mirrors the five-paragraph OPORD format. A warning order may include--

- The mission or nature of the operation (mission statement).
- Time and place for issuance of the operation (coordinating instructions).
- Who is participating in the operation (coordinating instructions).
- Time of the operation (timeline).

c. **STEP 3--MAKE A TENTATIVE PLAN**. The leader uses METT-TC to develop an estimate of the situation, which he will use as the basis for his tentative plan. This set of actions is referred to as the leader's mission analysis:

(1) **Conduct a detailed mission analysis.**

(a) *Concept and Intent.* Higher commanders' concept and intent two levels up. This information is found in paragraph 1b for two levels up and in paragraphs 2 and 3 for one higher.

(b) *Unit Tasks.* Tasks that are clearly stated in the order (Specified Tasks) or tasks that become apparent as the OPORD is analyzed (Implied Tasks).

EXAMPLE SPECIFIED TASKS

- Retain hill 545 to prevent envelopment of B Co.
- Provide one squad to the 81-mm platoon to carry ammo.
- Establish an OP vic GA124325 NLT 301500 NOV 89.

EXAMPLE IMPLIED TASKS

- Provide security during movement.
- Conduct resupply operations.
- Coordinate with adjacent units.

(c) *Unit Limitations.* The leader next determines all control measures or instructions in the OPORD that restrict his freedom of action; these are called limitations. In every operation, there are some limitations on the unit.

COMMON LIMITATIONS

Graphic control measures.
Cross the LD at 100030 OCT 94 (time).
MOPP4 in effect (uniform and environmental considerations).
ADA weapons status, tight; warning status, yellow (rules of engagement).

(d) *Mission-Essential Task(s).* After reviewing all the above factors, the leader identifies his mission-essential task(s). Failure to accomplish a mission-essential task results in the unit's failure to accomplish its primary purpose for that operation. The mission essential task should be found in the maneuver paragraph.

(e) *Restated Mission.* The restated mission statement becomes the focus for the remainder of the estimate process. This is a clear, concise statement of the mission essential task(s) to be accomplished by the unit and the purpose to be achieved. The mission statement will state WHO, WHAT (the task), WHEN (the critical time), WHERE (usually a grid coordinate), and WHY (the purpose the unit must achieve). Some examples of restated missions follow:

EXAMPLE RESTATED MISSIONS
(WHO) 1st Platoon attacks (WHAT) to seize (WHERE) HILL 482 vic NB 457371 (OBJ BLUE) (WHEN) NLT 09050OZ Dec 92 L 482 (WHY) to enable the company's main effort to destroy enemy command bunker.
(WHO) 1st Platoon, C Company defends (WHAT) to destroy from (WHERE) AB163456 to AB163486 to AB123486 to AB123456 (WHEN) NLT 28153OZ Oct 97 (WHY) to prevent enemy forces from enveloping B Company, 1-66 Infantry (L) from the south.

(2) Analyze the situation and develop a course of action. Each COA must be--suitable, acceptable, feasible, distinguishable, and complete.

(3) Upon developing a COA, the unit leader will assign C2 headquarters, complete generic task organization assigning all organic and attached elements, and prepare COA statement and sketch.

(4) With the restated mission from Step 1 to provide focus, the leader continues the estimate process using the remaining factors of METT-TC.

(a) What is known about the **ENEMY**?

Composition	This is an analysis of the forces and weapons that the enemy can bring to bear. Determine what weapons systems they have available, and what additional weapons and units are supporting him.
Disposition	The enemy's disposition is how he is arrayed on the terrain, such as in defensive positions, in an assembly area, or moving in march formation.
Strength	Percentage strength.
Recent Activities	Identify recent and significant enemy activities that may indicate future intentions.
Reinforcement Capabilities	Determine positions for reserves and estimated time to counterattack or reinforce.
Possible Courses of Action	Determine the enemy's possible COAs. This will help the leader determine how to best array his own forces against the enemy and fight the battle.

(b) How will **TERRAIN** and **WEATHER** affect the operation? Analyze terrain using OAKOC.

- **Observation and Fields of Fire**. Determine locations that provide the best observation and fields of fire along friendly and enemy avenues of approach, near the objective, and on key terrain. The analysis of fields of fire is mainly concerned with the ability to cover the terrain with direct fire.
- **Avenues of Approach**. Analyze and identify friendly and enemy avenues of approach. Identify avenues of approach en route to the objective, on and around the objective, and for possible enemy counterattack. Also, consider aerial and subterranean avenues.

Offensive Considerations (Friendly)	• How can these avenues support my movement? • What are the advantages/disadvantages of each? (Consider enemy, speed, cover, and concealment.) • What are the likely enemy counterattack routes?

Offensive Considerations (Enemy)	• How can the enemy use these approaches? • Which avenue is most dangerous? Least? (Prioritize each approach.) • Which avenues would support a counterattack?

- **Cover and Concealment**. The analysis of cover and concealment is often inseparable from the fields of fires and observation. Weapon positions must have both to be effective and to be survivable. Infantry units are capable of improving poor cover and concealment by digging in and camouflaging their positions. When moving, the terrain is used to provide cover and concealment.
- **Obstacles**. Identify the existing and reinforcing obstacles and hindering terrain that will affect mobility.
- **Key Terrain**. Key terrain is any location or area that the seizure, retention, or control of affords a marked advantage to either combatant. Using the map, aerial photos, and information already gathered, look for key terrain that dominates avenues of approach or the objective area. Next, look for decisive terrain that if held or controlled will have an extraordinary impact on the mission.

(5) Analyze courses of action (war-game). This analysis is conducted by war-gaming friendly courses of action against the enemy's most probable courses of action. The leader can--and should--war-game with his subordinates.

(6) Compare courses of action. The leader compares the COAs and selects the one that is most likely to accomplish the assigned mission. He considers the advantages and disadvantages for each COA. He also considers how the critical events impact on COAs.

(7) Make a decision. The leader selects the COA that he believes has the best chance of accomplishing the mission.

d. **STEP 4--START NECESSARY MOVEMENT**. The unit may need to begin movement while the leader is still planning or forward reconnoitering. This step may occur anytime during the TLP.

e. **STEP 5--RECONNOITER**. If time allows, the leader makes a personal reconnaissance. When time does not allow, the leader must make a map or aerial photo reconnaissance. Sometimes the leader must rely on others, such as scouts, to conduct the reconnaissance.

f. ***STEP 6--COMPLETE THE PLAN***. The leader completes his plan based on the reconnaissance and any changes in the situation.

g. **STEP 7--ISSUE THE COMPLETE ORDER**. Platoon and squad leaders normally issue oral operations orders to aid subordinates in understanding the concept of the mission. Leaders may require subordinates to repeat part of the order, demonstrate it on a terrain model, or sketch their understanding of the operation. Leaders should also quiz their Rangers to ensure that all Rangers understand the mission.

h. **STEP 8--SUPERVISE AND REFINE**. The leader supervises the unit's preparation for combat by conducting rehearsals and inspections.

(1) ***Rehearsals***. Rehearsals include the practice of having squad leaders brief their planned actions in execution sequence to the platoon leader. The leader should conduct rehearsals on terrain that resembles the actual ground, and in similar light conditions. At a minimum, the leader should conduct a backbrief (least preferred), full force rehearsal (most preferred), or reduced force rehearsal (key leaders), and a rock drill.

(a) *Purpose*. The leader uses rehearsals to--

- Practice essential tasks (improve performance).
- Reveal weaknesses or problems in the plan.
- Coordinate the actions of subordinate elements.
- Improve Ranger understanding of the concept of the operation (foster confidence in Rangers).

(b) *Times and Tasks*. The platoon may begin rehearsals of battle drills and other SOP items before the receipt of the operation order. Once the order has been issued, it can rehearse mission specific tasks. Some important tasks to rehearse include--

- Actions on the objective.
- Assaulting a trench, bunker, or building.
- Actions at the assault position.
- Breaching obstacles (mine and wire).
- Using special weapons or demolitions.
- Actions on unexpected enemy contact.

(c) *Types.*

- Backbrief.
  - -- Key leaders sequentially brief the actions required during operation.
  - -- Patrol leader controls.
  - -- Conducted twice: right after FRAGO (confirmation brief), and again after subordinates develop own plan.
- Reduced force.
  - -- Conducted when time is key constraint.
  - -- Conducted when security must be maintained.
  - -- Key leaders normally attend.
  - -- Mock-ups, sand tables, and small scale replicas used.
- Full force.
  - -- Most effective type.
  - -- First executed in daylight and open terrain.
  - -- Secondly conduct in same conditions as operation.
  - -- All Rangers participate.
  - -- May use force on force.
- Techniques.
  - -- Force on force.
  - -- Map (limited value and limited number of attendees).
  - -- Radio (cannot mass leaders; confirms communications).
  - -- Sand table or terrain model (key leaders; includes all control measures).
  - -- Rock drill (similar to sand table/terrain model; subordinates actually move themselves). Very effective for coordinating timing and sequence of events--especially when air assets are used.

(d) *Inspections.* Squad leaders should conduct initial inspections shortly after receipt of the warning order. The platoon sergeant spot-checks throughout the unit's preparation for combat. The platoon leader and platoon sergeant make a final inspection of--

- Weapons and ammunition.
- Uniforms and equipment.
- Mission-essential equipment.
- Soldier's understanding of the mission and their specific responsibilities.
- Communications.
- Rations and water.
- Camouflage.
- Deficiencies noted during earlier inspections.

2-2. **COMBAT INTELLIGENCE**. Gathering information is one of the most important aspects of conducting patrolling operations. The following details what information to collect and how to report it:

a. **Reports**. All information must be quickly, completely, and accurately reported. Use the SALUTE report format for reporting and recording information.

SIZE	– Seven enemy personnel
ACTIVITY	– Traveling SW
LOCATION	– GA123456
UNIT/UNIFORM	– OD uniforms with red six-point star on left shoulder
TIME	– 210200JAN99
EQUIPMENT	– Carry one machine gun and one rocket launcher

b. **Field Sketches**. Try to include a sketch with each report. Include only any aspects of military importance such as targets, objectives, obstacles, sector limits, or troop dispositions and locations (use symbols from FM 1-02). Use notes to explain the drawing, but they should not clutter the sketch.

c. **Captured Documents**. The leader collects documents and turns them in with his reports. He marks each document with the time and place of capture.

d. **Prisoners**. If prisoners are captured during a patrolling operation, they should be treated IAW the Geneva Convention and handled by the 5-S rule:

(1) Search
(2) Silence
(3) Segregate
(4) Safeguard
(5) Speed to rear

e. **Debriefs**. Immediately upon return from a mission, the unit is debriefed. The intelligence officer will generally have a unit-specific format for debriefing a patrol.

2-3. **WARNING ORDER**. Warning orders give subordinates advance notice of upcoming operations. This gives them time to prepare. A warning order must be brief, but complete. *A warning order only authorizes execution when it clearly says so.* (Figure 2-2 shows an example format; Figure 2-3 shows an example warning order.)

Figure 2-2. WARNING ORDER FORMAT

WARNING ORDER ________

*Roll call, pencil/pen/paper, RHB, map, protractor, leader's monitor, hold all questions till the end
References: Refer to higher headquarters' OPORD, and identify map sheet for operation.
Time Zone Used throughout the Order: (Optional)
Task Organization: (Optional; see paragraph 1c.)

1. **SITUATION** (Higher's OPORD para 1a[1]--[3])

a. **Enemy Forces**. Include significant changes in enemy composition, dispositions, and courses of action. Information not available for inclusion in the initial WARNO can be included in subsequent warning orders. (who, what, where)

b. **Friendly Forces**. (Optional) only address if essential to the WARNO.

(1) Higher commander's mission (who, what, where, why).
(2) Higher commander's intent. (Higher's [go to mapboard] OPORD para 1b[2], give task and purpose)

c. **Attachments and Detachments**. Initial task organization, only address major unit changes.

(1) Orient relative to each point on the compass.
(2) Box in the entire AO with grid lines.
(3) Trace each Zone using boundaries.
(4) Familiarize by identifying 3 natural (terrain) and 3 man-made features in each zone.
(5) Point out the enemy and friendly locations on the map board.

2. **MISSION**. State mission twice (who, what, when, where, and why) and concisely state task and purpose.

3. **EXECUTION**.

a. **Concept of Operation**. Provide as much information as available. The concept should describe the form of maneuver or defensive technique, critical events, decisive point of the operation and why it is decisive, task and purpose for the main and supporting efforts, purposes of the warfighting systems, such as engineers, fire support, intelligence etc), acceptable risk, and the desired endstate. After the desired endstate is stated, leaders should talk through the concept of the operation like a story. Use a simple sketch to assist in relaying the information.

b. **Tasks to Maneuver Units**. Provide specified tasks to subordinate units. These are tactical instructions on how to execute the mission for each element in task organization. Planning guidance consists of tasks assigned to subordinate units, special teams, and key individuals.

c. **Tasks to Combat Support Units**. See paragraph 3b.

d. **Coordinating Instructions**. Include any information available at that time. If you know it, then at least cover--

- Uniform and equipment common to all. Consider the factors of METT-TC and tailor the load for each Ranger.
- Time line. (State 4 W's and specified and implied times. Reverse plan. Use 1/3-2/3 rule.)
- CCIR.
- Risk guidance.
- Deception guidance.
- Specific priorities, in order of completion.
- Guidance on orders and rehearsals.
- Orders group meeting (attendees, location, and time).
- Earliest movement time and degree of notice.

4. **SERVICE SUPPORT**. Include any known logistics preparation for the operation.

a. Special equipment. Identify requirements and any coordination measures the unit needs to take to transfer the special equipment. (State the equipment you will use, need, or want for the mission.)

b. Transportation. Identify requirements and any coordination needed to pre-position assets. (State unit's means of infil/exfil.)

5. **COMMAND AND SIGNAL**.

a. **Command**. State the chain of command if different from unit SOP.

b. **Signal**. Identify current SOI edition, and pre-position signal assets to support operation.

- Give subordinates guidance on tasks to complete for preparation of the OPORD and the mission.
- Give time, place, and uniform of the OPORD.
- Give a time hack and ask for questions.

NOTES

Figure 2-3. EXAMPLE WARNING ORDER

WARNING ORDER (SQUAD)

ROLL CALL, CAMP DARBY SPECIAL (CDS) 1:50,000 MAP, PEN, PAPER, PENCIL, PROTRACTOR, RANGER HANDBOOK (RHB), HOLD ALL QUESTIONS, AND TLs MONITOR AND TASK-ORGANIZE

1. **SITUATION**.

a. **Enemy**.

WHO: The Aragon Liberation Front (ALF).
WHAT: Has been conducting aggressive offensive operations.
WHERE: In Zones A, B, C.

b. **Friendly**:

(1) **Higher's Mission**.

WHO: 2nd PLT, C CO.
WHAT: Conducts area ambushes to destroy enemy personnel and equipment.
WHERE: Zone D.
WHY: To prevent enemy logistical resupply.

(2) **Higher Commander's Intent**.

- Rapid infiltration into Zone D.
- Find, fix, and finish enemy forces in Zone D.
- Enemy personnel and equipment located and targeted.
- Squads ready to conduct future combat operations.

c. **Attachments/Detachments**. MG Tm #1/2 PLT 160630-190630.

2. **MISSION**.

WHO: 1st SQD, 2nd PLT
WHAT: Conducts a point ambush to destroy enemy personnel and equipment.
WHERE: GA 1805 7878.
WHEN: NLT 25 1830 DEC 2007.
WHY: To prevent enemy logistical resupply and gather PIR.

3. **EXECUTION**.

a. **Concept of Operation**. (Orient Rangers to sketch or terrain model). We are currently located at Camp Darby, GA 1962 7902, which is on a ridge. Our ground tactical plan will begin when we will move generally southwest to our ORP at Grid GA 187778. Terrain feature there is a draw for 1,000 meters, by foot and it will take 1 hour. Here will we finalize the preparing of M, W, and E. We will then move generally southwest to our objective at Grid GA 184774. Terrain feature there is a ridge for 200 meters, by foot, and it will take us 30 minutes to an hour. The reason for this is stealth on the objective while we occupy our positions.

b. **Tasks to Maneuver Units**.

(1) HQs: HQ is responsible for command and control and communications. Commo/HQ is 2nd in the OOM. M240 will be placed in a position to provide supporting fires in the kill zone. RTO will be the recorder en route and during actions on the objective. SL will be located where he can best control his squad and observe the kill zone. SL is responsible for dissemination of information.

(2) ATM: ATM will be 1st OOM/responsible for navigation, and for clearing and securing the far side at all linear danger areas. ATM is flank security for AOO, will provide early warning and seal off objective/1-2 man EPW, and search TM/1-2 man aid and litter TM/1-2 man demo TM/1 man (P), 1 man (A) ORP clearing TM/ 2-2 man flank security. TM for AOO/ 1 compassman, 1 pace man, or ATL will be the sec. TM LDR along the most likely avenue of approach.

(3) BTM: BTM 3RD OOM/responsible for securing the near side of all linear danger areas/BTM. Is assault for AOO. Must provide suppressive fire and assault thru kill zone/1-2 man EPW and search team/1-2 man aid and litter team/1-2 man demolitions team; 1 man (P); 1 man (A) ORP clearing team; 1-2 man S/O team; 1-2 man security team for reentry; 1 compassman; 1 pace man; and BTL. You will be the assault element leader for AOO.

c. **Tasks to Combat Support Units**. None.

d. **Coordinating Instructions**.

(1) Uniform and equipment common to all. At a minimum, each Ranger wears or carries--

(a) A ballistic helmet w/camouflaged band and luminous tape; clean and serviceable ACUs, camelback, black belt w/buckle, blackened combat boots, serviceable OD or black socks, serviceable OD/brown T-shirt, ID tags w/breakaway lanyard around neck (silenced). Right cargo pocket; (RHB, CDS [map], protractor, pencil, pen, and paper) waterproofed, left cargo pocket; patrol cap w/ luminous tape, ID card in right breast pocket, SL will carry sterile fire support overlay in left breast pocket.

(b) LCE, consisting of pistol belt, suspenders, 2 ammo pouches, seven 30-round magazines, one first aid dressing with pouch, one compass and pouch, two 1-quart canteens w/covers and cups, one whistle, and four M67 fragmentation grenades. All are tied down IAW 4th Bn SOP, except that the pace cord is tied to suspenders.

(c) Rucksack, complete; one set serviceable ACUs, two 2-quart canteens w straps; E-tool with cover; two pair socks; two brown T-shirts; weapons cleaning kit; one waterproof bag; a sewing kit; a patrol cap; MREs; poncho and liner; wet weather top; personal hygiene kit; one brown towel; and camouflage stick. RTO also packs and ties off the ANCD.

(d) Optional bug juice, 550 cord, foot powder, black tape, knife, all other items will be approved by SL 15 minutes prior to completion time of OPORD.

(2) Time schedule.

WHEN	WHAT	WHERE	WHO	REF
*0630	WARNING ORDER	BAY AREA	All	3D1
0700	INITIAL INSPECTION	BAY AREA	All	
*0730	Required Ammo/Supplies	CO TOC	BTL	4B1
*0745	P/U-Ammo/Supplies	CO TOC	BTL/Detail	4B1
0800	Test Fire	T/F AREA	All	
*0830	S-2/S-3 Coord	CO BAY AO	SL/RTO	3D1
*0830	Fire Spt Coord	BN TOC	SL/ RTO	3A2
0850	Adj Unit Coord	BN TOC	ATL/compass man	3D1
*0900	Enter Net	BAY AREA	RTO	5B1
0930	Complete Plan	BAY AREA	All	
*1000	Issue OPORD	BAY AREA	All	3D1
1300	Rehearsal	BAY AREA	All	
1400	Final Inspection	BAY AREA	All	
1430	Conduct Move	BAY AREA	All	
1800	In ORP	GA	All	
2000	In Position	GA 184774	All	
*2300	Mission Complete	GA 186874	All	3D1
*0200	Reenter FFU	GA	All	
0500	S-2 Debrief	BN TOC	All	
* Specified time				

4. **SERVICE AND SUPPORT**.

a. **Special Equipment**.

(1) HQs: One hand-held radio and extra batteries; one AN/PRC-119 radio and extra batteries, one ANCD, one AN/PSN-11 and extra batteries, one M-22 binos, one AN/PVS-14 and extra batteries; one sheets overlay material, three sheets carbon paper; two HE smoke, two red smoke, one M240-B MG complete w/tripod, flex mount, T&E, and Pintle, one AN/PVS-4 and extra batteries, one AN/PVS-7D, and extra batteries.

(2) ATM: One hand-held radio and extra batteries, one M-24 binos, two AN/PVS-14 and extra batteries, two HE smoke, two yellow smoke, one thermite grenade, two gags, two flex cuffs, two sandbags, one CLS bag, one pole-less litter, two M18A1 claymore mines, one AT-4, one LAW, one M-203 vest, two M-249 SAW pouches, one small metal flashlight, and extra batteries secured to rifle.

(3) BTM: One hand-held radio and extra batteries, one M-24 binos, two AN/PVS 14 and extra batteries, two HE smoke, two yellow smoke, one thermite grenade; two each gags, flex cuffs, and sandbags; one CLS bag, one pole-less Litter, two M18A1 claymore mines, one AT-4, one LAW, one M-203 vest, two M249 SAW pouches, one small metal flashlight, and extra batteries secured to rifle.

b. **Ammo**. (per weapon system).

M16/M4	M249	M203	M240	SL AND TLs
210 Rounds 5.56-MM Ball	625 Rounds 5.56-MM (4:1 Mix)	16 Rounds 40-MM HEDP 5 Rounds 40-MM Shot 4 Rounds 40-MM Illumination 2 Rounds 40-MM TP-T	825 Rounds 7.62-MM (4:1 Mix)	42 Rounds 5.56-MM Tracer

c. **Transportation**. NA

5. **COMMAND AND SIGNAL**.

a. **Command**.

PERSONNEL	POS	SOC
HQ/SUPPORT		
BARNETT	SL	1
MARENGO	RTO	4
CASELY	MG	11
WADE	AG	12
ATM/SECURITY		
TRUITT	ATL	3
SILCOX	GRN	5
WISE	R	6
EDMUNDS	R	7
BTM/ASSAULT		
SPENCER	BTL	2
CHRISTENSEN	GRN	8
RUFUS	R	9
ROBINSON	R	10

SUBORDINATE GUIDANCE:	
ATL:	Write paragraph 1 of squad OPORD, and for drawing all sketches, terrain models, routes, and nonsterile fire support overlays.
BTL:	You are second in the chain of command and in charge at all times during my absence. Write paragraph 4 of squad OPORD Write the supply, DX, and ammo lists, draw and issue all items, ensure that everyone conducts a test fire, and all equipment is tied down IAW 4th RTBn SOP, and update the squad status card and hand receipt.
RTO:	Write paragraph 5 of squad OPORD, ensure all radios are operational with proper freqs loaded, and ensure we enter the net on time.
TL:	TLs: You will update the Warning Order board with all the correct information. As a task is accomplished, you will line it out. Also, post your succession of command (SOC), duty positions, and job descriptions (specialty teams and key individuals). Come and see me for further guidance at the conclusion of this WARNO.

b. **Signal**.

TACTICAL FREQUENCY MEDEVAC FREQUENCY CHALLENGE PASSWORD RUNNING PASSWORD NUMBER COMBO	62,000 35,000 BAYONET MILLETT RANGER 7
TIME, PLACE, AND UNIFORM FOR OPORD. TIME HACK. ANY QUESTIONS?	

NOTES

2-4. **OPERATIONS ORDER**. An operations order (OPORD) is a directive issued by a leader to his subordinates in order to effect the coordinated execution of a specific operation. A five-paragraph format (shown below) is used to organize the briefing, to ensure completeness, and to help subordinate leaders understand and follow the order. Use a terrain model or sketch along with a map to explain the order. When possible, such as in the defense, give the order while observing the objective. The platoon/squad leader briefs his OPORD orally off notes that follow the five-paragraph format. Before the issuance of the OPORD, the leader ensures that the following resources are in place: pencil, pen, paper; RHB; map; protractor; leader's monitor. Then he calls roll and says "*Please hold all questions till the end.*" Figure 2-4 shows an example OPORD format. Figure 2-5 shows an OPORD shell.

Figure 2-4. EXAMPLE OPORD FORMAT

OPERATIONS ORDER

[Plans and orders normally contain a code name and are numbered consecutively within a calendar year.]

References: [The heading of the plan or order includes a list of maps, charts, datum, or other related documents the unit will need to understand the plan or order. The user need not reference the SOP, but may refer to it in the body of the plan or order. He refers to a map by map series number (and country or geographic area, if required), sheet number and name, edition, and scale, if required. "*Datum*" refers to the mathematical model of the earth that applies to the coordinates on a particular map. It is used to determine coordinates. Different nations use different datum for printing coordinates on their maps. The datum is usually referenced in the marginal information of each map.]

Time zone used throughout the order: The time zone used throughout the order (including annexes and appendixes) is the time zone applicable to the operation. Operations across several time zones use Zulu time.

Task organization: Describe the allocation of forces to support the commander's concept. Task organization may be shown in one of two places: preceding paragraph one, or in an annex, if the task organization is long and complicated.

**Weather and light data, and general forecast (Only discuss what will affect your patrol. Note effects on friendly and enemy personnel and equipment.):*

High
Moonrise
Sunrise
Low
Moonset
Sunset
Wind Speed
Moon Phase
BMNT
Wind Direction
Percent Illumination
EENT

**This is the information the leader received from higher's OPORD.*

Terrain: Make a statement concerning the effects terrain will have on both friendly and enemy forces in the area of operation using the OAKOC format.

1. **SITUATION**.

a. **Enemy Forces**. The enemy situation in higher headquarters' OPORD (paragraph 1a) is the basis for this, but the leader refines this to provide the detail required by his subordinates.

(1) Enemy's composition, disposition, strength.
(2) Recent activities.
(3) Known/suspected locations and capabilities.
(4) Describe the enemy's most likely and most dangerous course of action on a map.

b. **Friendly Forces**. This information is in paragraphs 1b, 2, and 3 of higher headquarters' OPORD.

(1) State the mission of the Higher Unit (2 levels up).

(2) State the mission of the Higher Unit (1 level up).

(3) State intent 2 levels up.

(4) State locations of units to the left, right, front, and rear. State those units' tasks and purposes; and say how those units will influence your unit's patrol.

(a) Show other units locations on map board.

(b) Include statements about the influence each of the above patrols will have on your mission, if any.

- Obtain this information from higher's OPORD. It gives each leader an idea of what other units are doing and where they are going. This information is in paragraph 3a (1) (execution, concept of the operation, maneuver).
- Also include any information obtained when the leader conducts adjacent unit coordination.

c. **Attachments and Detachments**. Do not repeat information already listed under Task Organization. Try to put all information in the Task Organization. However, when not in the Task Organization, list units that are attached or detached to the headquarters that issues the order. State when attachment or detachment is to be effective if different from when the OPORD is effective, such as on order, on commitment of the reserve). Use the term "*remains attached*" when units will be or have been attached for some time.

2. **MISSION**. State the mission derived during the planning process. There are no subparagraphs in a mission statement. Include the 5 W's: Who, What (task), Where, When, and Why (purpose).

- Make it a clear and concise statement and read it twice.
- Go to map and point out the exact location of the OBJ and point out the unit's present location.

3. **EXECUTION**. State the commander's intent.

a. **Concept of Operation**. The concept statement should be concise and clear. It describes how the unit will accomplish its mission from start to finish. The number of subparagraphs, if any, is based on what the leader considers appropriate for the level of leadership and complexity of the operation. Generally, the concept should describe the form of maneuver or defensive technique and critical events; identify what the decisive point of the operation is and why; describe the task and purpose for the main effort and supporting efforts; state the purpose of the warfighting functions, such as engineers, fire support, or intelligence; define acceptable risk; and identify the desired endstate. After stating the desired endstate, leaders should talk through the concept of the operation like a story, using a simple sketch to help them convey the information. Figure 1 shows the six warfighting functions (WFF):

Figure 1. Warfighting functions.

- Fire support
- Movement and Maneuver
- Protection
- Command and Control
- Intelligence
- Sustainment (formerly called "CSS")

(1) ***Maneuver***. The maneuver paragraph details the mechanics of the operations. This subparagraph specifically addresses all subordinate units and attachments by name, giving each its mission in the form of a task and purpose. The main effort's mission must be supported by the subordinate unit's missions. Actions on the objective will comprise the majority of this paragraph and should include a detailed plan for engagement/disengagement criteria, an alternate plan in the event of compromise or unplanned movement of enemy forces, and a withdrawal plan.

(2) ***Fires***. State scheme of fires to support the overall concept. This paragraph should state which maneuver unit has priority of fires. Fires should be planned using the PLOT-CR format (purpose, location, observer, trigger, communication method, resources). A target list worksheet and overlay are referenced here, if applicable. Specific targets are discussed and pointed out on the terrain model (see Chapter 3, Fire Support).

(3) ***Casualty Evacuation***.

(a) *Plan*. State a detailed CASEVAC plan during each phase of the operation. Include CCP locations, tentative extraction points, and methods of extraction.

b. **Tasks to Maneuver Units**. Clearly state the missions or tasks for each maneuver unit that reports directly to the headquarters issuing the order. List the units in the task organization, including reserves. Use a separate subparagraph for each maneuver unit. Only state the tasks that are necessary for comprehension, clarity, and emphasis. Place tactical tasks that affect two or more units in Coordinating Instructions (subparagraph 3d). Platoon leaders task their subordinate squads. Those squads may be tasked to provide any of the following special teams: reconnaissance and security, assault, support, aid and litter, EPW and search, clearing, and demolitions. Detailed instructions may also be given to the platoon sergeant, RTO, compassman, and paceman.

(1) ***Tasks to Combat Support Units***. Add subparagraphs here only as necessary. List CS units in subparagraphs from the task organization. Use CS subparagraphs to list only those specific tasks that CS units must accomplish and that are not specified or implied elsewhere. Include organization for combat if not clear from task organization.

c. **Coordinating Instructions**. This is always the last subparagraph under paragraph 3. List only the instructions that apply to two or more units, and which are seldom covered in unit SOPs. Refer the user to an annex for more complex instructions. The following information is required:

1. Time schedule (state time, place, uniform, and priority of rehearsals, backbriefs, inspections, and movement).

2. Give the commander's critical information requirements (CCIR), which includes priority intelligence requirements (PIR), essential elements of friendly information (EEFI), and friendly force information requirements (FFIR):

a. PIR is intelligence that the commander must have for planning and decision making.

b. EEFI are critical aspects of friendly operations that, if known by the enemy, would compromise, lead to failure, or limit success of the operation.

c. FFIR include what the commander needs to know about friendly forces available for the operation. It can include personnel status, ammunition status, and leadership capabilities.

3. Risk reduction control measures. These are measures unique to the operation. They supplement the unit SOP and can include mission-oriented protective posture, operational exposure guidance, vehicle recognition signals, and fratricide prevention measures.

4. Rules of engagement (ROE).

5. Environmental considerations.

6. Force protection

7. Movement plan. Use terrain model, sketch, or both. State azimuths, directions, and grid coordinates. Use names of real Rangers on the sketches.

a. Sketch out the order of movement, the formation, and the movement technique.

b. Sketch actions at halts (long and short).

c. Routes (use terrain model):

1) Orient the terrain model (North, South, East, and West).

2) Identify the grid lines, both North-South and East-West.

3) Brief the legend.

4) Box in the AO with distinguishable natural or man-made terrain features such as roads, ridgelines, rivers, and streams.

5) Brief all the routes depicted on the terrain model from start to finish using specific azimuths and distance.

a) The primary route from the start point to the ORP, from the ORP to the objective, and from the objective to the patrol base while describing the terrain over which the squad/platoon will travel. (Include near/far side rally points for danger areas.)

b) The alternate route from the start point to the ORP, from the ORP to the objective, and from the objective to the patrol base while describing the terrain over which the squad/platoon will travel. (Include near/far side rally points for danger areas.)

6) Brief the fire support plan (if not given in fires paragraph). When covering fires the leader should cover the following (PLOT-CR):

- *Purpose.*
- *Location and terrain feature to the targets.*
- *Observer.*
- *Trigger.*
- *Communication method.*
- *Resources.*

7) State where the sterile fire support overlay is being carried.

a) Rally points and actions at all rally points (include grid location and terrain reference, and use sketch).

b) Actions at danger areas. This is a general plan for unknown linear, small open areas and large open areas; specific plan for all known danger areas that unit will encounter along the route (use sketch).

c) Actions on enemy contact; battle drills (use sketch).

4. **SERVICE SUPPORT**. Address service support in the areas shown below as needed to clarify the service support concept. Subparagraphs can include:

a. **General**. Refer to the SOPs that govern the sustainment operations of the unit. Provide current and proposed company trains locations, casualty, and damaged equipment collection points and routes.

b. **Materiel and Services**.

(1) ***Supply***.

Class I--Rations plan.
Class III--Petroleum.
Class V--Ammunition.
Class VII--Major end items.
Class VIII--Medical.
Class IX--Repair parts.
Distribution Methods.

(2) ***Transportation***.

(3) ***Services***. (Laundry and showers.)

(4) ***Maintenance***. (Weapons and equipment.)

(a) *Medical Evacuation and Hospitalization.* Method of evacuating dead and wounded, friendly and enemy personnel. Include priorities and location of CCP.

(b) *Personnel Support.* Method of handling EPWs and designation of the EPW collection point.

5. **COMMAND AND SIGNAL**.

This paragraph states where command and control facilities and key leaders are located during the operation. Use the OPORD shell to save time and organize your thoughts while preparing an OPORD.

a. **Command**.

- Location of the higher unit commander and CP.
- Location of key personnel and CP during each phase of the operation.

- Succession of command.
- Adjustments to the patrol SOP.

b. **Signal**.
- SOI index in effect.
- Methods of communication in priority.
- Pyrotechnics and signals, to include arm and hand signals (demonstrate).
- Code words such as OPSKEDs.
- Challenge and password (use behind friendly lines).
- Number combination (use forward of friendly lines).
- Running password.
- Recognition signals (near/far and day/night).

c. **Special Instructions to RTOs**.

d. **Actions after Issuance of OPORD--**

(1) Issue annexes.
(2) Give time hack.
(3) ASK for questions.

Figure 2-5. OPORD SHELL

OPORD

DTG:

REFERENCE:

TASK ORGANIZATION:

1ST PLT	2ND PLT	TM CONTROL:

1. **SITUATION**.

BMNT: SR: SS: EENT: MR: MS: HI: LO: ILLUM:

a. **Enemy**.

Terrain	(1) Higher mission: Cdr's intent:	Terrain
End state	(2) Units to N, S, E, W.	End state

2. **MISSION**:

3. **EXECUTION**: a. **Concept of Operations**.

Cdr's intent:

(ME): T: P:	(SE1): T: P:	(SE2): T: P:	(SE3): T: P:	(SE4): T: P:	(SE5): T: P:

b. **Maneuver**

PHASE 1:	PHASE 2:
ME:	ME:
SE1:	SE1:
SE2:	SE2:
SE3:	SE3:
SE4:	SE4:
SE5:	SE5:
FIRES:	FIRES:
MTRS:	MTRS:
C2:	C2:
PHASE 3:	PHASE 4:
ME:	ME:
SE1:	SE1:
SE2:	SE2:
SE3:	SE3:
SE4:	SE4:
SE5:	SE5:
FIRES:	FIRES:
MTRS:	MTRS:
C2:	C2:

c. **Tasks to Maneuver Units**.

(1) Risk-reduction control measures.
Rules of engagement.
(2) Environmental considerations.
(3) Non-Airborne personnel.

(ME):	(SE1):	(SE2):
(SE3):	(SE4):	(SE5):

d. **Tasks to Subordinate Units**

Mortars

Coordinating instructions:

PIR:

EEFI:

FFIR:

Timeline.

Coordinating Instructions.

WHEN	WHO	WHAT	WHERE

4. **SERVICE SUPPORT**.

a. **General**.

b. **Material and Services**.

(1) Supply:

Class I.

Class III:

Class V:

Class VIII:

Class IX:

(2) LOGSTAT.

(3) The chain of command is responsible to ensure that ammo lasts between resupplies.

c. **Medical Evacuation**.

(1) Air MEDEVAC is on call through [this] frequency. JP capable; use standard 9-line, 24-hr coverage.

(2) Establish friendly CCPs in AAs. Adjust CCPs as mission progresses.

(3) Troop medical clinic located vic [this location].

(4) Precedence for evacuation of WIA is US military, host nation military, host nation civilian, and enemy military.

(5) Precedence for evacuation of KIA is same. Evacuation of KIA will be as per mission constraints.

d. **Personnel**.

(1) Method of handling EPWs:

ANNEXES:

5. **COMMAND AND SIGNAL**.

a. **Command**.

(1) TF CP:

(2) Succession of Command:

b. **Signal**.

(1) Challenge and Password:

(2) Number Combination:

(3) Running Password:

(4) Recognition:

(a) Far:

(b) Near:

(7) Code word:

(8) OPSKED:

(9) Reports. SALUTE to higher on enemy contact.

2-5. **FRAGMENTARY ORDER**. A FRAGO is an abbreviated form of an operations order, usually issued on a day-to-day basis that eliminates the need for restating information contained in a basic operations order. It is issued after an OPORD to change or modify that order or to execute a branch or sequel to that order. Figure 2-6 shows an annotated FRAGO format:

Figure 2-6. ANNOTATED FRAGO FORMAT

FRAGMENTARY ORDER____________

Time Zone referenced throughout order:

Task Organization:

Weather and Light Data:

High:	BMNT:	Moonrise:
Low:	Sunrise:	Moonset:
Wind Speed:	Sunset:	% Illum:
Wind Direction:	EENT:	
Forecast:		

Terrain (changes that will effect operation in new area of operations):

OAKOC: Observation and fields of fire, avenues of approach, key terrain, observation, and cover and concealment.

1. **SITUATION**. (Brief changes from base OPORD specific to this day's operation.)

a. **Enemy Situation**.

(1) Composition, disposition and strength.

(2) Capabilities.

(3) Recent activities.
(4) Most likely COA.

b. **Friendly Situation**.
(1) Higher mission.
(2) Adjacent patrols task/purpose.
(3) Adjacent patrol objective/route (if known).

2. **MISSION** (Who, what [task], when, where, why [purpose]—from higher HQ's maneuver paragraph).

3. **EXECUTION**

a. **Concept of Operation**. (Explain how unit will accomplish mission in general terms. Identify mission essential task, designate the main effort and how the supporting efforts support the main effort.)

(1) ***Maneuver***. (Assign task/purpose and discuss actions on the objective in detail. Use a sketch or terrain model to brief.)

(2) ***Fires***. (In support of today's mission, portions that do not change are not briefed.) PLOT-CR:
(a) Purpose.
(b) Location.
(c) Observer.
(d) Trigger.
(e) Communication method.
(f) Resource.

(3) ***CASEVAC***.

b. **Tasks to Maneuver Units**. (List unique tasks that apply to each squad.)

c. **Coordinating Instructions**.

(1) ***Timeline***.
(a) Hit time:
(b) ORP time:
(c) Movement time from PB:
(d) Final inspection:
(e) Rehearsals:
(f) FRAGO complete:

(2) ***Movement Plan***. (Use a sketch to brief.)
(a) Routes (primary and alternate):
(b) OOM:
(c) Formations:
(d) Movement technique:

(3) ***PIR***. (Specific to this mission):

(4) ***Rehearsal Plan***.

(5) ***Patrol Base Plan***. (If not IAW SOP):
(a) Teams:
(b) Occupation plan:
(c) Operations plan (security plan, alert plan, Black and Gold):
(d) Priorities of work:

(6) ***Air Assault Plan***. (Provided with higher's FRAGO, if applicable):
(a) Number, type, and ACL aircraft:
(b) PZ grid, DOL, and PZ posture time:
(c) Load time/lift time/flight time:
(d) Number of lifts and composition:
(e) Air checkpoints en route:
(f) LZ grid/DOL/LZ time:
(g) Actions after getting off aircraft:
(h) Actions on contact on LZ:

(7) ***Linkup Plan***. (If applicable):
(a) Time of linkup:
(b) Location of linkup site:
(c) Stationary element:
(d) Moving element:
(e) Rally points:
(f) Actions at linkup point:
(g) Near/far recognition signals (day and night):

4. **SUPPORT**. (Only cover changes from base order that apply for today's mission).
a. **Material and Services**.
(1) Any changes in classes of supply:
(2) Resupply plan:
(3) Water resupply plan:
(4) Aerial resupply plan (if applicable) :
(5) Truck plan:
(6) Maintenance issues specific to plan:
b. **Medical Evacuation Plan**. (Specific to mission):
(1) CCP point and markings:
(2) Aid and litter duties (if not SOP):

5. **COMMAND AND SIGNAL**.
a. **Command**.
(1) Location of Company CP:
(2) Location of PL:
(3) Location of key leaders:
(4) Succession of command:
b. **Signal**.
(1) Location of radios:
(a) During movement:
(b) During actions on the objective:
(2) SOI in effect:

(a) Bn command freq:	Co CP call sign:
(b) Bn MEDEVAC freq:	PL call sign:
(c) Retrans freq:	PSG call sign:
(d) Company freq:	1 SL call sign:
(e) Platoon freq:	2 SL call sign:
	3 SL call sign:
	WSL call sign:

(3) Pyro signals in use:
(4) Running password/number combination:
(5) OPSKEDs in effect:
(a) The field FRAGO should take no more than 40 minutes to issue, with 30 minutes for the target. The proposed planning guide is as follows:

Paragraphs 1 and 2:	5 minutes
Paragraph 3:	20 to 30 minutes
Paragraphs 4 and 5:	5 minutes

(b) The FRAGO should focus on actions on the objective. The PL may use subordinates to prepare para 1, 4, 5 and routes and fires for the FRAGO. It is acceptable for subordinates to brief the portions of the FRAGO they prepare.
(c) Use of sketches and a terrain model are critical to allow rapid understanding of the operation/FRAGO.
(d) Rehearsals are critical as elements of the constrained planning model. The FRAGO used in conjunction with effective rehearsals reduces preparation time and allows the PL more time for movement and recon.
(e) Planning in a field environment will necessarily reduce the amount of time leaders have for in-depth mission planning. The TLP gives leaders a framework to plan missions and produce orders when time is short.

2-6. **ANNEXES**. Operation order annexes are necessary to complete the plan and to provide greater clarity and understanding during complex or critical aspects of the operation. Information issued in annex form include the: aerial re-supply, truck movement, air assault, patrol base, small boat, linkup, and stream crossing annexes. Annexes are prepared only if the subject is not addressed thoroughly enough in the OPORD; brevity remains the standard. Annexes are always issued after the operation order.

AIR MOVEMENT ANNEX

1. **SITUATION**.
   a. **Enemy Situation**.
      (1) Enemy air capability.
      (2) Enemy ADA capability.
      (3) Include in weather: % Illum, illum angle, NVG window, ceiling, and visibility.
   b. **Friendly Situation**.
      (1) Unit(s) supporting operation.
      (2) Friendly ADA status.
2. **MISSION**.
3. **EXECUTION**.
   a. **Concept of Operation**.
   b. **Subunit Missions**.
   c. **Coordinating instructions**.
      (1) ***Pickup Zone***.
         (a) Name/Number/
         (b) Coordinates/
         (c) Load Time/
         (d) Takeoff Time/
         (e) Markings/
         (f) Control/
         (g) Landing Formation/
         (h) Approach/Departure Direction.
         (i) Alternate PZ Name/Number.
         (j) Penetration Points.
         (k) Extraction Points.
      (2) ***Landing Zone***.
         (a) Name/Number.
         (b) Coordinates.
         (c) H-Hour.
         (d) Markings.
         (e) Control.
         (f) Landing Formation/Direction.
         (g) Alt LZ Name/Number.
         (h) Deception Plan.
         (i) Extraction LZ.
      (3) ***Laager Site***.
         (a) Communications.
         (b) Security Force.
      (4) ***Flight Routes and Alternates***.
      (5) ***Abort Criteria***.
      (6) ***Down Aircraft/Crew*** (designated area of recovery [DAR]).
      (7) ***Special Instructions***.
      (8) ***Cross-FLOT Considerations***.
      (9) ***Aircraft Speed***.

(10) ***Aircraft Altitude***.
(11) ***Aircraft Crank Time***.
(12) ***Rehearsal Schedule/Plan***.
(13) ***Actions on Enemy Contact*** (en route and on the ground).

4. **SERVICE SUPPORT**.
   a. **Forward Area Refuel/Rearm Points**.
   b. **Classes I, III, and V** (specific).

5. **COMMAND AND SIGNAL**.
   a. **Command**.
   b. **Signal**.
      (1) Air/ground call signs and frequencies.
      (2) Air/ground emergency code.
      (3) Passwords/number combinations.
      (4) Fire net/Quickfire net.
      (5) Time zone.
      (6) Time hack.

AERIAL RESUPPLY ANNEX

1. **SITUATION**.
   a. **Enemy Forces** (include weather).
   b. **Friendly Forces**.
   c. **Attachments and Detachments**.

2. **MISSION**.

3. **EXECUTION**.
   a. **Concept of Operation**.
      (1) ***Maneuver***.
      (2) ***Fires***.
   b. **Tasks to Combat Units**.
      (1) ***Command and Control***.
      (2) ***Security***.
      (3) ***Marking***.
      (4) ***Recovery/Transport***.
   c. **Tasks to Combat Support Units**.
   d. **Coordinating Instructions**.
      (1) Flight Route.
         (a) General.
         (b) Checkpoints.
         (c) Communication checkpoint (CCP).
            - Marking of CCP.
            - Report time.
         (d) Heading from CCP.
      (2) Landing/Drop Zone.
         (a) Location.
            - Primary.
            - Alternate.
         (b) Marking.
            - Near.
            - Far.

(3) Drop Information.
   (a) Date/time of resupply (and alternates).
   (b) Code letter at DZ/LZ.
   (c) Length of DZ in seconds or dimensions of LZ.
   (d) Procedures for turning off DZ/LZ.
   (e) Formation, altitude, and air speed.
      - En route.
      - At DZ/LZ.
(4) Actions on enemy contact during resupply.
(5) Abort Criteria: En route and at DZ/LZ.
(6) Actions at DZ/LZ.
   - Rehearsals.

4. **SERVICE SUPPORT**.

5. **COMMAND AND SIGNAL**.
   a. **Command**.
      (1) Location of patrol leader.
      (2) Location of assistant patrol leader.
      (3) Location of members not involved in resupply.
   b. **Signal**.
      (1) Air to ground call-signs and frequencies (primary and alternate).
      (2) Long range visual signals.
      (3) Short range visual signals.
      (4) Emergency procedures and signals.
      (5) Air drop communication procedures.
      (6) Code words.

PATROL BASE ANNEX

1. **SITUATION**
   a. Enemy Forces
   b. Friendly Forces
   c. Attachments and Detachments

2. **MISSION**.

3. **EXECUTION**.
   a. **Concept of Operation**.
      (1) ***Maneuver***.
      (2) ***Fires***.
   b. **Tasks to Combat Units**.
      (1) ***Teams***.
         - Security
         - Recon
         - Surveillance
         - LP/OPs
      (2) ***Individuals***.
   c. **Tasks to Combat Support Units**.
   d. **Coordinating instructions**.
      (1) Occupation plan.
      (2) Operations plan.
         - Security Plan.
         - Alert Plan.

- Priority of work.
- Evacuation plan.
- Alternate patrol base (used when primary is unsuitable or compromised).

4. **SERVICE SUPPORT**.
   a. Water plan.
   b. Maintenance plan.
   c. Hygiene plan.
   d. Messing plan.
   e. Rest plan.

5. **Command and Signal**.
   a. **Command**.
      (1) Location of patrol leader.
      (2) Location of assistant patrol leader.
      (3) Location of patrol CP.
   b. **Signal**.
      (1) Call signs and frequencies.
      (2) Code words.
      (3) Emergency signals.

SMALL BOAT ANNEX

1. **SITUATION**.
   a. **Enemy Forces**.
      (1) **Weather**.
         (a) Tide.
         (b) Surf.
         (c) Wind.
      (2) **Terrain**.
         (a) River width.
         (b) River depth and water temperature.
         (c) Current.
         (d) Vegetation.
      (3) **Identification, Location, Activity, and Strength**.
   b. **Friendly Forces** (unit furnishing support).
   c. **Attachments and Detachments**.
   d. **Organization for Movement**.

2. **MISSION**.

3. **EXECUTION**.
   a. **Concept of Operation**.
      (1) ***Maneuver***.
      (2) ***Fires***.
   b. **Tasks to Combat Units**.
      (1) Security.
      (2) Tie-down teams.
         (a) Load equipment.
         (b) Secure equipment.
      (3) Designation of coxswains and boat commanders.
      (4) Selection of navigator(s) and observer(s).

c. **Coordinating Instructions**.

(1) Formations and order of movement.
(2) Route and alternate route.
(3) Method of navigation.
(4) Actions on enemy contact.
(5) Rally points.
(6) Embarkation plan.
(7) Debarkation plan.
(8) Rehearsals.
(9) Time schedule.

4. **SERVICE SUPPORT**.

a. **Ration Plan**.

b. **Weapons and Ammunition**.

c. **Uniform and Equipment**.

(1) Method of distribution of paddles and life jackets.
(2) Disposition of boat, paddles and life jackets upon debarkation.

5. **COMMAND AND SIGNAL**.

a. **Command**.

(1) Location of patrol leader.
(2) Location of assistant patrol leader.

b. **Signal**.

(1) Signals used between boats and in boats.
(2) Code words.

STREAM CROSSING ANNEX

1. **SITUATION**.

a. **Enemy Forces**.

(1) ***Weather***.
(2) ***Terrain***.

(a) River width.
(b) River depth and water temperature.
(c) Current.
(d) Vegetation.
(e) Obstacles.

(3) ***Enemy*** (location, identification, activity).

b. **Friendly Forces**.

c. **Attachments and Detachments**.

2. **MISSION**.

3. **EXECUTION**.

a. **Concept of Operation**.

(1) ***Maneuver***.
(2) ***Fires***.

b. **Tasks to Combat Units**.

(1) ***Elements***.
(2) ***Teams***.
(3) ***Individuals***.

c. **Tasks to Combat Support Units**.

d. **Coordinating Instructions**.

(1) Crossing procedure/techniques.
(2) Security.
(3) Order of crossing.
(4) Actions on enemy contact.
(5) Alternate plan.
(6) Rallying points.
(7) Rehearsal plan.
(8) Time schedule.

4. **SERVICE SUPPORT**.

5. **COMMAND AND SIGNAL**.

a. **Command**.

(1) Location of patrol leader.
(2) Location of assistant patrol leader.
(3) Location of CP.

b. **Signal**.

(1) Emergency signals.
(2) Signals.

TRUCK ANNEX

1. **SITUATION**.

a. **Enemy Forces**.

b. **Friendly Forces**.

c. **Attachments and Detachments**.

2. **MISSION**.

3. **EXECUTION**.

a. **Concept of Operation**.

(1) ***Maneuver***.
(2) ***Fires***.

b. **Tasks to Combat Units**.

c. **Tasks to Combat Support Units**.

d. **Coordinating Instructions**.

(1) Time of departure and return.
(2) Loading plan and order of movement.
(3) Route (primary and alternate).
(4) Air Guards.
(5) Actions on enemy contact (vehicle ambush) during movement, loading, and downloading.
(6) Actions at the detrucking point.
(7) Rehearsals.
(8) Vehicle speed, separation, and recovery plan.
(9) Broken vehicle instructions.

4. **SERVICE SUPPORT**.

5. **COMMAND AND SIGNAL**.
   a. **Command**. Location of PL and PSG
   b. **Signal**.
      (1) Radio call signs and frequencies
      (2) Code words

2-7. **COORDINATION CHECKLISTS**. The following checklists are items that a platoon/squad leader must check when planning for a combat operation. In some cases, he will coordinate directly with the appropriate staff section, in most cases this information will be provided by the company commander or platoon leader. The platoon/squad leader, to keep him from overlooking anything that may be vital to his mission, may carry copies of these checklists:

Figure 2-7. INTELLIGENCE COORDINATION CHECKLIST

INTELLIGENCE COORDINATION CHECKLIST

In this coordination, the leader is informed of any changes in the situation as given in the operation order of mission briefing. He must keep himself constantly updated to ensure the plan is sound:

1. Identification of enemy unit.
2. Weather and light data.
3. Terrain update.
   a. Aerial photos.
   b. Trails and obstacles not on map.
4. Known or suspected enemy locations.
5. Weapons.
6. Probable course of action.
7. Recent enemy activities.
8. Reaction time of reaction forces.
9. Civilians on the battlefield.
10. Update to CCIR.

Figure 2-8. OPERATIONS COORDINATION CHECKLIST

OPERATIONS COORDINATION CHECKLIST

This coordination occurs with the platoon leader/company commander so that the platoon/squad leader can confirm his mission and operational plan, receive any last-minute changes to his mission or plan, and to update his subordinates or issue a FRAGO, if required:

1. Mission backbrief.
2. Identification of friendly units.
3. Changes in the friendly situation.
4. Route selection, LZ/PZ/DZ selection.
5. Linkup procedures.
6. Transportation/movement plan.

7. Resupply (in conjunction with S4).
8. Signal plan.
9. Departure and re-entry of forward units.
10. Special equipment requirements.
11. Adjacent units operating in the area of operations.
12. Rehearsal areas.
13. Method of insertion/extraction.

Figure 2-9. FIRE SUPPORT COORDINATION CHECKLIST

FIRE SUPPORT COORDINATION CHECKLIST

The platoon/squad leader will normally coordinate the following with the platoon forward observer (FO):

1. Mission backbrief.
2. Identification of supporting unit.
3. Mission and objective.
4. Route to and from the objective (include alternate routes).
5. Time of departure and expected time of return.
6. Unit target list (from fire plan).
7. Type of available support (artillery, mortar, naval gunfire, and aerial support to include Army, Navy, and Air Force) and their locations.
8. Ammunition available (to include different fuses).
9. Priority of fires.
10. Control measures.
    a. Checkpoints.
    b. Boundaries.
    c. Phase lines.
    d. Fire support coordination measures.
    e. Priority targets (target list).
    f. RFA (restrictive fire area).
    g. RFL (restrictive fire line).
    h. NFA (no-fire areas).
    i. Precoordinated authentication.
11. Communication (include primary and alternate means, emergency signals and code words).

Figure 2-10. COORDINATION WITH FORWARD UNIT CHECKLIST

COORDINATION WITH FORWARD UNIT CHECKLIST

A platoon/squad that requires foot movement through a friendly forward unit must coordinate with that unit's commander for a safe and orderly passage. If no time and place has been designated for coordination with the forward unit, the platoon/squad leader should set a time and place when he coordinates with the S-3. He must talk with someone at the forward unit who has the authority to commit that unit in assisting the platoon/squad during departure. Coordination entails a two-way exchange of information:

1. Identification (yourself and your unit).
2. Size of platoon/squad.
3. Time(s) and place(s) of departure and return, location(s) of departure point(s), ERRP, and detrucking points.
4. General area of operations.
5. Information on terrain and vegetation.
6. Know or suspected enemy positions or obstacles.
7. Possible enemy ambush sites.
8. Latest enemy activity.
9. Detailed information on friendly positions such as crew-served weapons, FPF.

10. Fire and barrier plan.
   a. Support the unit can furnish. How long and what can they do?
      (1) Fire support.
      (2) Litter teams.
      (3) Navigational signals and aids.
      (4) Guides.
      (5) Communications.
      (6) Reaction units.
      (7) Other.
   b. Call signs and frequencies.
   c. Pyrotechnic plan.
   d. Challenge and password, running password, number combination.
   e. Emergency signals and code words.
   f. If the unit is relieved, pass the information to the relieving unit.
   g. Recognition signals.

Figure 2-11. ADJACENT UNIT COORDINATION CHECKLIST

ADJACENT UNIT COORDINATION CHECKLIST

Immediately after the operation order of mission briefing, the platoon/squad leader should check with other platoon/squad leaders who will be operating in the same areas. If the leader is not aware of any other units operating is his area, he should check with the S3 during the operations coordination. The S3 can help arrange this coordination if necessary. The platoon/squad leaders should exchange the following information with other units operating in the same area:

1. Identification of the unit.
2. Mission and size of unit.
3. Planned times and points of departure and reentry.
4. Route(s).
5. Fire support and control measures.
6. Frequencies and call signs.
7. Challenge and password, running password, number combination.
8. Pyrotechnic plan.
9. Any information that the unit may have about the enemy.
10. Recognition signals.

Figure 2-12. REHEARSAL AREA COORDINATION CHECKLIST

REHEARSAL AREA COORDINATION CHECKLIST

This coordination is conducted with the platoon leader/company commander to facilitate the unit's safe, efficient and effective use of rehearsal area prior to its mission:

1. Identification of your unit.
2. Mission.
3. Terrain similar to objective site.
4. Security of the area.
5. Availability of aggressors.
6. Use of blanks, pyrotechnics, and ammunition.
7. Mock-ups available.
8. Time the area is available (preferably when light conditions approximate light conditions of patrol).
9. Transportation.
10. Coordination with other units using area.

Figure 2-13. ARMY AVIATION COORDINATION CHECKLIST

ARMY AVIATION COORDINATION CHECKLIST

This coordination is conducted with the platoon leader/company commander and/or S3 Air to facilitate the time, detailed and effective use of aviation assets as they apply to your tactical mission:

1. **SITUATION**.
   a. **Enemy Situation**.
      (1) Enemy air capability.
      (2) Enemy ADA capability.
      (3) Include in Weather: Percent illum, illum angle, NVG window, ceiling, and visibility.
   b. **Friendly Situation**.
      (1) Unit(s) supporting operation, Axis of movement/corridor/routes.
      (2) Friendly ADA status.

2. **MISSION**.

3. **EXECUTION**.
   a. **Concept of the Operation**. Overview of what requesting unit wants to accomplish with the air assault/air movement.
   b. **Tasks to Combat Units**.
      (1) Infantry.
      (2) Attack aviation.
   c. **Tasks to Combat Support Units**.
      (1) Artillery.
      (2) Aviation (lift).
   d. **Coordinating Instructions**.
      (1) ***Pickup Zone***.
      - Direction of landing.
      - Time of landing/flight direction.
      - Location of PZ/alternate PZ.
      - Loading procedures.
      - Marking of PZ (panel, smoke, SM, lights).
      - Flight route planned (SP, ACP, RP).
      - Formations: PZ, en route, LZ.
      - Code words:
        -- PZ secure (prior to landing), PZ clear (lead bird and last bird).
        -- Alternate PZ (at PZ, en route, LZ), names of PZ/alt PZ.
      - TAC air/artillery.
      - Number of pax per bird and for entire lift.
      - Equipment carried by individuals.
      - Marking of key leaders.
      - Abort criteria (PZ, en route, LZ).

      (2) ***Landing Zone***.
      - Direction of landing.
      - False insertion plans.
      - Time of landing (LZ time).
      - Location of LZ and Alternate LZ.
      - Marking of LZ (panel, smoke, SM, lights).
      - Formation of landing.
      - Code words, LZ name, alternate LZ name.
      - TAC air/artillery preparation, fire support coordination.
      - Secure LZ or not?

4. **SERVICE AND SUPPORT**.
   a. Number of aircraft per lift and number of lifts.
   b. Refuel/rearm during mission or not?
   c. Special equipment/aircraft configuration for weapons carried by unit personnel.
   d. Bump plan.
5. **COMMAND AND SIGNAL**.
   a. Frequencies, call signs and code words.
   b. Locations of air missions commander ground tactical commander, and air assault task force commander.

Figure 2-14. VEHICULAR MOVEMENT COORDINATION CHECKLIST

VEHICULAR MOVEMENT COORDINATION CHECKLIST

This is coordinated with the supporting unit through the platoon sergeant/first sergeant to facilitate the effective, detailed, and efficient use of vehicular support and/or assets:

1. Identification of the unit.
2. Supporting unit identification.
3. Number and type of vehicles and tactical preparation.
4. Entrucking point.
5. Departure/loading time.
6. Preparation of vehicles for movement.
   a. Driver responsibilities.
   b. Platoon/squad responsibilities.
   c. Special supplies/equipment required.
7. Availability of vehicles for preparation/rehearsals/inspection (time and location).
8. Routes.
   a. Primary.
   b. Alternate.
   c. Checkpoints.
9. Detrucking points.
   a. Primary.
   b. Alternate.
10. March internal/speed.
11. Communications (frequencies, call signs, codes).
12. Emergency procedures and signals.

2-8. DOCTRINAL TERMS

TACTICAL TASK--A clearly defined, measurable activity accomplished by individuals and organizations. Tasks are specific activities that contribute to the accomplishment of encompassing missions or other requirements. A task should be definable, measurable, and decisive (achieve the purpose).

Actions by Friendly Force		**Effects on Enemy Force**
Assault	Follow and Assume	Block
Attack-by-Fire	Follow and Support	Canalize
Breach	Linkup	Contain
Bypass	Occupy	Defeat
Clear	Reconstitution	Destroy
Combat Search and Rescue	Reduce	Disrupt
Consolidation and Reorganization	Retain	Fix
Control	Secure	Interdict
Counterreconnaissance	Seize	Isolate
Disengagement	Support-by-Fire	Neutralize
Exfiltrate	Suppress	Penetrate
		Turn

PURPOSE (in order to)--The desired or intended result of the tactical operation stated in terms related to the enemy or the desired situation. The why of the mission statement. The most important component of the mission statement.

Allow	Divert	Prevent
Cause	Enable	Protect
Create	Envelop	Support
Deceive	Influence	Surprise
Deny	Open	

OPERATION--A military action or the carrying out of a military action to gain the objectives of any battle or campaign.

Types of operations include:	Forms of Offensive Maneuver:	Types of Defensive Techniques:	Other Operations:
Movement to Contact	Envelopment	Area Defense	Reconnaissance Ops
Search and Attack	Frontal Attack	Mobile Defense	Security Ops
Attack	Infiltration	Retrograde Ops	Information Ops
Ambush	Penetration	Delay	Combined Arms
Demonstration	Turning Movement	Withdrawal	Breach
Feint		Retirement	Passage of Lines
Raid			Relief in Place
Spoiling Attack			River Crossing Ops
Exploitation			Troop Movement
Pursuit			Admin
			Approach March
			Road March

2-9. **TERRAIN MODEL**. The terrain model (Figure 2-15) is an outstanding visual means used during the planning process to communicate the patrol routes and also detailed actions on the objective. At a minimum, the model is required to during planning to display routes to the objective and to highlight prominent terrain features the patrol will encounter during movement. A second terrain model is usually constructed of the objective area, enlarged to a sufficient size and detail to brief the patrol's actions on the objective.

a. **Checklist**.

(1) North-seeking arrow.
(2) Scale.
(3) Grid lines.
(4) Objective location.
(5) Exaggerated terrain relief, water obstacles.
(6) Friendly patrol locations.
(7) Targets (indirect fires, including grid and type of round).
(8) Routes, primary and alternate.
(9) Planned RPs (ORP, L/URP, RP).
(10) Danger areas (roads, trails, open areas).
(11) Legend.
(12) Blowup of objective area.

b. **Construction**. The following are some field-expedient techniques to aid in terrain model construction:

(1) Use a 3 x 5 card, MRE box, or piece of paper to label the objective or key sites.

(2) Grid lines can be made using string from the guts of 550 cord or colored tape. (Grid lines are identified by writing numbers on small pieces of paper.)

(3) Trees and vegetation are replicated by using moss, green or brown spray paint, pine needles, crushed leaves, or cut up grass.

(4) Water is designated by blue chalk, blue spray paint, blue yarn, tin foil, or MRE creamer.

(5) North seeking arrows are made from sharpened twigs, pencils, or colored yarn.

(6) Enemy positions are designated using red yarn, M-16 rounds, toy Rangers, or poker chips.

(7) Friendly positions such as security elements, support by fire, and assault elements are made using M-16 rounds, toy Rangers, poker chips, small MRE packets of sugar and coffee, or pre-printed acetate cards.

(8) Small pieces of cardboard or paper can identify target reference points and indirect fire targets. Ensure grids are shown for each point.

(9) Breach, support by fire, and assault positions are made using colored yarn or string so that these positions can be easily identified.

(10) Bunkers and buildings are constructed using MRE boxes or tongue depressors/sticks.

(11) Perimeter wire is constructed from a spiral notebook.

(12) Key phase lines are constructed with colored string or yarn.

(13) Trench lines are replicated by colored tape or yarn, by digging a furrow and coloring it with colored chalk or spray paint.

Note: All symbols used on the terrain model must be clearly identified in the legend.

Figure 2-15. TERRAIN MODEL

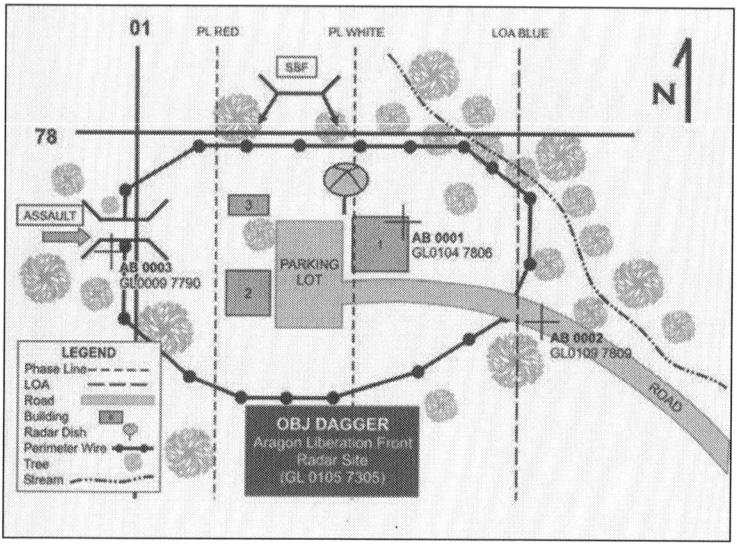

NOTES

Chapter 3
FIRE SUPPORT

Indirect fire support can greatly increase the combat effectiveness and survivability of any Infantry unit. The ability to plan for and effectively use this asset is a task that every Ranger and small unit leader should master. Fire support assets can assist a unit suppress, fix, destroy, or neutralize the enemy. Employment of indirect fire support should be considered throughout every offensive and defensive operation. This chapter discusses planning, tasks, capabilities, risk estimate distances, target overlays, close air support, elements and sequence of call for fire, and example call-for-fire transmissions.

3-1. **PLANNING**. Planning is the continual process of selecting targets on which fires are prearranged to support a phase of the commander's plan. The principles of planning follow:

- Consider what the commander wants to do.
- Plan early and continuously.
- Exploit all available targeting assets.
- Use all available lethal and nonlethal fire support means.
- Use the lowest echelon able to furnish effective support.
- Observe all fires.
- Use the most effective fire support asset available.
- Provide adequate fire support.
- Avoid unnecessary duplication.
- Provide for safety of friendly forces and installations.
- Provide for flexibility.
- Furnish the type of fire support requested.
- Consider the airspace.
- Provide rapid and effective coordination.
- Keep all fire support informed.

3-2. **TASKS**.

a. **All Operations**.

(1) Locate targets.
(2) Integrate all available assets.
(3) Destroy, neutralize, or suppress all enemy direct and indirect fire systems.
(4) Provide illumination and smoke.
(5) Provide fires in support of JAAT and SEAD missions.
(6) Deliver scatterable mines.
(7) Prepare for future operations.
(8) Provide positive clearance of fires.

b. **Offensive Operations**.

(1) Support the movement to contact, chance contact.
(2) Soften enemy defenses before the attack by arranging short, violent preparations, where required.
(3) Provide support during the attack by attacking high payoff targets.
(4) Plan for deep and flanking fires.
(5) Plan fires during consolidation.
(6) Provide counterfires.

c. **Defensive Operations**.

(1) Execute all plans as the commander intends.
(2) Disorganize, delay, and weaken the enemy before the attack.
(3) Provide counterfire.
(4) Provide fires in support of planned engagement areas (EAs).
(5) Attack high payoff targets.
(6) Plan fires in support of barriers and obstacle plans.
(7) Plan for deep, flanking, and rear fires.

(8) Provide fires in support of counterattacks.

(9) Plan final protective fire (FPFs).

3-3. **CAPABILITIES**. Tables 3-1 and 3-2 show capabilities of field artillery and mortars.

Table 3-1. CAPABILITIES OF FIELD ARTILLERY

WEAPON	MAX RANGE (meters)	MIN RANGE (meters)	MAX RATE Rds per Min	Burst Radius	SUSTAINED RATE Rds per Min
105-mm Howitzer M119, Towed	14,000 m	0	6 for 2 min	35 m	3 Rds for 30 min, then 1 rd per min
155-mm Howitzer M198, Towed	18 m,100 m 30,000 m (RAP)	0	4 for 3 min 2 for 30 min	50 m	1 rd per min Temp Dependent
155-mm Howitzer M109A6 SP	18 m, 100 m 30,000 m (RAP)	0	4 for 3 min	50 m	1 for 60 min 0.5

Table 3-2. CAPABILITIES OF MORTARS

WPN	MUNITIONS AVAILABLE	MAX HE RANGE	MIN RANGE (meters)	MAX RATE (Rds per Min)	BURST RADIUS	SUSTAINED RATE (Rds per Min)
60mm	HE, WP, Illum	3,500 m	70 m (HE)	30 for 4 min	30 m	20
81mm	HE, WP, Illum	5,600 m	70 m (HE)	25 for 2 min	38 m	8
120mm	HE, Smoke, Illum	7,200 m	180 m (HE)	15 for 1 min	60 m	5

DANGER

DANGER CLOSE

1. **When the target is within 600 meters of any friendly troops (for mortars and field artillery), announce DANGER CLOSE in the Method of Engagement portion of the call for fire.**
2. **When adjusting 5-inch or smaller naval guns on targets within 750 meters, announce DANGER CLOSE. For larger naval guns, announce DANGER CLOSE for targets within 1,000 meters. Failure to adhere to this guidance can result in fratricide.**
3. **Avoid making corrections using the bracketing method of adjustment, because doing so can cause serious injury or death. Use only the creeping method of adjustment during DANGER CLOSE missions. Make corrections of no more than 100 meters by creeping the rounds to the target.**

3-4. **RISK ESTIMATE DISTANCES**. RED applies to combat only. Minimum safe distances (Table 3-3) apply to training IAW AR 350-1. RED takes into account the bursting radius of particular munitions and the characteristics of the delivery system and associates this combination with a percentage representing the likelihood of becoming a casualty, that is, the percentage of risk. RED is defined as the minimum distance friendly troops can approach the effects of friendly fires without suffering appreciable casualties of 0.1% PI or higher.

WARNING

[Commanders] use risk-estimate distance formulas to determine acceptable risk levels in combat only. Specifically, use them to identify the risk to your Rangers at various distances from their targets. Risk estimate distances apply only in combat. In training, use minimum safe distances (MSD).

a. **Casualty Criterion**. The casualty criterion is the 5-minute assault criterion for a prone soldier in winter clothing and helmet. Physical incapacitation means that a soldier is physically unable to function in an assault within a 5-minute period after an attack. A PI value of less than 0.1 percent can be interpreted as being less than or equal to one chance in one thousand.

Table 3-3. RISK ESTIMATE DISTANCES FOR MORTARS AND CANNON ARTILLERY

System	Risk Estimate Distances (Meters)					
	10 % PI			0.1 % PI		
	1/3 range	2/3 range	Max range	1/3 range	2/3 range	Max range
60-mm mortar	60	65	65	100	150	175
81-mm mortar	75	80	80	165	185	230
120-mm mortar	100	100	100	150	300	400
105-mm howitzer	85	85	90	175	200	275
155-mm howitzer	100	100	125	200	280	450
155-mm DPICM	150	180	200	280	300	475

b. **Risk**. Using echelonment of fires within the specified RED for a delivery system requires the unit to assume some risks. The maneuver commander determines by delivery system how close he will allow fires to fall in proximity to his forces. The maneuver commander makes the decision for this risk level, but he relies heavily on the FSO's expertise.

3-5. **TARGET OVERLAYS**.

a. Figure 3-1 shows contents of fire support overlay.

Figure 3-1. CONTENTS OF FIRE SUPPORT OVERLAY

- Unit and official capacity of person making overlay.
- Date the overlay was prepared.
- Map Sheet Number.
- Effective Period of Overlay (DTG).
- Priority target.
- ORP Location.
- Call signs and frequencies. (PRI/ALT)
- Routes--Primary/Alternate.
- Phase Lines/Checkpoints used by the patrol.
- Spares.
- Index marks to position overlay on map.
- Objective.
- Target Symbols.
- Description, location and remarks column, complete.

b. Using a PLOT-CR checklist helps ensure the leader's fire support plan is complete:
   (1) Purpose--Purpose of the planned indirect fires.
   (2) Location--Planned targets with an 8-digit grid (minimum).
   (3) Observer--Planned observer to see the impact of the rounds and adjust.
   (4) Trigger--Method to initiate fires.
   (5) Communication--Method to communicate between the observer and the supporting unit.
   (6) Resources--Planned allocated resource for each target.

c. Sterile overlay includes---
   (1) Index marks to position overlay on map.
   (2) Target symbols.

d. Target overlay symbols (Figure 3-2) include--
   (1) Point target:
   (2) Linear target:
   (3) Circular target:

Figure 3-2. TARGET OVERLAY SYMBOLS--POINT (TOP), LINEAR (CENTER), AND CIRCULAR (BOTTOM)

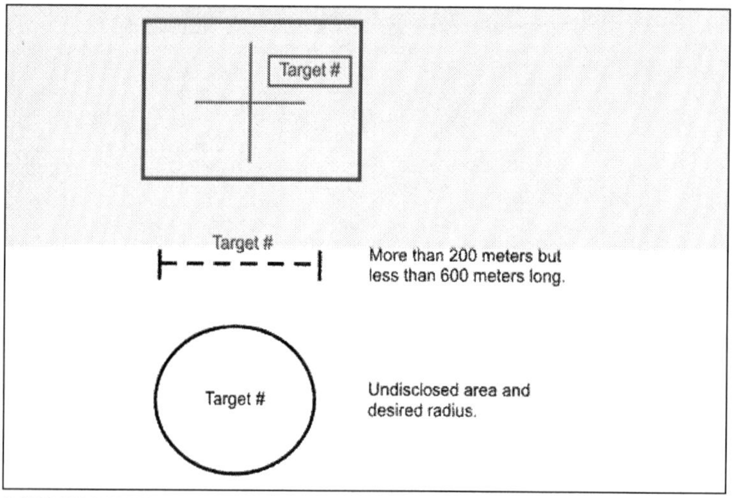

3-6. **CLOSE AIR SUPPORT**. The two types of close air support requests are planned and immediate. Planned requests are processed by the Army chain to corps for approval. Immediate requests are initiated at any level and processed by the battalion S-3 FSO, and air liaison officer.

a. Format for requesting immediate CAS:
   (1) Observer identification.
   (2) Warning order (request close air).
   (3) Target location (grid).

(4) Target description. (Description must include, as a minimum: type and number of targets, activity or movement; point or area targets, include desired results on target and time on target.

(5) Location of friendly forces.

(6) NAV details (elevation).

(7) Threats--ADA, small arms, and so on.

(8) Hazards--friendly aircraft in area.

(9) Wind direction.

b. Close air support capabilities are shown in Table 3-4.

Table 3-4. CLOSE AIR SUPPORT

AIRCRAFT	SERVICE	CHARACTERISTICS
A-10 *	AF	Specialized CAS aircraft; subsonic; typical load 6,000 lbs, max load 16,000 lbs; 30-mm gun.
F-16 *	AF	A multirole aircraft; complements the F-4 and F-15 in an air-to-air role. Most accurate ground delivery system in the inventory; supersonic; typical load 6,000 lbs, max load 10,600 lbs.
F-18 *	N/MC	A multirole fighter scheduled to replace the F-4. A wide variety of air-to-surface weapons. Typical load 7,000 lbs, max load 17,000 lbs, 20-mm gun mounted in the nose; air-to-air missiles.
AC-130	AF/R	A specialized CAS/RACO aircraft, propeller driven. Two models: The "A" model has two 40-mm guns, two 20-mm guns and two 7.62-mm miniguns. The similar "H" model has no 7.62-mm miniguns, and has a 105-mm howitzer in place of one of the 40-mm guns. Both models have advanced sensors and a target-acquisition system, including FLIR and low light TV. Very accurate. Vulnerable to enemy air defense systems, so must operate in a low-threat environment.
*FM capability.		

3-7. **CALL FOR FIRE**.

a. **Observer's Identification - Call Signs**.

b. **Warning Order**.

(1) ***Type of mission***.

(a) Adjust fire.

(b) Fire for effect.

(c) Suppress.

(d) Immediate suppression/immediate smoke.

(2) ***Size of element to Fire for Effect***. When the observer does not specify what size element to fire, the battalion FDC will decide.

(3) ***Method of Target Location***.

(a) Polar plot.

(b) Shift from a known point.

(c) Grid.

(4) **Location of Target**.

(a) *Grid Coordinate - 6 Digit*. 8 digit if greater accuracy is required.

(b) *Shift from a Known Point*. Send observer-target (OT) direction:

- Mils (nearest 10).
- Degrees.
- Cardinal direction.
- Send lateral shift (right/left, nearest 10 mils).
- Send range shift (add/drop, nearest 100 mils).
- Send vertical shift (up/down). Use only if it exceeds 35 mils (nearest 5 mils).

(c) *Polar Plot*. Send direction. (Nearest 10 mils)

- Send distance (nearest 100 mils).
- Send vertical shift (nearest 5 mils).

(5) ***Description of Target***.

(a) Type.

(b) Activity.

(c) Number.

(d) Degree of protection.

(e) Size and shape (length/width or radius).

(6) ***Method of Engagement, Type of Adjustment***. When the observer does not request a specific type of fire control adjustment, area fire is issued.

(a) Area fire--moving target.

(b) Precision fire--point target.

(7) ***Danger Close***. Use when friendly troops are within--

(a) 600 mils for mortars.

(b) 600 mils for artillery.

(c) 750 mils for naval guns 5 inches or smaller.

(8) ***Mark***. Use to orient observer or to indicate targets.

(9) ***Trajectory***.

(a) Low angle (standard).

(b) High angle (mortars or if requested).

(10) ***Ammunition***. Use HE quick used unless observer specifies otherwise.

(a) Projectile (HE, illum, ICM, smoke, and so on).

(b) Fuse (quick, time, and so on).

(c) Volume of fire (observer may request the number of rounds to be fired).

(11) ***Distribution***.

(a) 100 mil sheaf (standard).

(b) Converged sheaf (used for small hard targets).

(c) Special sheaf (any length, width, and attitude).

(d) Open sheaf (separate bursts).

(e) Parallel sheaf (linear target).

(12) ***Method of Fire and Control***.

(a) *Method of Fire*. Specific guns and a specific interval between rounds. Normally adjust fire, one gun is used with a 5-second interval between rounds.

(b) *Method of Control*.

- AT MY COMMAND - FIRE--Remains in effect until observer announces, "*Cancel at my command.*"
- CANNOT OBSERVE--Observer can't see the target.
- TIME ON TARGET--Observer tells FDC when he wants the rounds to impact.
- CONTINUOUS ILLUMINATION. Calculated by the FDC, otherwise observer indicates interval between rounds in seconds.
- COORDINATED ILLUMINATION. Observer may order the interval between illum and HE shells.
- CEASE LOADING--Stop loading rounds.
- CHECK FIRING--Stop firing at once.
- CONTINUOUS FIRE--Load and fire as fast as possible.
- REPEAT--Fire another round(s) with or without adjustments.

(13) ***Authentication***. Challenge and reply.

(14) ***Message to Observer***.

(a) Battery(ies) to fire for effect.

(b) Adjusting battery.

(c) Changes to the initial call for fire.

(d) Number of rounds (per tube) to be fired for effect.

(e) Target numbers.

(f) Additional information:

- Time of flight. Moving target mission.
- Probable error in range. 38 meters or greater (normal mission).
- Angle T. 500 mils or greater.

c. **Correction of Errors**. When FDC has made an error when reading back he fire support data, the observer announces "*CORRECTION*" and transmits the correct data in its entirety.

d. **Call-for-Fire Transmissions**. Example call-for-fire transmissions are shown in Table 3-5.

Table 3-5. EXAMPLE CALL-FOR-FIRE TRANSMISSIONS.

GRID MISSION	
OBSERVER	**FIRING UNIT**
F24, this is J42, ADJUST FIRE, OVER.	J42, this is F24, AJUST FIRE, OUT.
GRID WM180513, DIRECTION 0530, OVER.	GRID WM180513, DIRECTION 0530, OUT.
Infantry platoon dug in, OVER.	Infantry platoon dug in, OUT.
	SHOT OVER.
SHOT OUT.	
	SPLASH, OVER.
SPLASH OUT.	
End of mission, 15 casualties, platoon dispersed, OVER.	End of mission, 15 casualties, platoon dispersed, OUT.
SHIFT FROM KNOWN POINT	
OBSERVER	**FIRING UNIT**
J42, THIS IS F24, ADJUST FIRE, SHIFT AB1001, OVER.	F24, THIS IS J42, ADJUST FIRE, SHIFT AB1001, OUT.
DIRECTION 2420, RIGHT 400, ADD 400, OVER.	DIRECTION 2420, RIGHT 400, ADD 400, OUT.
FIVE T-72 TANKS AT POL SITE, OVER.	FIVE T-72 TANKS AT POL SITE, AUTHENTICATE JULIET NOVEMBER, OVER.
I AUTHENTICATE TANGO, OVER.	
	SHOT, OVER.
SHOT OUT.	SPLASH, OVER.
SPLASH OUT.	
END OF MISSION, 2 TANKS DESTROYED, 3 IN WOODLINE, OVER	END OF MISSION, 2 TANKS DESTROYED, 3 IN WOODLINE, OUT

POLAR	
OBSERVER	**FIRING UNIT**
J42, this is F24, adjust fire, polar OVER.	F24, this is J42, adjust fire, polar, OUT
Direction 2300, distance 4,000, OVER.	Direction 2300, distance 4,000, OUT.
Infantry platoon dug in, OVER	Infantry platoon dug in, OUT
	Shot OVER.
Shot OUT.	Splash, OVER.
Splash OUT.	
End of mission, 15 casualties, platoon dispersed, OVER.	End of mission, 15 casualties, platoon dispersed, OUT.

NOTES

Chapter 4
MOVEMENT

To survive on the battlefield, stealth, dispersion, and security must be enforced in all tactical movements. The leader must be skilled in all movement techniques.

4-1. **FORMATIONS**. Movement formations are comprised of elements and Soldiers in relation to each other. Fire teams, squads and platoons use several movement formations. Formations provide the leader control based on a METT-TC analysis. Leaders position themselves where they can best command and control formations. The formations below allow the fire team leader to lead by example, "*Follow me, and do as I do.*" All Soldiers in the team must be able to see their leader. The formations in Figure 4-1 reflect fire team formations. Squad formations are very similar with more Soldiers. Squads can operate in lines and files similar to fire teams. When squads operate in wedges or in echelon, the fire teams use those formations and simply arrange themselves in column or one team behind the other. Squads may also use the vee, with one team forming the lines of the vee, and the SL at the apex for command and control. Platoons work on the same basic formations as the Squads. When operating as a platoon, the platoon leader must carefully select the location for his machine guns in the movement formation.

Figure 4-1. FORMATIONS

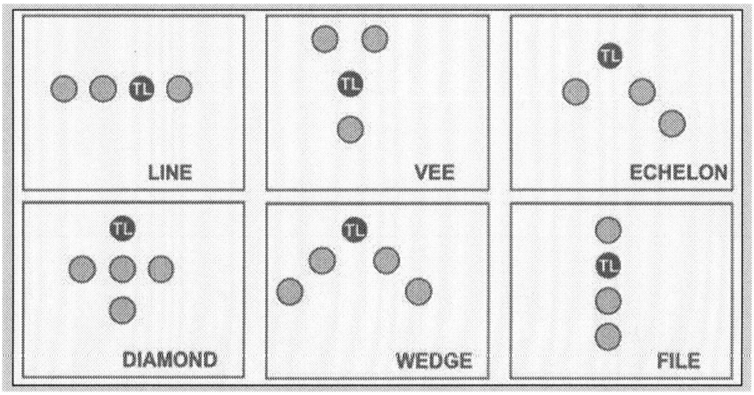

a. **Techniques**. A movement technique is the method a unit uses to traverse terrain. There are three movement techniques: traveling, traveling overwatch, and bounding overwatch. The selection of a movement technique is based on the likelihood of enemy contact and the need for speed. Factors to consider for each technique are control, dispersion, speed, and security. Movement techniques are not fixed formations. They refer to the distances between Soldiers, teams, and squads that vary based on METT-TC. Soldiers must be able to see their fire team leaders. The platoon leader should be able to see his lead squad leader. Leaders control movement with hand-and-arm signals and use radios only when needed.

b. **Standards**.

(1) Unit moves on designated route or arrives at specified location IAW OPORD maintaining accountability of all assigned/attached personnel.

(2) Unit uses movement formation and technique ordered by the leader based on METT-TC.

(3) Leaders remain oriented (within 200 meters) and follow planned route unless METT-TC dictates otherwise.

(4) Unit maintains 360 degree security and remains 100% alert during movement.

(5) Unit maintains 360 degree security and a minimum of 75% security during halts.

(6) If contact with the enemy is made, it is made with the smallest element possible.

(7) Control measures are used during movement such as head counts, rally points, or phase lines.

c. **Fundamentals**.

(1) ***Land Navigation***. Mission accomplishment depends on successful land navigation. The patrol should use stealth and vigilance to avoid chance contact. Designate a primary and alternate compass and pace man per patrol.

Note: The point man will not be tasked to perform compass or pace duties. The point man's sole responsibility is forward security for the element.

(2) ***Avoidance of Detection***. Patrols must use stealth, and use the cover and concealment of the terrain to its maximum advantage. Whenever possible, move during limited visibility in order to maximize technological advantages gained by night vision devices and to hinder the enemy's ability to detect the patrol. Exploit the enemy's weaknesses, and attempt to time movements to coincide with other operations that distract the enemy. The enemy threat and terrain determines which of the three movement techniques will be used:

(a) Fire teams maintain visual contact, but the distance between them is such that the entire patrol does not become engaged if contact is made. Fire teams can spread their formations as necessary to gain better observation to the flanks. Although widely spaced, men retain their relative position in their wedge and follow their team leader. Only in extreme situations should the file be used.

(b) The lead squad must secure the front along with assuming responsibility for navigation. For a long movement, the PL may rotate the lead squad responsibilities. The fire team/squad in the rear is charged with rear security.

(c) Vary movement techniques to meet the changing situation.

(d) With the exception of fire team leaders, leaders move inside their formations where they can maintain the best control.

(3) ***Security***. The patrol must use both active and passive security measures constantly. Assign subunits responsibility for security at danger areas, patrol bases, and most importantly in the objective area.

(4) ***Fire Support***. Plan fire support (mortars, artillery, tactical air, attack helicopter, naval gunfire).

(5) ***Three-Dimensional Battlefield***. 360 Degree Security is achieved through high and low security. Within a fire team, squad, and so on, the leader must assign appropriate sectors of fire to their subordinate in order to ensure all aspects of the battlefield are covered. This includes trees, multiple storied structures, tunnels, sewers, ditches.

d. **Movement Techniques**.

(1) The traveling technique is used when enemy contact is not likely but speed is necessary.

(2) The traveling overwatch technique is used when enemy contact is possible.

(3) The bounding overwatch technique is used when enemy contact is likely, or when crossing a danger area.

e. **Traveling**. In the traveling technique, the distance between individuals is about 10 meters with 20 meters between squads. It has the following characteristics:

(1) More control than traveling overwatch but less than bounding overwatch.

(2) Minimum dispersion.

(3) Maximum speed.

(4) Minimum security.

f. **Traveling Overwatch**. The traveling overwatch technique is the basic movement technique. The distance between individuals is about 20 meters, between teams about 50 meters.

(1) In platoon traveling overwatch, the lead squad must be far enough ahead of the rest of the platoon to detect or engage any enemy before the enemy observes or fires on the main body. However, it must be close enough to be supported by the platoon's small arm's fires. This is normally between 50 to 100 meters, depending on terrain, vegetation, and light and weather conditions.

(2) In a column formation, only the lead squad should use the traveling overwatch; however, if greater dispersion is desired, all squads may use it.

(3) In other formations, all squads use traveling overwatch unless the platoon leader specifies otherwise.

(4) Traveling overwatch has the following characteristics:

- Good control.
- Good dispersion.
- Good speed.
- Good security forward.

g. **Bounding Overwatch**. In the bounding overwatch technique (Figure 4-2), the distance between men remains about 20 meters. The distance between teams and squads varies.

(1) The squad or platoon has a bounding element and an overwatch element. The bounding element moves while the overwatch element occupies an overwatch position that can cover the route of the bounding element by fire. Each bound is within supporting range of the overwatch element.

(2) There are two types of bounding, successive and alternating (Figure 4-2). Successive is nothing more than one squad moving to a position, then the overwatching squad moving to a position generally online with the first squad. Alternating bounding is when one squad moves into position, then the overwatching squad moves to a position in front of the first squad.

Figure 4-2. SQUAD BOUNDING OVERWATCH

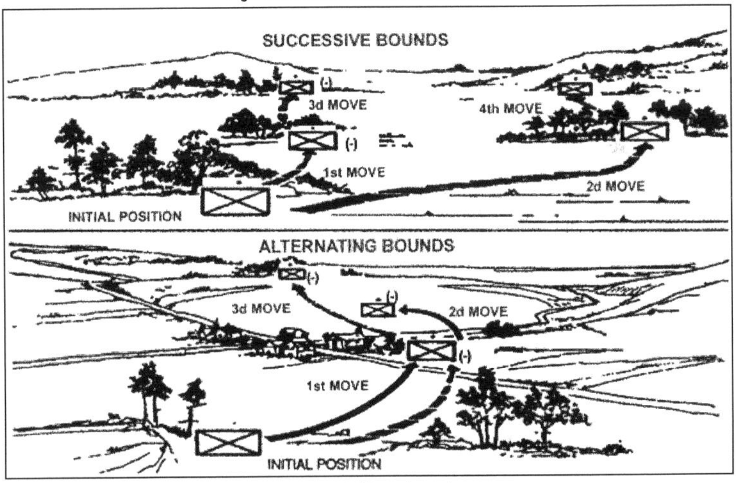

(3) The length of a bound depends on the terrain, visibility, and control.

(4) Before a bound, the leader gives the following instructions to his subordinates:

- Direction of the enemy if known
- Position of overwatch elements
- Next overwatch position
- Route of the bounding element
- What to do after the bounding element reaches the next position
- How the elements receive follow-on orders

(5) The characteristics of bounding overwatch are:

- Maximum control
- Maximum dispersion
- Minimum speed
- Maximum security

h. **Platoon Bounding Overwatch** (Figure 4-3).

(1) ***Method One***. When platoons use bounding overwatch, one squad bounds and one squad overwatches; the third squad awaits orders. Forward observers stay with the overwatching squad to call for fire. Platoon leaders normally stay with the overwatching squad who use machine guns and attached weapons to support the bounding squad.

(2) ***Method Two***. Another way is to have one squad use bounding overwatch and have the other two squads use traveling or traveling overwatch technique.

Figure 4-3. PLATOON BOUNDING OVERWATCH

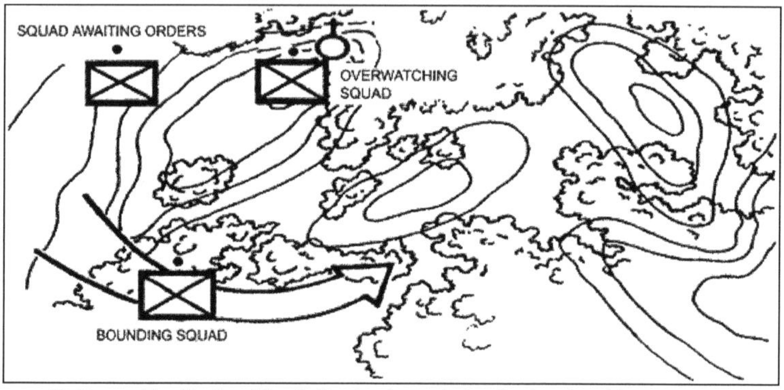

(3) ***Movement Considerations***. When deciding where to move the bounding element, consider--

- Where the enemy is likely to be.
- The mission.
- The routes to the next overwatch position.
- The weapons ranges of the overwatching unit.
- The responsiveness of the rest of the unit.
- The fields of fire at the next overwatch position.

4-2. **TACTICAL MARCHES**. Platoons conduct two types of marches with the company: foot marches and motor (road) marches.

a. **Purpose/General**. A successful foot march is when troops arrive at their destination at the prescribed time, physically able to execute their tactical mission.

b. **Standard**.

(1) The unit crosses the start point and release point at the time specified in the order.

(2) The unit follows the prescribed route, rate of march, and interval without deviation unless required otherwise by enemy action or higher headquarters action.

c. **Fundamentals**.

(1) Effective control.

(2) Detailed planning.

(3) Rehearsals.

d. **Considerations**.

(1) ***METT-TC***.

- Mission ....Task and purpose.
- Enemy ....Intentions, capabilities, and course of action.
- Terrain and weather.................Road condition/trafficability, and visibility.

- Troops/equipment....................Condition of Soldiers and their loads, numbers and types of weapons and radios.
- Time..Start time, release time, rate of march, time available.
- Civilians ....Movement through populated areas, refugees, OPSEC.

(2) ***Task Organization***:
- Security....................................Advance and trail teams.
- Main body ....Two remaining line squads and weapons squad.
- Headquarters...........................Command and control.
- Control measures.

(3) ***Start Point and Release Point*** (given by higher):
(a) Checkpoints.............................At checkpoints report to higher and use checkpoint to remain oriented.
(b) Rally or rendezvous points ......Used when elements become separated.
(c) Location of leaders ....Where they can best control their elements.
(d) Commo plan ....Locations of radios, frequencies, call signs, and OPSKEDs.
(e) Dispersion between Soldiers--
- 3 to 5 meters / day.
- 1 to 3 meters / night.

(4) ***March Order***. May be issued as an OPORD, FRAGO, or annex to either (must use operational overlay or strip map)"
(a) Formations and order of movement
(b) Route of march........................Assembly area, start point, release point, rally points, checkpoints, break/halt points.
(c) Start point time, release point time, and rate of march.
(d) March interval for squads, teams, and individuals.
(e) Actions on enemy contact........Air and ground.
(f) Actions at halts.
(g) Fires ....Detailed plan of fire support for the march.
(h) Water supply plan.
(i) MEDEVAC plan.

e. **Duties and Responsibilities**.

(1) ***Platoon Leader***:
- Before ....Issues warning order and FRAGO; inspects and supervises.
- During ....Ensures unit makes movement time; ensures interval is maintained and that unit remains oriented; maintains security; checks condition of men; and enforces water discipline and field sanitation.
- After ....Ensures men are prepared to accomplish their mission, supervises SLs, and ensures medical coverage is provided to men as needed.

(2) ***Platoon Sergeant***:
- Before ....Assists PL, makes recommendations, and enforces uniform and packing lists.
- During ....Controls stragglers, helps platoon leader maintain proper interval and security.
- At Halts ....Enforces security, ensures welfare of men, enforces field sanitation and litter discipline, and enforces use of preventive medicine.
- After..Coordinates for water, rations, and medical supplies. Recovers casualties.

(3) ***Squad Leaders***:
- Before ....Provides detailed instruction to TLs, inspects boots and socks for serviceability and proper fit, inspects adjustment of equipment, inspects to ensure canteens are full, and ensures equal distribution of loads.
- During ....Controls squad, maintains proper interval between men and equipment, enforces security, remains oriented.

- At Halts....................................Ensures security is maintained, provides men for water resupply as detailed. Physically checks the men in his squad, ensures they drink water, and that they change socks as necessary. Rotates heavy equipment. (Units should plan this in detail to avoid confusion before, during, and after halts.)
- After..Occupies squad sector of assembly area, conducts foot inspection, reports condition of men to PL, and prepares men to accomplish the mission.

(4) ***Security Squad***:
- Lead Team.
  - -- Serves as point element for platoon.
  - -- Recons route to SP.
  - -- Calls in checkpoints.
  - -- Provides early warning.
  - -- Maintains rate of march.
  - -- Moves 10 to 20 meters in front of main body.
- Trail Team.
  - -- Provides rear security.
  - -- Moves 10 to 20 meters behind main body.

(5) ***Medic***:
- Assesses and treats march casualties.
- Advise chain of command on evacuation and transportation requirements of casualties.

(6) ***Individual***:
- Maintains interval.
- Follows TL's examples.
- Relays hand and arm signals; remains alert during movement and at halts.
- Remains alert during movement and at halts.

4-3. **MOVEMENT IN LIMITED VISIBILITY CONDITIONS**. During hours of limited visibility, a platoon will use surveillance, target acquisition, and night observation (STANO) devices to enhance their effectiveness. Leaders must be able to control, navigate, maintain security, and move during limited visibility.

a. **Control**. When visibility is poor, the following methods aid in control.

(1) Use of night vision devices.
(2) Leaders move closer to the front.
(3) Platoon reduces speed.
(4) Use of luminescent tape on equipment.
(5) Reduce intervals between men and elements.
(6) Conducts headcount regularly.

b. **Navigation**. While navigating during limited visibility the same techniques are used as during the day; however, leaders exercise more care to keep the patrol oriented.

c. **Security**.

(1) Enforce strict noise and light discipline.
(2) Use radio-listening silence.
(3) Use camouflage.
(4) Use terrain to avoid detection by enemy surveillance or night vision devices.
(5) Make frequent listening halts; Conduct SLLS (Stop, Look, Listen, Smell).
(6) Mask the sounds of movement when possible. (Rain, wind, and flowing water will mask the sounds of movement).

d. **Rally Points**. Plan actions to be taken at rally points in detail. All elements must maintain communications at all time. There are two techniques for actions at rally points:

(1) ***Minimum Force***: Patrol members assemble at the rally point, and the senior leader assumes command. When the minimum force (designated in the OPORD) is assembled and organized the patrol will continue the mission.

(2) ***Time Available***: The senior leader determines if the patrol has enough time remaining to accomplish the mission.

e. **Actions at Halts**. During halts, post security and cover all approaches into the sector with key weapons.

(1) ***Short Halt***. Typically 1-2 minutes long. Soldiers seek immediate cover and concealment and take a knee. Leaders assign sectors of fire.

(2) ***Long Halt***. More than 2 minutes. Soldiers assume the prone position behind cover and concealment. (Ensure soldiers have clear fields of fire.) Leaders assign sectors of fire.

4-4. **DANGER AREAS**. A danger area is any place on a unit's route where the leader determines his unit may be exposed to enemy observation or fire. Some examples of danger areas are open areas, roads and trails, built-up areas, enemy positions, and natural and man-made obstacles. Bypass danger areas whenever possible.

a. **Standards**.

(1) The unit prevents the enemy from surprising the main body.
(2) The unit moves all personnel and equipment across the danger area.
(3) The unit prevents decisive engagement by the enemy.

b. **Fundamentals**.

(1) Designate near and far side rally points.
(2) Secure near side, left and right flank, and rear security.
(3) Recon and secure the far side.
(4) Cross the danger area.
(5) Plan for fires on all known danger areas.

c. **Technique for Crossing Danger Areas**.

(1) ***Linear Danger Area*** (LDA, Figure 4-4) ***Actions for a Squad***:

STEP 1: The alpha team leader (ATL) observes the linear danger area and sends the hand and arm signal to the SL who determines to bound across.

STEP 2: SL directs the ATL to move his team across the LDA far enough to fit the remainder of the squad on the far side of the LDA. Bravo team moves to the LDA to the right or left to provide an overwatch position prior to A team crossing.

STEP 3: SL receives the hand and arm signal that it is safe to move the rest of the squad across (B team is still providing overwatch).

STEP 4: SL moves himself, radio operator, and B team across the LDA. (A team provides overwatch for squad missions.)

STEP 5: A team assumes original azimuth at SL's command or hand and arm signal.

Figure 4-4. LINEAR DANGER AREA

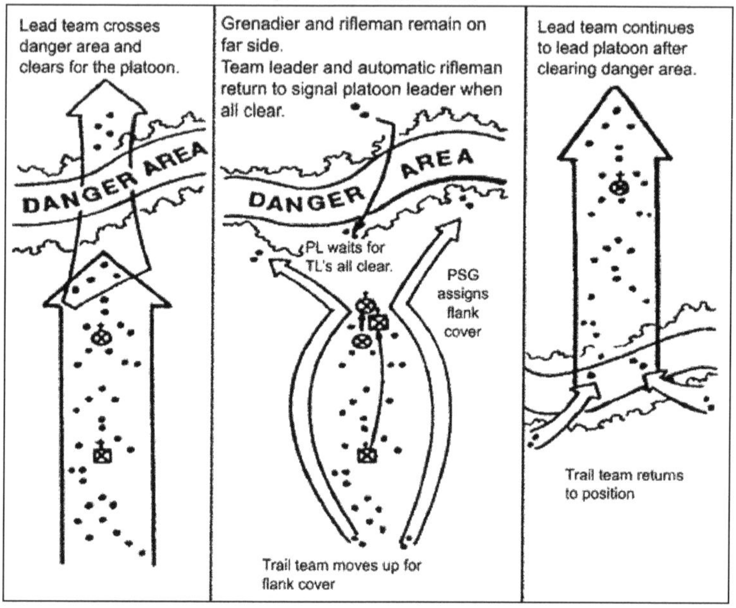

(2) ***LDA Crossing for a Platoon***:

(a) The lead squad halts the platoon, and signals danger area.

(b) The platoon leader moves forward to the lead squad to confirm the danger area and decides if current location is a suitable crossing site.

(c) The platoon leader confirms danger area/crossing site and establishes near and far side rally points.

(d) On the platoon leader's signal, the A team of the lead squad establishes an overwatch position to the left of the crossing site. Prior to crossing, the compassman with the lead two squads confirm azimuth and pace data.

(e) B team of the lead squad establishes an overwatch position to the right of the crossing site.

(f) Once overwatch positions are established, the platoon leader gives the second squad in movement the signal to bound across by fire team.

(g) Once across, the squad is now lead in movement and continues on azimuth

(h) Once Stop, Look, Listen, and Smell (SLLS) is conducted, squad leader signals platoon leader all clear.

- Day time—hand and arm signal such as thumbs up.
- Night time—Clandestine signal such as IR, red lens.

(i) Platoon leader receives all clear and crosses with radio operator, FO, WSL, and 2 gun teams.

(j) Once across, PL signals the 3rd squad in movement to cross.

(k) PSG with medic and one gun team crosses after 2nd squad is across (sterilizing central crossing site).

(l) PSG signals security squad to cross at their location.

Notes: Platoon leader will plan for fires at all known LDA crossing sites.
Nearside security in overwatch will sterilize signs of the patrol .

(3) ***Danger Area (Small/Open)***

(a) The lead squad halts the platoon and signals danger area.

(b) The PL moves forward to the lead squad to confirm the danger area.

(c) The platoon leader confirms danger area and establishes near and far side rally points.

(d) The PL designates lead squad to bypass danger area using the detour-bypass method.

(e) Paceman suspends current pace count and initiates an interim pace count. Alternate pace/compass man moves forward and offsets compass 90 degrees left or right as designated and moves in that direction until clear of danger area.

(f) After moving set distance (x-meters as instructed by PL). Lead squad assumes original azimuth, and primary pace man resumes original pace.

(g) After the open area, the alternate pace/compass man offsets his compass 90 degrees left or right and leads the platoon/squad the same distance (x-meters) back to the original azimuth.

(4) ***Danger Areas (Series)***: A series of danger areas is two or more danger areas within an area that can be either observed or covered by fire.

- Double linear danger area (use linear danger area technique and cross as one LDA).
- Linear/small open danger area (use by-pass/contour technique. Figure 4-5).
- Linear/large open danger area (use platoon wedge when crossing).

Note: A series of danger areas is crossed using the technique which provides the most security.

(5) ***Danger Area (Large)***:

(a) Lead squad halts the platoon, and signals danger area.

(b) The platoon leader moves forward with RTO and FO and confirms danger area.

(c) The platoon leader confirms danger area and establishes near and far side rally points.

(d) PL designates direction of movement.

(e) PL designates change of formation as necessary to ensure security.

Notes: Before point man steps into danger area, PL and FO adjust targets to cover movement.
If far side of danger area is within 250 meters, PL establishes overwatch, and designates lead squad to clear wood line on far side.

Figure 4-5. SMALL OPEN AREA

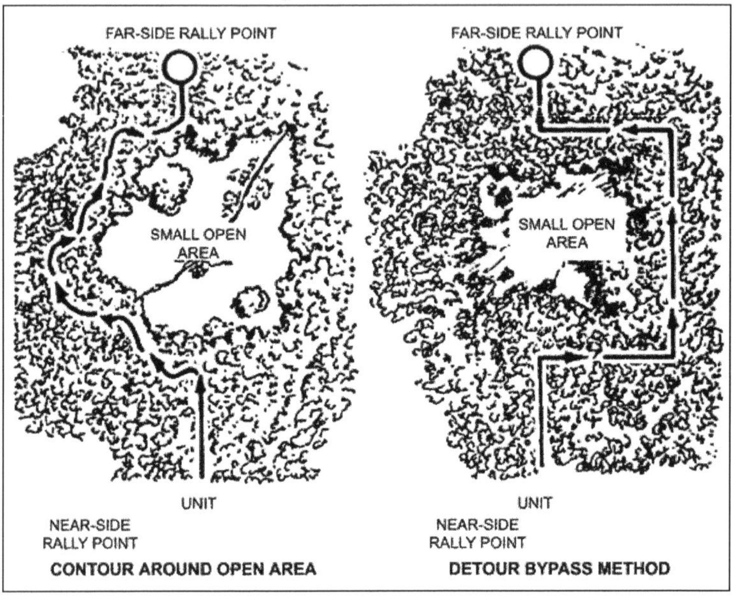

NOTES

Chapter 5
PATROLS

Infantry platoons and squads primarily conduct two types of patrols: reconnaissance and combat. This chapter describes the principles of patrolling, planning considerations, types of patrols, supporting tasks, patrol base, and movement to contact (FM 7-8, FM 3-0, and FM 101-5-1). In this chapter, the terms "*element*" and "*team*" refer to the squads, fire teams, or buddy teams that perform the tasks as described.

5-1. **PRINCIPLES**. All patrols are governed by five principles:

a. **Planning**. Quickly make a simple plan and effectively communicate it to the lowest level. A great plan that takes forever to complete and is poorly disseminated isn't a great plan. Plan and prepare to a realistic standard and rehearse everything.

b. **Reconnaissance**. Your responsibility as a Ranger leader is to confirm what you think you know, and to learn that which you do not already know.

c. **Security**. Preserve your force as a whole. Every Ranger and every rifle counts; anyone could be the difference between victory and defeat.

d. **Control**. Clear understanding of the concept of the operation and commander's intent, coupled with disciplined communications, to bring every man and weapon available to overwhelm the enemy at the decisive point.

e. **Common Sense**. Use all available information and good judgment to make sound, timely decisions.

5-2. **PLANNING**. This paragraph provides the planning considerations common to most patrols. It discusses task organization, initial planning and coordination, completion of the plan, and contingency planning.

a. **Task Organization**. A patrol is a mission, not an organization. To accomplish the patrolling mission, a platoon or squad must perform specific tasks, for example, secure itself, cross danger areas, recon the patrol objective, breach, support, or assault. As with other missions, the leader tasks elements of his unit in accordance with his estimate of the situation, identifying those tasks his unit must perform and designating which elements of his unit will perform which tasks. Where possible, in assigning tasks, the leader should maintain squad and fire team integrity. The chain of command continues to lead its elements during a patrol. Squads and fire teams may perform more than one task in an assigned sequence; others may perform only one task. The leader must plan carefully to ensure that he has identified and assigned all required tasks in the most efficient way. Elements and teams for platoons conducting patrols include--

(1) Elements common to all patrols:

(a) *Headquarters Element*. The headquarters consists of the platoon leader (PL), RTO, platoon sergeant (PSG), FO, RTO, and medic. It may include any attachments that the PL decides that he or the PSG must control directly.

(b) *Aid and Litter Team*. Aid and litter teams are responsible for buddy aid and evacuation of casualties.

(c) *Enemy Prisoner of War Team*. EPW teams control enemy prisoners using the five S's and the leader's guidance.

(d) *Surveillance Team*. The surveillance team keeps watch on the objective from the time that the leader's reconnaissance ends until the unit deploys for actions on the objective. They then rejoin their parent element.

(e) *En Route Recorder*. Part of the HQ element, maintains communications with higher and acts as the recorder for all CCIR collected during the mission.

(f) *Compass Man*. The compass man assists in navigation by ensuring the patrol remains on course at all times. Instructions to the compass man must include initial and subsequent azimuths. As a technique, the compass man should preset his compass on the initial azimuth before the unit moves out, especially if the move will be during limited visibility conditions. The platoon or squad leader should also designate an alternate compass man.

(g) *Point/Pace Man*. As required, the PL designates a primary and alternate point man and a pace man for the patrol. The pace man aids in navigation by keeping an accurate count of distance traveled. The point man selects the actual route through the terrain, guided by the compass man or team leader. In addition, the point man also provides frontal security.

(2) Elements common to all combat patrols:

(a) *Assault Element*. The assault element seizes and secures the objective and protects special teams as they complete their assigned actions on the objective.

(b) *Security Element*. The security element provides security at danger areas, secures the ORP, isolates the objective, and supports the withdrawal of the rest of the patrol once actions on the objective are complete. The security element may have separate security teams, each with an assigned task or sequence of tasks.

(c) *Support Element*. The support element provides direct and indirect fire support for the unit. Direct fires include machine guns, medium and light antiarmor weapons, small recoilless rifles. Indirect fires available may include mortars, artillery, CAS, and organic M203 weapon systems.

(d) *Demolition Team*. Demolition teams are responsible for preparing and detonating the charges to destroy designated equipment, vehicles, or facilities on the objective.

(e) *EPW and Search Teams*. The assault element may provide two-man (buddy teams) or four-man (fire team) search teams to search bunkers, buildings, or tunnels on the objective. These teams will search the objective or kill zone for any PIR that may give the PL an idea of the enemy concept for future operations. Primary and alternate teams may be assigned to ensure enough prepared personnel are available on the objective.

(f) *Breach Element*. The breach team conducts initial penetration of enemy obstacles to seize a foothold and allow the patrol to enter an objective. This is typically done IAW METT-TC and the steps outlined in the "*Conduct an Initial Breach of a Mined Wire Obstacle*" battle drill in Chapter 6 of this Handbook.

(3) Elements common to all reconnaissance patrols:

(a) *Reconnaissance Team*. Reconnaissance teams reconnoiter the objective area once the security teams are in position. Normally these are two-man teams (buddy teams) to reduce the possibility of detection.

(b) *Reconnaissance and Security Teams*. R&S teams are normally used in a zone reconnaissance, but may be useful in any situation when it is impractical to separate the responsibilities for reconnaissance and security.

(c) *Security Element*. When the responsibilities of reconnaissance and security are separate, the security element provides security at danger areas, secures the ORP, isolates the objective, and supports the withdrawal of the rest of the platoon once the recon is complete. The security element may have separate security teams, each with an assigned task or sequence of tasks.

b. **Initial Planning and Coordination**. Leaders plan and prepare for patrols using the troop-leading procedures and the estimate of the situation, as described in Chapter 2. Through an estimate of the situation, leaders identify required actions on the objective (mission analysis) and plan backward to departure from friendly lines and forward to reentry of friendly lines. Because patrolling units act independently, move beyond the direct-fire support of the parent unit, and operate forward of friendly units, coordination must be thorough and detailed. Coordination is continuous throughout planning and preparation. PLs use checklists to preclude omitting any items vital to the accomplishment of the mission.

(1) ***Coordination with Higher Headquarters***. This coordination includes Intelligence, Operations, and Fire Support IAW Chapter 2, *Coordination Checklists* (**page 2-34**). This initial coordination is an integral part of Step 3 of Troop-Leading Procedures, *Make a Tentative Plan*.

(2) ***Coordination with Adjacent Units***. The leader also coordinates his unit's patrol activities with the leaders of other units that will be patrolling in adjacent areas at the same time, IAW Chapter 2-7, *Coordination Checklists* (**page 2-34**).

c. **Completion of Plan**. As the PL completes his plan, he considers--

(1) ***Specified and Implied Tasks***. The PL ensures that he has assigned all specified tasks to be performed on the objective, at rally points, at danger areas, at security or surveillance locations, along the route(s), and at passage lanes. These make up the maneuver and tasks to maneuver units subparagraphs of the Execution paragraph.

(2) ***Key travel and Execution Times***. The leader estimates time requirements for movement to the objective, leader's reconnaissance of the objective, establishment of security and surveillance, completion of all assigned tasks on the objective, and passage through friendly lines. Some planning factors are--

- Movement: Average of 1 Kmph during daylight hours in Woodland Terrain; Average limited visibility ½ Kmph. Add additional time for restrictive, or severely restrictive terrain such as mountains, swamps, or thick vegetation.
- Leader's recon: NLT 1.5 hr.
- Establishment of security and surveillance: 0.5 hr.

(3) ***Primary and Alternate Routes***. The leader selects primary and alternate routes to and from the objective. The return routes should differ from the routes to the objective. The PL may delegate route selection to a subordinate, but is ultimately responsible for the routes selected.

(4) ***Signals***. The leader should consider the use of special signals. These include hand-and-arm signals, flares, voice, whistles, radios, and infrared equipment. Primary and alternate signals must be identified and rehearsed so that all Soldiers know their meaning.

(5) ***Challenge and Password Forward of Friendly Lines***. The challenge and password from the unit's ANCD must not be used beyond the FLOT.

(a) *Odd-Number System*. The leader specifies an odd number. The challenge can be any number less than the specified number. The password will be the number that must be added to it to equal the specified number, for example, the number is 7, the challenge is 3, and the password is 4.

(b) *Running Password*. ANCDs may also designate a running password. This code word alerts a unit that friendly Soldiers are approaching in a less than organized manner and possibly under pressure. The number of Soldiers approaching follows the running password. For example, if the running password is "*Ranger*," and five friendly Soldiers are approaching, they would say "*Ranger five*."

(6) ***Location of Leaders***. The PL considers where he and the PSG and other key leaders are located during each phase of the mission. The PL positions himself where he can best control the actions of the patrol. The PSG is normally located with the assault element during a raid or attack to help the PL control the use of additional assaulting squads, and will assist with securing the OBJ. The PSG will locate himself at the CCP to facilitate casualty treatment and evacuation. During a reconnaissance mission, the PSG will stay behind in the ORP to facilitate the transfer of Intel to the higher headquarters, and control the recon elements movement into and out of the ORP.

(7) ***Actions on Enemy Contact***. Unless required by the mission, the unit avoids enemy contact. The leader's plan must address actions on chance contact at each phase of the patrol mission. The unit's ability to continue will depend on how early contact is made, whether the platoon is able to break contact successfully (so that its subsequent direction of movement is undetected), and whether the unit receives any casualties because of the contact. The plan must address the handling of seriously wounded Soldiers and KIAs. The plan must also address the handling of prisoners who are captured because of chance contact and are not part of the planned mission.

(8) ***Contingency Plans***. The leader leaves his unit for many reasons throughout the planning, coordination, preparation, and execution of his patrol mission. Each time the leader departs the patrol main body, he must issue a five-point contingency plan to the leader left in charge of the unit. The contingency plan is described by the acronym GOTWA, as follows. The patrol leader will additionally issue specific guidance stating what tasks are to be accomplished in the ORP in his absence:

- G: Where the leader is GOING.
- O: OTHERS he is taking with him.
- T: TIME he plans to be gone.
- W: WHAT to do if the leader does not return in time.
- A: The unit's and the leader's ACTIONS on chance contact while the leader is gone.

(9) ***Rally Points***. The leader considers the use and location of rally points. A rally point is a place designated by the leader where the unit moves to reassemble and reorganize if it becomes dispersed. Soldiers must know which rally point to move to at each phase of the patrol mission should they become separated from the unit. They must also know what actions are required there and how long they are to wait at each rally point before moving to another.

(a) *Criteria*. Rally points must be--

- Easily identifiable in daylight and limited visibility.
- Show no signs of recent enemy activity.
- Covered and concealed.
- Away from natural lines of drift and high-speed avenues of approach.
- Defendable for short periods of time.

(b) *Types*. The most common types of rally points are initial, en route, objective, and near-and-far-side rally points.

(10) ***Objective Rally Point***. The ORP is typically 200 to 400m from the objective, or at a minimum, one major terrain feature away. Actions at the ORP include--

- Conduct SLLS and pinpoint location.
- Leaders Recon of the Objective.
- Issuing a FRAGO, if needed.
- Making final preparations before continuing operations, for example, recamouflaging, preparing demolitions, lining up rucksacks for quick recovery. Preparing EPW bindings, first aid kits, litters, and inspecting weapons.
- Accounting for Soldiers and equipment after actions at the objective are complete.
- Reestablishing the chain of command after actions at the objective are complete.
- Disseminating information from reconnaissance, if contact was not made.

(11) ***Leader's reconnaissance of the objective***. The plan must include a leader's reconnaissance of the objective once the platoon or squad establishes the ORP. Before departing, the leader must issue a 5-point contingency plan. During his

reconnaissance, the leader pinpoints the objective, selects reconnaissance, security, support, and assault positions for his elements, and adjusts his plan based on his observation of the objective. Each type of patrol requires different tasks during the leader's reconnaissance. The platoon leader will bring different elements with him. (These are discussed separately under each type of patrol). The leader must plan time to return to the ORP, complete his plan, disseminate information, issue orders and instructions, and allow his squads to make any additional preparations. During the Leader's Reconnaissance for a Raid or Ambush, the PL will leave surveillance on the OBJ.

(12) ***Actions on the objective***. Each type of patrol requires different actions on the objective. Actions on the objective are discussed under each type of patrol.

5-3. **RECONNAISSANCE PATROLS**. Recon patrols are one of the two types of patrols. provide timely and accurate information on the enemy and terrain. They confirm the leader's plan before it is executed. Units on reconnaissance operations collect specific information (*priority intelligence requirements* [PIR]) or general information (*information requirements* [IR]) based on the instructions from their higher commander. The two types of recon patrols discussed here are area and zone. This section discusses the fundamentals of reconnaissance, task standards for the two most common types of recon, and actions on the objective for those types of recon.

a. **Fundamentals of Reconnaissance**. In order to have a successful area reconnaissance, the platoon leader must apply the fundamentals of the reconnaissance to his plan during the conduct of the operation.

(1) ***Gain all required information***. The parent unit tells the patrol leader (PL) what information is required. This is in the form of the IR and PIR. The platoon's mission is then tailored to what information is required. During the entire patrol, members must continuously gain and exchange all information gathered, but cannot consider the mission accomplished unless all PIR has been gathered.

(2) ***Avoid detection by the enemy***. A patrol must not let the enemy know that it is in the objective area. If the enemy knows he is being observed, he may move, change his plans, or increase his security measures. Methods of avoiding detection are--

(a) Minimize movement in the objective area (area reconnaissance).

(b) Move no closer to the enemy than necessary.

(c) If possible, use long-range surveillance or night vision devices.

(d) Camouflage, stealth, noise, and light discipline.

(e) Minimize radio traffic.

(3) ***Employ security measures***. A patrol must be able to break contact and return to the friendly unit with what information is gathered. If necessary, break contact and continue the mission. Security elements are emplaced so that they can overwatch the reconnaissance elements and suppress the enemy so the reconnaissance element can break contact.

(4) ***Task organization***. When the platoon leader receives the order, he analyzes his mission to ensure he understands what must be done. Then he task organizes his platoon to best accomplish the mission IAW METT-TC. Recons are typically squad-sized missions.

b. **Task Standards**.

(1) ***Area Reconnaissance***. The area recon patrol collects all available information on PIR and other intelligence not specified in the order for the area. The patrol completes the recon and reports all information by the time specified in the order. The patrol is not compromised.

(2) ***Zone Reconnaissance***. The zone recon patrol determines all PIR and other intelligence not specified in the order for its assigned zone. The patrol reconnoiters without detection by the enemy. The patrol completes the recon and reports all information by the time specified in the order.

c. **Actions on the Objective, Area Reconnaissance** (Figure 5-1).

(1) The element occupies the ORP as discussed in the section on occupation of the ORP. The RTO calls in spare for occupation of ORP. The leader confirms his location on map while subordinate leaders make necessary perimeter adjustments.

(2) The PL organizes the platoon in one of two ways: separate recon and security elements, or combined recon and security elements.

(3) The PL takes subordinates leaders and key personnel on a leader's recon to confirm the objective and plan.

(a) Issues a 5-point contingency plan before departure.

(b) Establishes a suitable release point that is beyond sight and sound of the objective if possible, but that is definitely out of sight. The RP should also have good rally point characteristics.

(c) Allow all personnel to become familiar with the release point and surrounding area.

(d) Identifies the objective and emplaces surveillance. Designates a surveillance team to keep the objective under surveillance. Issues a contingency plan to the senior man remaining with the surveillance team. The surveillance team is positioned with one man facing the objective, and one facing back in the direction of the release point.

(e) Takes subordinate leaders forward to pinpoint the objective, emplace surveillance, establish a limit of advance, and choose vantage points.

(f) Maintains commo with the platoon throughout the leader's recon.

(4) The PSG maintains security and supervises priorities of work in the ORP.

(a) Reestablishes security at the ORP.

(b) Disseminates the PLs contingency plan.

(c) Oversees preparation of recon personnel (personnel recamouflaged, NVDs and binos prepared, weapons on safe with a round in the chamber).

(5) The PL and his recon party return to the ORP.

(a) Confirms the plan or issues a FRAGO.

(b) Allows subordinate leaders time to disseminate the plan.

(6) The patrol conducts the recon by long-range observation and surveillance if possible.

(a) R&S elements move to observation points that offer cover and concealment and that are outside of small-arms range.

(b) Establishes a series of observation posts (OP) if information cannot be gathered from one location.

(c) Gathers all PIR using the SALUTE format.

(7) If necessary, the patrol conducts its recon by short-range observation and surveillance.

(a) Moves to an OP near the objective.

(b) Passes close enough to the objective to gain information.

(c) Gathers all PIR using the SALUTE format.

(8) R&S teams move using a technique such as the cloverleaf method to move to successive OP's. In this method, R&S teams avoid paralleling the objective site, maintain extreme stealth, do not cross the limit of advance, and maximize the use of available cover and concealment.

(9) During the conduct of the recon, each R&S team will return to the release point when any of the following occurs:

- They have gathered all their PIR.
- They have reached the limit of advance.
- The allocated time to conduct the recon has elapsed.
- Contact has been made.

(10) At the release point, the leader will analyze what information has been gathered and determine if he has met the PIR requirements.

(11) If the leader determines that he has not gathered sufficient information to meet the PIR requirements, or if the information he and the subordinate leader gathered differs drastically, he may have to send R&S teams back to the objective site. In this case, R&S teams will alternate areas of responsibilities. For example, if one team reconnoitered from the 6 – 3 – 12, then that team will now recon from the 6 – 9 – 12.

(12) The R&S element returns undetected to the ORP by the specified time.

(a) Disseminates information to all patrol members through key leaders at the ORP, or moves to a position at least one terrain feature or one kilometer away to disseminate. To disseminate, the leader has the RTO prepare three sketches of the objective site based on the leader's sketch and provides the copies to the subordinate leaders to assist in dissemination.

(b) Reports any information requirements and/or any information requiring immediate attention to higher headquarters, and departs for the designated area.

(13) If contact is made, move to the release point. The recon element tries to break contact and return to the ORP, secure rucksacks, and quickly move out of the area. Once they have moved a safe distance away, the leader will inform higher HQ of the situation and take further instructions from them.

(a) While emplacing surveillance, the recon element withdraws through the release point to the ORP, and follows the same procedures as above.

(b) While conducting the reconnaissance, the compromised element returns a sufficient volume of fire to allow them to break contact. Surveillance can fire an AT-4 at the largest weapon on the objective. All elements will pull off the objective and move to the release point. The senior man will quickly account for all personnel and return to the ORP. Once in the ORP, follow the procedures previously described.

Figure 5-1. ACTIONS ON THE OBJECTIVE, AREA RECONNAISSANCE

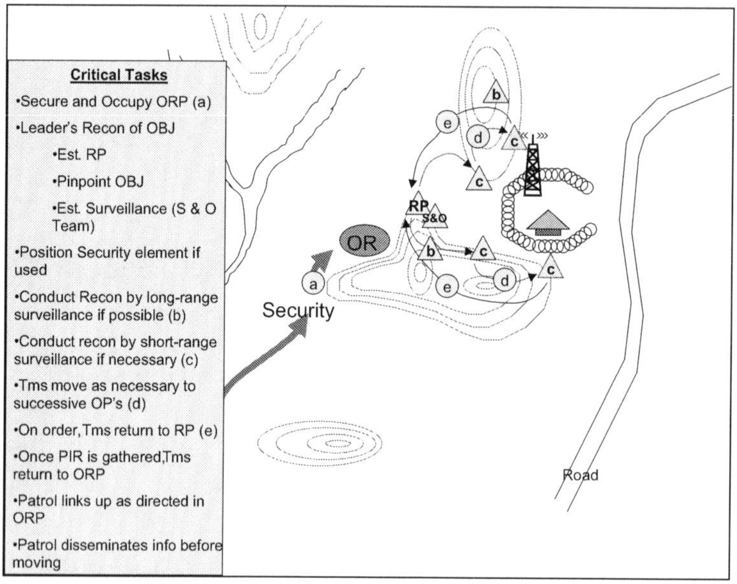

d. **Actions on the Objective, Zone Reconnaissance**.

(1) The element occupies the initial ORP as discussed in the section occupation of the ORP. The radio operator calls in spare for occupation of ORP. The leader confirms his location on map while subordinate leaders make necessary perimeter adjustments.

(2) The recon team leaders organize their recon elements.

(a) Designate security and recon elements.

(b) Assign responsibilities (point man, pace man, en route recorder, and rear security), if not already assigned.

(c) Designates easily recognizable rally points.

(d) Ensure local security at all halts.

(3) The patrol recons the zone.

(a) Moves tactically to the ORPs.

(b) Occupies designated ORPs.

(c) Follows the method designated by the PL (fan, converging routes, or box method, Figure 5-2).

Figure 5-2. COMPARISON OF ZONE RECONNAISSANCE METHODS

FAN METHOD	*CONVERGING ROUTES METHOD*	*BOX METHOD*
• Uses a series (fan) of ORPs. • Patrol establishes security at first ORP. • Each recon element moves from ORP along a different fan-shaped route. Route overlaps with that of other recon elements. This ensures recon of entire area. • Leader maintains reserve at ORP. • When all recon elements return to ORP, PL collects and disseminates all info before moving to next ORP.	• PL selects routes from ORP thru zone to a linkup point at the far side of the zone from the ORP. Each recon element moves and recons along a specified route. They converge (link up) at one time and place.	• PL sends recon elements from the first ORP along routes that form a box. He sends other elements along routes throughout the box. All teams link up at the far side of the box from the ORP.

(d) The recon teams reconnoiter.

- During movement, the squad will gather all PIR specified by the order.
- Recon team leaders will ensure sketches are drawn or digital photos are taken of all enemy hard sites, roads, and trails.
- Return to the ORP, or link up at the rendezvous point on time.
- When the squad arrives at new rendezvous point or ORP, the recon team leaders report to the PL with all information gathered.

(e) The PL continues to control the recon elements.

- PL moves with the recon element that establishes the linkup point.
- PL changes recon methods as required.
- PL designates times for the elements to return to the ORP or to linkup.
- PL collects all information and disseminates it to the entire patrol. PL will brief all key subordinate leaders on information gathered by other squads, establishing one consolidated sketch if possible, and allow team leaders time to brief their teams.
- PL and PSG account for all personnel.

(f) The patrol continues the recon until all designated areas have been reconned, and returns undetected to friendly lines.

5-4. **COMBAT PATROLS**. Combat patrols are the second type of patrol. Combat patrols are further divided into raids and ambushes Units conduct combat patrols to destroy or capture enemy Soldiers or equipment; destroy installations, facilities, or key points; or harass enemy forces. Combat patrols also provide security for larger units. This section describes overall combat patrol planning considerations, task considerations for each type of combat patrol, and finally actions on the objective for each type.

a. **Planning Considerations (General)**. In planning a combat patrol, the PL considers the following:

(1) ***Tasks to Maneuver Units***. Normally the platoon headquarters element controls the patrol on a combat patrol mission. The PL makes every try to maintain squad and fire team integrity as he assigns tasks to subordinates units.

(a) The PL must consider the requirements for assaulting the objective, supporting the assault by fire, and security of the entire unit throughout the mission.

- For the assault on the objective, the PL considers the required actions on the objective, the size of the objective, and the known or presumed strength and disposition of the enemy on and near the objective.
- The PL considers the weapons available, and the type and volume of fires required to provide fire support for the assault on the objective.
- The PL considers the requirement to secure the platoon at points along the route, at danger areas, at the ORP, along enemy avenues of approach into the objective, and elsewhere during the mission.
- The PL will also designate engagement/disengagement criteria.

(b) The PL assigns additional tasks to his squads for demolition, search of EPWs, guarding of EPWs, treatment and evacuation (litter teams) of friendly casualties, and other tasks required for successful completion of patrol mission (if not already in the SOP).

(c) The PL determines who will control any attachments of skilled personnel or special equipment.

(2) ***Leader's Reconnaissance of the Objective***. In a combat patrol, the PL has additional considerations for the conduct of his reconnaissance of the objective from the ORP.

(a) *Composition of the leader's reconnaissance party.* The platoon leader will normally bring the following personnel.

- Squad leaders to include the weapons squad leader.
- Surveillance team.
- Forward observer.
- Security element (dependent on time available).

(b) *Conduct of the leader's reconnaissance.* In a combat patrol, the PL considers the following additional actions in the conduct of the leader's reconnaissance of the objective.

- The PL designates a release point approximately half way between the ORP and this objective. The PL posts the surveillance team. Squads and fire teams separate at the release point, and then they move to their assigned positions.
- The PL confirms the location of the objective or kill zone. He notes the terrain and identifies where he can emplace claymores to cover dead space. Any change to his plan is issued to the squad leaders (while overlooking the objective if possible).
- If the objective is the kill zone for an ambush, the leader's reconnaissance party should not cross the objective; to do so will leave tracks that may compromise the mission.
- The PL confirms the suitability of the assault and support positions and routes from them back to the ORP.
- The PL issues a five-point contingency plan before returning to the ORP.

b. **Ambush**.

(1) ***Planning Considerations***. An ambush is a surprise attack from a concealed position on a moving or temporarily halted target. Ambushes are classified by category--hasty or deliberate; type--point or area; and formation--linear or L-shaped. The leader uses a combination of category, type, and formation in developing his ambush plan. The key planning considerations include--

(a) Coverage of entire kill zone by fire.

(b) METT-TC.

(c) Use of existing or reinforcing obstacles, including Claymores, to keep the enemy in the kill zone.

(d) Security teams (typically equipped with hand-held antitank weapons such as AT-4 or LAW; Claymores; and various means of communication.

(e) Protect the assault and support elements with claymores or explosives.

(f) Use security elements or teams to isolate the kill zone.

(g) Assault through the kill zone to the limit of advance (LOA). (The assault element must be able to move quickly through its own protective obstacles.)

(h) Time the actions of all elements of the platoon to preclude loss of surprise. In the event any member of the ambush is compromised, he may immediately initiate the ambush.

(i) When the ambush must be manned for a long time, use only one squad to conduct the entire ambush and determining movement time of rotating squads from the ORP to the ambush site.

(2) ***Categories***.

(a) *Hasty Ambush*. A unit conducts a hasty ambush when it makes visual contact with an enemy force and has time to establish an ambush without being detected. The actions for a hasty ambush must be well rehearsed so that Soldiers know what to do on the leader's signal. They must also know what action to take if the unit is detected before it is ready to initiate the ambush.

(b) *Deliberate Ambush*. A deliberate ambush is conducted at a predetermined location against any enemy element that meets the commander's engagement criteria. The leader requires the following detailed information in planning a deliberate ambush: size and composition of the targeted enemy, and weapons and equipment available to the enemy.

(3) ***Types***.

(a) Point ambush. In a point ambush, Soldiers deploy to attack an enemy in a single kill zone.

(b) Area ambush. In an area, Soldiers deploy in two or more related point ambushes.

(4) ***Formations***. (Figure 5-3).

(a) *Linear Ambush*. In an ambush using a linear formation, the assault and support elements deploy parallel to the enemy's route. This positions both elements on the long axis of the kill zone and subjects the enemy to flanking fire. This formation can be used in close terrain that restricts the enemy's ability to maneuver against the platoon, or in open terrain provided a means of keeping the enemy in the kill zone can be effected.

(b) *L-Shaped Ambush*. In an L-shaped ambush, the assault element forms the long leg parallel to the enemy's direction of movement along the kill zone. The support element forms the short leg at one end of and at right angles to the assault element. This provides both flanking (long leg) and enfilading fires (short leg) against the enemy. The L-shaped ambush can be used at a sharp bend in a trail, road, or stream. It should not be used where the short leg would have to cross a straight road or trail.

Figure 5-3. AMBUSH FORMATIONS

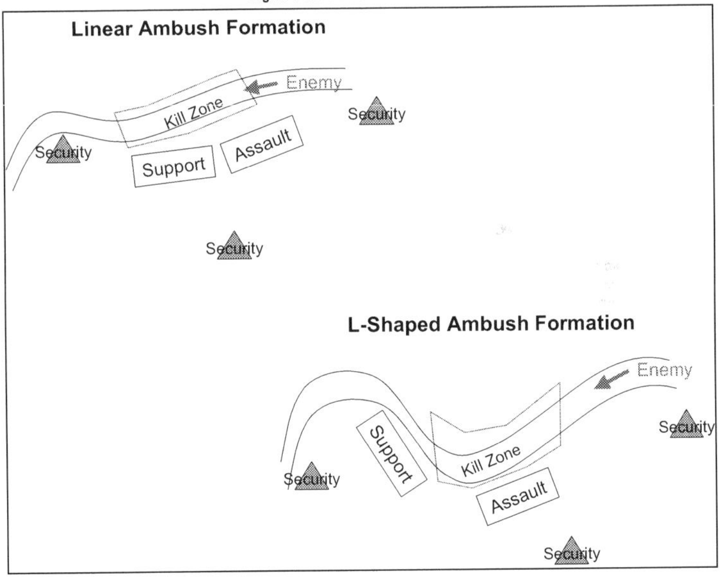

c. **Hasty Ambush**.

(1) ***Task Standards***. The platoon moves quickly to concealed positions. The ambush is not initiated until the majority of the enemy is in the kill zone. The unit does not become decisively engaged. The platoon surprises the enemy. The patrol captures, kills, or forces the withdrawal of all of the enemy within the kill zone. On order, the patrol withdraws all personnel and equipment in the kill zone from observation and direct fire. The unit does not become decisively engaged by follow-on elements. The platoon continues follow-on operations.

(2) ***Actions on the Objective (Hasty Ambush)***. See Figure 5-4.

(a) Using visual signals, any Soldier alerts the unit that an enemy force is in sight. The Soldier continues to monitor the location and activities of the enemy force until his team or squad leader relieves him, and gives the enemy location and direction of movement.

(b) The platoon or squad halts and remains motionless.

- The PL gives the signal to conduct a hasty ambush, taking care not to alert the enemy of the patrol's presence.
- The leader determines the best nearby location for a hasty ambush. He uses arm-and-hand signals to direct the unit members to covered and concealed positions.

(c) The leader designates the location and extent of the kill zone.

d) Teams and squads move silently to covered and concealed positions, ensuring positions are undetected and have good observation and fields of fire into the kill zone.

(e) Security elements move out to cover each flank and the rear of the unit. The leader directs the security elements to move a given distance, set up, and then rejoin the unit on order or, after the ambush (the sound of firing ceases). At squad level, the two outside buddy teams normally provide flank security as well as fires into the kill zone. At platoon level, fire teams make up the security elements.

(f) The PL assigns sectors of fire and issues any other commands necessary such as control measures.

(g) The PL initiates the ambush, using the greatest casualty-producing weapon available, when the largest percentage of enemy is in the kill zone. The PL--

- Controls the rate and distribution of fire.
- Employs indirect fire to support the ambush.
- Orders cease fire.
- (If the situation dictates) Orders the patrol to assault through the kill zone.

(h) The PL designates personnel to conduct a hasty search of enemy personnel and process enemy prisoners and equipment.

(i) The PL orders the platoon to withdraw from the ambush site along a covered and concealed route.

(j) The PL gains accountability, reorganizes as necessary, disseminates information, reports the situation, and continues the mission as directed.

d. **Deliberate (Point/Area) Ambush**.

(1) ***Task Standards***. The ambush is emplaced NLT the time specified in the order. The patrol surprises the enemy and engages the enemy main body. The patrol kills or captures all enemy in the kill zone and destroys equipment based on the commander's intent. The patrol withdraws all personnel and equipment from the objective, on order, within the time specified in the order. The patrol obtains all available PIR from the ambush and continues follow-on operations.

(2) ***Actions on the Objective (Deliberate Ambush).*** See Figure 5-5.

(a) The PL prepares the patrol for the ambush in the ORP.

(b) The PL prepares to conduct a leader's reconnaissance. He--

- Designates the members of the leader's recon party (typically includes squad leaders, surveillance team, FO, and possibly the security element.
- Issues a contingency plan to the PSG.

Figure 5-4. ACTIONS ON THE OBJECTIVE -- HASTY AMBUSH

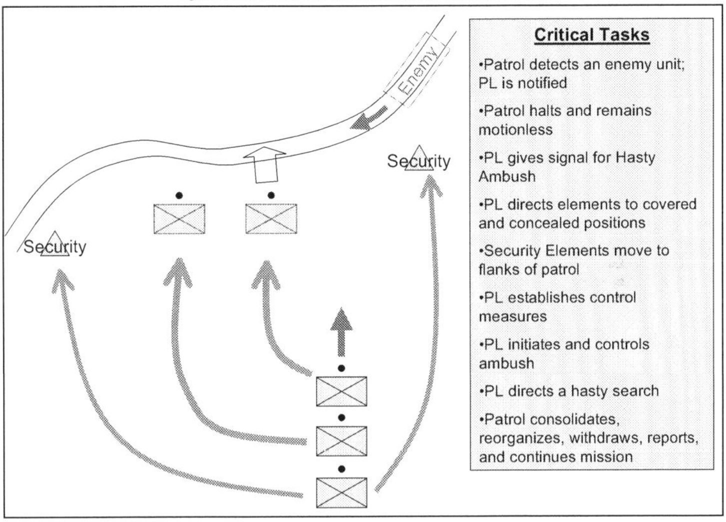

(c) The PL conducts his leader's reconnaissance. He--

- Ensures the leader's recon party moves undetected.
- Confirms the objective location and suitability for the ambush.
- Selects a kill zone.
- Posts the surveillance team at the site and issues a contingency plan.
- Confirms suitability of assault and support positions, and routes from them to the ORP.
- Selects the position of each weapon system in the support-by-fire position, and then designates sectors of fire.
- Identifies all offensive control measures to be used. Identifies the PLD, the assault position, LOA, any boundaries or other control measures. If available, the PL can use infrared aiming devices to identify these positions on the ground.

(d) The PL adjusts his plan based on info from the reconnaissance. He--

- Assigns positions.
- Designates withdrawal routes.

(e) The PL confirms the ambush formation.

(f) The security team(s) occupy first, securing the flanks of the ambush site, and providing early warning. The security element must be in position before the support and assault elements move forward of the release point. A security team remains in the ORP if the patrol plans to return to the ORP after actions on the objective. If the ORP is abandoned, a rear security team should be emplaced.

(g) Support element leader assigns sectors of fire. He--

- Emplaces claymores and obstacles as designated.
- Identifies sectors of fire and emplaces limiting stakes to prevent friendly fires from hitting other elements.
- Overwatches the movement of the assault element into position.

(h) Once the support element is in position, or on the PLs order, the assault element--

- Departs the ORP and moves into position.
- Upon reaching the PLD, the assault element transitions from the movement formation to the battle formation.
- Identifies individual sectors of fire as assigned by the PL. Emplaces aiming stakes or uses metal-to-metal contact with the machine gun tripods to prevent fratricide on the objective.
- Emplaces claymores to help destroy the enemy in the kill zone.
- Camouflages positions.

(i) The security element spots the enemy and notifies the PL, and reports the direction of movement, size of the target, and any special weapons or equipment carried. The security element also keeps the platoon leader informed if any enemy forces are following the lead force.

(j) The PL alerts other elements, and determines if the enemy force is too large, or if the ambush can engage the enemy successfully.

(k) The PL initiates the ambush using the highest casualty-producing device. He may use a command-detonated claymore. He must also plan a backup method for initiating the ambush, in case his primary means fails. This should also be a casualty-producing device such as his individual weapon. He passes this information to all Rangers, and practices it during rehearsals.

(l) The PL ensures that the assault and support elements deliver fire with the heaviest, most accurate volume possible on the enemy in the kill zone. In limited visibility, the PL may use infrared lasers to further define specific targets in the kill zone.

(m) Before assaulting the target, the PL gives the signal to lift or shift fires.

(n) The assault element--

- Assaults before the remaining enemy can react.
- Kills or captures enemy in the kill zone.
- Uses individual movement techniques or bounds by fire teams to move.
- Upon reaching the limit of advance, halts and establishes security. If needed, it reestablishes the chain of command and remains key weapon systems. All Soldiers will load a fresh magazine or drum of ammunition using the buddy system. ACE reports will be submitted through the chain of command. The PL will submit an initial contact report to higher.

(o) The PL directs special teams (EPW search, aid and litter, demo) to accomplish their assigned task once the assault element has established its LOA.

- Once the kill zone had been cleared, collect and secure all EPWs and move them out of the kill zone before searching bodies. Coordinate for an EPW exchange point to link up with higher to extract all EPWs and treat them IAW the five S's.
- Search from one side to the other and mark bodies that have been searched to ensure the area is thoroughly covered. Units should use the "*clear out, search in*" technique, clear from the center of the objective out ensuring the area is clear of all enemy combatants; then search all enemy personnel towards the center of the objective. Search all dead enemy personnel using two-man search techniques.
  -- As the search team approaches a dead enemy Soldier, one-man guards while the other man searches. First, he kicks the enemy weapon away.
  -- Second, he rolls the body over (if on the stomach) by lying on top and when given the go ahead by the guard (who is positioned at the enemy's head), the searcher rolls the body over on him. This is done for protection in case the enemy Soldier has a grenade with the pin pulled underneath him.
  -- The searchers then conduct a systematic search of the dead Soldier from head to toe removing all papers and anything new (different type rank, shoulder boards, different unit patch, pistol, weapon, or NVD). They note if the enemy has a fresh or shabby haircut and the condition of his uniform and boots. They note the radio frequency, and then they secure the SOI, maps, documents, and overlays.
  -- Once the body has been thoroughly searched, the search team will continue in this manner until all enemy personnel in and near the kill zone have been searched.
- Identify, collect, and prepare all equipment to be carried back or destroyed.
- Evacuate and treat friendly wounded first, then enemy wounded, time permitting.

- The demolition team prepares dual-primed explosives or incendiary grenades and awaits the signal to initiate. This is normally the last action performed before the unit departs the objective and may signal the security elements to return to the ORP.
- Actions on the objective with stationary assault line; all actions are the same with the exception of the search teams. To provide security within the teams to the far side of the kill zone during the search, they work in three-Ranger teams. Before the search begins, the Rangers move all KIAs to the near side of the kill zone.

(p) If enemy reinforcements try to penetrate the kill zone, the flank security will engage to prevent the assault element from being compromised.

(q) The platoon leader directs the unit's withdrawal from the ambush site:

- Elements normally withdraw in the reverse order that they established their positions.
- The elements may return to the RP or directly to the ORP, depending on the distance between elements.
- The security element of the ORP must be alert to assist the platoon's return to the ORP. It maintains security for the ORP while the rest of the platoon prepares to leave.
- If possible, all elements should return to the location at which they separated from the main body. This location should usually be the RP.

(r) The PL and PSG direct actions at the ORP, to include accountability of personnel and equipment and recovery of rucksacks and other equipment left at the ORP during the ambush.

(s) The platoon leader disseminates information, or moves the platoon to a safe location (no less than one kilometer or one terrain feature away from the objective) and disseminates information.

(t) As required, the PL and FO execute indirect fires to cover the platoon's withdrawal.

Figure 5-5. ACTIONS ON THE OBJECTIVE -- DELIBERATE AMBUSH

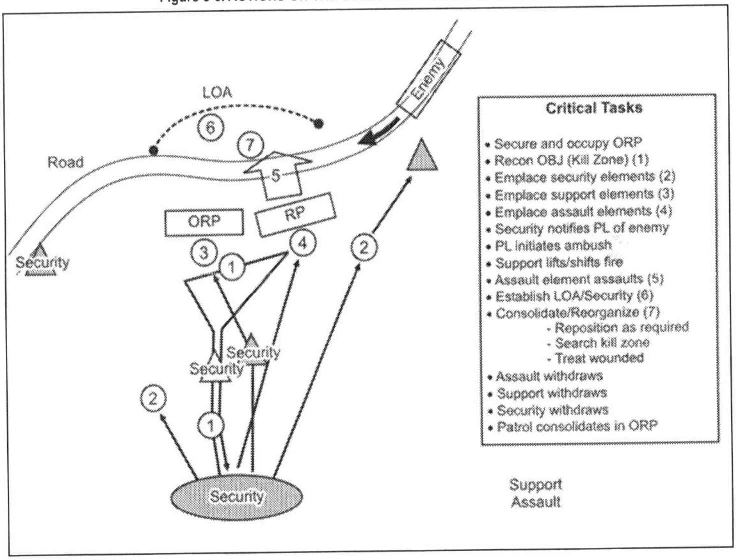

e. **Perform Raid**. The patrol initiates the raid NLT the time specified in the order, surprises the enemy, assaults the objective, and accomplishes its assigned mission within the commander's intent. The patrol does not become decisively engaged en route to the objective. The patrol obtains all available PIR from the raid objective and continues follow-on operations.

(1) ***Planning Considerations***. A Raid is a form of attack, usually small scale, involving a swift entry into hostile territory to secure information, confuse the enemy, or destroy installations followed by a planned withdrawal. Squads do not conduct raids. The sequence of platoon actions for a raid is similar to those for an ambush. Additionally, the assault element of the platoon may have to conduct a breach of an obstacle. It may have additional tasks to perform on the objective such as demolition of fixed facilities. Fundamentals of the raid include--

- Surprise and speed. Infiltrate and surprise the enemy without being detected.
- Coordinated fires. Seal off the objective with well-synchronized direct and indirect fires.
- Violence of action. Overwhelm the enemy with fire and maneuver.
- Planned withdrawal. Withdraw from the objective in an organized manner, maintaining security.

(2) ***Actions on the Objective (Raid)***. See Figure 5-6.

(a) The patrol moves to and occupies the ORP IAW the patrol SOP. The patrol prepares for the leader's recon.

(b) The PL, squad leaders, and selected personnel conduct a leader's recon.

- PL leaves a five point contingency plan with the PSG.
- PL establishes the RP, pinpoints the objective, contacts the PSG to prep men, weapons, and equipment, emplaces the surveillance team to observe the objective, and verifies and updates intelligence information.
- Leader's recon verifies location of and routes to security, support, and assault positions.
- Security teams are brought forward on the leader's reconnaissance and emplaced before the leader's recon leaves the RP
- Leaders conduct the recon without compromising the patrol.
- Leaders normally recon support by fire position first, then the assault position.

(c) The PL confirms, denies, or modifies his plan and issues instructions to his squad leaders.

- Assigns positions and withdrawal routes to all elements.
- Designates control measures on the objective (element objectives, lanes, limits of advance, target reference points, and assault line).
- Allows SLs time to disseminate information, and confirm that their elements are ready.

(d) Security elements occupy designated positions, moving undetected into positions that provide early warning and can seal off the objective from outside support or reinforcement.

(e) The support element leader moves the support element to designated positions. The support element leader ensures his element can place well-aimed fire on the objective.

(f) The PL moves with the assault element into the assault position. The assault position is normally the last covered and concealed position before reaching the objective. As it passes through the assault position the platoon deploys into its assault formation; that is, its squads and fire teams deploy to place the bulk of their firepower to the front as they assault the objective.

- Makes contact with the surveillance team to confirm any enemy activity on the objective.
- Ensures that the assault position is close enough for immediate assault if the assault element is detected early.
- Moves into position undetected, and establish local security and fire control measures.

(g) Element leaders inform the PL when their elements are in position and ready.

(h) The PL directs the support element to fire.

(i) Upon gaining fire superiority, the PL directs the assault element to move towards the objective.

- Assault element holds fire until engaged, or until ready to penetrate the objective.
- PL signals the support element to lift or shift fires. The support element lifts or shifts fires as directed, shifting fire to the flanks of targets or areas as directed in the FRAGO.

(j) The assault element attacks and secures the objective. The assault element may be required to breech a wire obstacle. As the platoon, or its assault element, moves onto the objective, it must increase the volume and accuracy of fires. Squad leaders assign specific targets or objectives for their fire teams. Only when these direct fires keep the enemy suppressed can the rest of the unit maneuver. As the assault element gets closer to the enemy, there is more emphasis on suppression and less on maneuver. Ultimately, all but one fire team may be suppressing to allow that one fire team to break into the enemy position. Throughout the assault, Soldiers use proper individual movement techniques, and fire teams retain their basic shallow wedge formation. The platoon does not get "*on-line*" to sweep across the objective.

- Assault element assaults through the objective to the designated LOA.
- Assault element leaders establish local security along the LOA, and consolidate and reorganize as necessary. They provide ACE reports to the PL and PSG. The platoon establishes security, operates key weapons, provides first aid, and prepares wounded Soldiers for MEDEVAC. They redistribute ammunition and supplies, and they relocate selected weapons to alternate positions if leaders believe that the enemy may have pinpointed them during the attack. They adjust other positions for mutual support. The squad and team leader provide ammunition, casualty, and equipment (ACE) reports to the platoon leader. The PL/PSG reorganizes the patrol based on the contact.
  -- On order, special teams accomplish all assigned tasks under the supervision of the PL, who positions himself where he can control the patrol.
  -- Special team leaders report to PL when assigned tasks are complete.

(k) On order or signal of the PL, the assault element withdraws from the objective. Using prearranged signals, the assault line begins an organized withdrawal from the objective site, maintaining control and security throughout the withdrawal. The assault element bounds back near the original assault line, and begins a single file withdrawal through the APL's choke point. All Rangers must move through the choke point for an accurate count. Once the assault element is a safe distance from the objective and the headcount is confirmed, the platoon can withdraw the support element. If the support elements were a part of the assault line, they withdraw together, and security is signaled to withdraw. Once the support is a safe distance off the objective, they notify the platoon leader, who contacts the security element and signals them to withdraw. All security teams link up at the release point and notify the platoon leader before moving to the ORP. Personnel returning to the ORP immediately secure their equipment and establish all-round security. Once the security element returns, the platoon moves out of the objective area as soon as possible, normally in two to three minutes.

- Before withdrawing, the demo team activates demo devices and charges.
- Support element or designated personnel in the assault element maintain local security during the withdrawal.
- Leaders report updated accountability and status (ACE report) to the PL and PSG.

(l) Squads withdraw from the objective in the order designated in the FRAGO to the ORP.

- Account for personnel and equipment.
- Disseminate information.
- Redistribute ammunition and equipment as required.

(m) The PL reports mission accomplishment to higher and continues the mission.

- Reports raid assessment to higher.
- Informs higher of any IR/PIR gathered.

NOTES

Figure 5-6. ACTIONS ON THE OBJECTIVE -- RAID

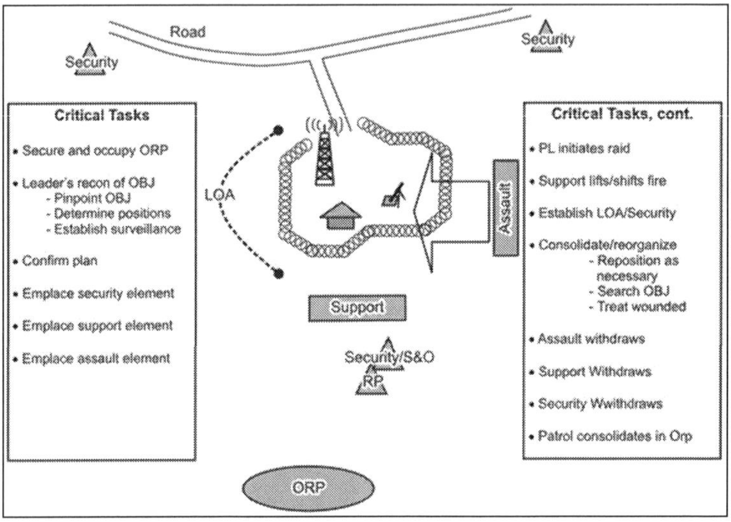

5-5. **SUPPORTING TASKS**. This section covers Linkup, Patrol Debriefing, and Occupation of an ORP.

a. **Linkup**. A linkup is a meeting of friendly ground forces. Linkups depend on control, detailed planning, communications, and stealth.

(1) ***Task Standard***. The units link up at the time and place specified in the order. The enemy does not surprise the main bodies. The linkup units establish a consolidated chain of command.

(2) ***Linkup Site Selection***. The leader identifies a tentative linkup site by map reconnaissance, other imagery, or higher headquarters designates a linkup site. The linkup site should have the following characteristics:

- Ease of recognition.
- Cover and concealment.
- No tactical value to the enemy.
- Location away from natural lines of drift.
- Defendable for a short period of time.
- N=Multiple access and escape routes.

(3) ***Execution***. Linkup procedure begins as the unit moves to the linkup point. The steps of this procedure are:

(a) The stationary unit performs linkup actions.

- Occupies the linkup rally point (LRP) NLT the time specified in the order.
- Establishes all-around security, establishes commo, and prepares to accept the moving unit.
- The security team clears the immediate area around the linkup point. It then marks the linkup point with the coordinated recognition signal. The security team moves to a covered and concealed position and observes the linkup point and immediate area around it.

(b) The moving unit--
- Performs linkup actions.
- The unit reports its location using phase lines, checkpoints, or other control measures.
- Halts at a safe distance from the linkup point in a covered and concealed position (the linkup rally point).

(c) The PL and a contact team--
- Prepare to make physical contact with the stationary unit.
- Issue a contingency plan to the PSG.
- Maintain commo with the platoon; verify near and far recognition signals for linkup (good visibility and limited visibility).
- Exchange far and near recognition signals with the linkup unit; conduct final coordination with the linkup unit.

(d) The stationary unit--
- Guides the patrol from its linkup rally point to the stationary unit linkup rally point.
- Linkup is complete by the time specified in the order.
- The main body of the stationary unit is alerted before the moving unit is brought forward.

(e) The patrol continues its mission IAW the order.

(4) ***Coordination Checklist***. The PL coordinates or obtains the following information from the unit that his patrol will link up with:
- Exchange frequencies, call signs, codes, and other communication information.
- Verify near and far recognition signals.
- Exchange fire coordination measures.
- Determine command relationship with the linkup unit; plan for consolidation of chain of command.
- Plan actions following linkup.
- Exchange control measures such as contact points, phase lines, contact points, as appropriate.

b. **Debrief**. Immediately after the platoon or squad returns, personnel from higher headquarters conduct a thorough debrief. This may include all members of the platoon or the leaders, RTOs, and any attached personnel. Normally the debriefing is oral. Sometimes a written report is required. Information on the written report should include--
- Size and composition of the unit conducting the patrol.
- Mission of the platoon such as type of patrol, location, and purpose.
- Departure and return times.
- Routes. Use checkpoints, grid coordinates for each leg or include an overlay.
- Detailed description of terrain and enemy positions that were identified.
- Results of any contact with the enemy.
- Unit status at the conclusion of the patrol mission, including the disposition of dead or wounded Soldiers.
- Conclusions or recommendations.

c. **Objective Rally Point**. The ORP is a point out of sight, sound, and small arms range of the objective area. It is normally located in the direction that the platoon plans to move after completion of actions on the objective. The ORP is tentative until the objective is pinpointed.

(1) ***Occupation of the ORP***. See Figure 5-7.

(a) The patrol halts beyond sight and sound of the tentative ORP (200-400m in good visibility, 100-200m in limited visibility).

(b) The patrol establishes a security halt IAW the unit SOP.

(c) After issuing a five-point contingency plan to the PSG, the PL moves forward with a recon element to conduct a leader's recon of the ORP.

(d) For a squad-sized patrol, the PL moves forward with a compass man and one member of each fire team to confirm the ORP.
- After physically clearing the ORP location, the PL leaves two Rangers at the 6 o'clock position facing in opposite directions.
- The PL issues a contingency plan and returns with the compass man to guide the patrol forward.
- The PL guides the patrol forward into the ORP, with one team occupying from 3 o'clock through 12 o'clock to 9 o'clock, and the other occupying from 9 o'clock through 6 o'clock to 3 o'clock.

(e) For a platoon-size patrol, the PL, RTO, WSL, three ammo bearers, a team leader, a SAW gunner, and riflemen go on the leaders recon for the ORP and position themselves at 10, 2, and 6 o'clock.

- The first squad in the order of march is the base squad, occupying from 10 to 2 o'clock.
- The trail squads occupy from 2 to 6 o'clock and 6 to 10 o'clock, respectively.
- The patrol headquarters element occupies the center of the triangle.

(2) ***Actions in the ORP***. The unit prepares for the mission in the ORP. Once the leader's recon pinpoints the objective, the PSG generally lines up rucks IAW unit SOP in the center of the ORP.

Figure 5-7. OCCUPATION OF THE ORP

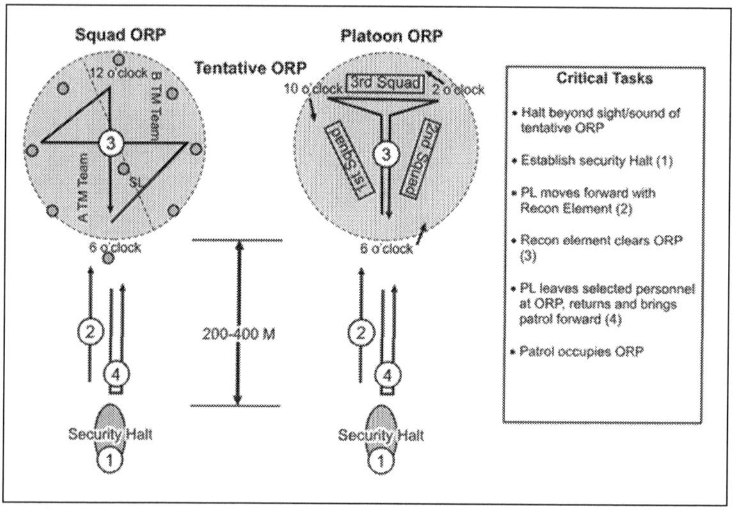

d. **Patrol Base**. A patrol base is a security perimeter that is set up when a squad or platoon conducting a patrol halts for an extended period. Patrol bases should not be occupied for more than a 24-hour period (except in emergency). A patrol never uses the same patrol base twice.

(1) ***Use***. Patrol bases are typically used--

- To avoid detection by eliminating movement.
- To hide a unit during a long detailed reconnaissance.
- To perform maintenance on weapons, equipment, eat and rest.
- To plan and issue orders.
- To reorganize after infiltrating on an enemy area.
- To establish a base from which to execute several consecutive or concurrent operations.

(2) ***Site Selection***. The leader selects the tentative site from a map or by aerial reconnaissance. The site's suitability must be confirmed and secured before the unit moves into it. Plans to establish a patrol base must include selecting an alternate patrol base site. The alternate site is used if the first site is unsuitable or if the patrol must unexpectedly evacuate the first patrol base.

(3) ***Planning Considerations***. Leaders planning for a patrol base must consider the mission and passive and active security measures. A patrol base (PB) must be located so it allows the unit to accomplish its mission.

- Observation posts and communication with observation posts.
- Patrol or platoon fire plan.
- Alert plan.
- Withdrawal plan from the patrol base to include withdrawal routes and a rally point, rendezvous point, or alternate patrol base.
- A security system to make sure that specific Soldiers are awake at all times.
- Enforcement of camouflage, noise, and light discipline.
- The conduct of required activities with minimum movement and noise.
- Priorities of Work.

(4) ***Security Measures***.

- Select terrain the enemy would probably consider of little tactical value.
- Select terrain that is off main lines of drift.
- Select difficult terrain that would impede foot movement, such as an area of dense vegetation, preferably bushes and trees that spread close to the ground.
- Select terrain near a source of water.
- Select terrain that can be defended for a short period and that offers good cover and concealment.
- Avoid known or suspected enemy positions.
- Avoid built-up areas.
- Avoid ridges and hilltops, except as needed for maintaining communications.
- Avoid small valleys.
- Avoid roads and trails.

(5) ***Occupation***. See Figure 5-8.

(a) A PB is reconned and occupied in the same manner as an ORP, with the exception that the platoon will typically plan to enter at a 90 degree turn. The PL leaves a two-man OP at the turn, and the patrol covers any tracks from the turn to the PB.

(b) The platoon moves into the PB. Squad-sized patrols will generally occupy a cigar-shaped perimeter; platoon-sized patrols will generally occupy a triangle-shaped perimeter.

(c) The PL and another designated leader inspect and adjust the entire perimeter as necessary.

(d) After the PL has checked each squad sector, each SL sends a two-man R&S team to the PL at the CP. The PL issues the three R&S teams a contingency plan, reconnaissance method, and detailed guidance on what to look for (enemy, water, built-up areas or human habitat, roads, trails, or possible rally points).

(e) Where each R&S team departs is based on the PLs guidance. The R&S team moves a prescribed distance and direction, and reenters where the PL dictates.

- Squad-sized patrols do not normally send out an R&S team at night.
- R&S teams will prepare a sketch of the area to the squad front if possible.
- The patrol remains at 100 % alert during this recon.
- If the PL feels the patrol was tracked or followed, he may elect to wait in silence at 100 % alert before sending out R&S teams.
- The R&S teams may use methods such as the "*I*," the "*Box*," or the "*T*." Regardless of the method chosen, the R&S team must be able to provide the PL with the same information.
- Upon completion of R&S, the PL confirms or denies the patrol base location, and either moves the patrol or begins priorities of work.

(6) ***Passive (Clandestine) Patrol Base (Squad)***.

- The purpose of a passive patrol base is for the rest of a squad or smaller size element.
- Unit moves as a whole and occupies in force.
- Squad leader ensures that the unit moves in at a 90-degree angle to the order of movement.
- A claymore mine is emplaced on route entering patrol base.
- Alpha and Bravo teams sit back-to-back facing outward, ensuring that at least one individual per team is alert and providing security.

(7) ***Priorities of Work (Platoon and Squad)***. Once the PL is briefed by the R&S teams and determines the area is suitable for a patrol base, the leader establishes or modifies defensive work priorities in order to establish the defense for the patrol base. Priorities of work are not a laundry list of tasks to be completed; to be effective, priorities of work must consist of a task, a given time, and a measurable performance standard. For each priority of work, a clear standard must be issued to guide the element in the successful accomplishment of each task. It must also be designated whether the work will be controlled in a centralized or decentralized manner. Priorities of work are determined IAW METT-TC. Priorities of Work may include, but are not limited to the following tasks:

(a) *Security (Continuous).*

- Prepare to use all passive and active measures to cover all of the perimeter all of the time, regardless of the percentage of weapons used to cover that all of the terrain.
- Readjust after R&S teams return, or based on current priority of work (such as weapons maintenance).
- Employ all elements, weapons, and personnel to meet conditions of the terrain, enemy, or situation.
- Assign sectors of fire to all personnel and weapons. Develop squad sector sketches and platoon fire plan.
- Confirm location of fighting positions for cover, concealment, and observation and fields of fire. SLs supervise placement of aiming stakes and claymores.
- Only use one point of entry and exit, and count personnel in and out. Everyone is challenged IAW the unit SOP.
- Hasty fighting positions are prepared at least 18 inches deep (at the front), and sloping gently from front to rear, with a grenade sump if possible.

(b) *Withdrawal Plan.* The PL designates the signal for withdrawal, order of withdrawal, and the platoon rendezvous point and/or alternate patrol base.

(c) *Communication (Continuous).* Commo must be maintained with higher headquarters, OP's, and within the unit. May be rotated between the patrol's RTOs to allow accomplishment of continuous radio monitoring, radio maintenance, act as runners for PL, or conduct other priorities of work.

(d) *Mission Preparation and Planning.* The PL uses the patrol base to plan, issue orders, rehearse, inspect, and prepare for future missions.

(e) *Weapons and Equipment Maintenance.* The PL ensures that machine guns, weapon systems, commo equipment, and night vision devices (as well as other equipment) are maintained. These items are not disassembled at the same time for maintenance (no more than 33 percent at a time), and weapons are not disassembled at night. If one machine gun is down, then security for all remaining systems is raised.

(f) *Water Resupply.* The PSG organizes watering parties as necessary. The watering party carries canteens in an empty rucksack or duffel bag, and must have commo and a contingency plan prior to departure.

(g) *Mess Plan.* At a minimum, security and weapons maintenance are performed prior to mess. Normally no more than half the platoon eats at one time. Rangers typically eat 1 to 3 meters behind their fighting positions.

- Rest/sleep plan management. The patrol conducts rest as necessary to prepare for future operations.
- Alert Plan and Stand-to. The PL states the alert posture and the stand-to time. He develops the plan to ensure all positions are checked periodically, OP's are relieved periodically, and at least one leader is always alert. The patrol typically conducts stand-to at a time specified by unit SOP (such as 30 minutes before and after BMNT or EENT).
- Resupply. Distribute or cross-load ammunition, meals, equipment, etc.
- Sanitation and personal hygiene. The PSG and medic ensure a slit trench is prepared and marked. All Soldiers will brush teeth, wash face, shave, wash hands, armpits, groin, feet, and darken (brush shine) boots daily. The patrol will not leave trash behind.

Figure 5-8. PATROL BASE

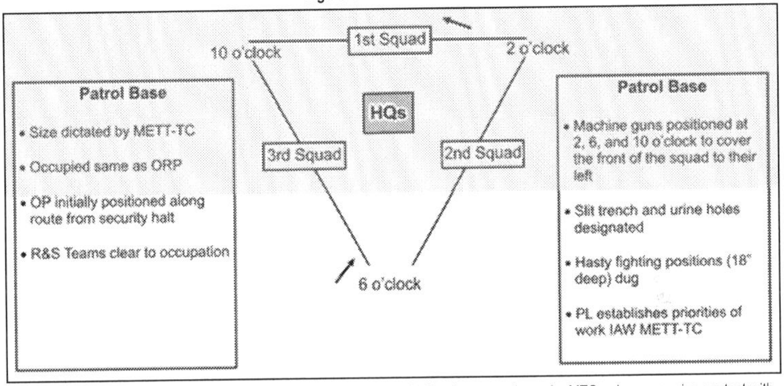

5-6. **MOVEMENT TO CONTACT**. The MTC is one of the five types of offensive operations. An MTC gains or regains contact with the enemy. Once contact is made, the unit develops the situation. Normally a platoon conducts an MTC as part of a larger force.

a. **Techniques**. The two techniques of conducting a movement to contact are search and attack and approach march.

(1) ***Search and Attack (S&A)***. This technique is used when the enemy is dispersed, is expected to avoid contact, disengage or withdraw, or you have to deny his movement in an area. The search and attack technique involves the use of multiple platoons, squads, and fire teams coordinating their actions to make contact with the enemy. Platoons typically try to find the enemy and then fix and finish him. They combine patrolling techniques with the requirement to conduct hasty or deliberate attacks once the enemy has been found.

(a) *Planning Considerations*.

- The factors of METT-TC.
- The requirement for decentralized execution.
- The requirement for mutual support.
- The length of operations.
- Minimize Soldier's load to improve stealth and speed.
- Resupply and MEDEVAC.
- Positioning key leaders and equipment.
- Employment of key weapons.
- Requirement for patrol bases.
- Concept for entering the zone of action.
- The concept for linkups while in contact.

(b) *Critical Performance Measures*.

- The platoon locates the enemy without being detected.
- Once engaged, fixes the enemy in position and maneuvers against the enemy.
- Maintains security throughout actions to avoid being flanked.

(2) ***Approach March***. The concept of the approach march is to make contact with the smallest element, allowing the commander the flexibility of destroying or bypassing the enemy. A platoon uses the approach march method as part of a larger unit. It can be tasked as the advance guard, move as part of the main body, or provide flank or rear security for the company or battalion. They may also receive on-order missions as part of the main body.

(a) *Fundamentals*. These basics are common to all movements to contact.

- Make enemy contact with smallest element possible.
- Rapidly develop combat power upon enemy contact.

- Provide all-round security for the unit.
- Support higher unit's concept.
- Reports all information rapidly and accurately and strives to gain and maintain contact with the enemy.
- Requires decentralized execution.

(b) *Planning Considerations.* The following issues should be considered heavily for MTC operations:

- Factors of METT-TC.
- Reduced Soldier's load.

(c) *Critical Performance Measures.*

- PL selects the appropriate movement formation based on likelihood of enemy contact.
- Maintains contact, once contact is made, until ordered to do otherwise.

b. **Task Standards**. The platoon moves NLT the time specified in the order, the platoon makes contact with the smallest element possible, and the main body is not surprised by the enemy. Once the platoon makes contact, it maintains contact. The platoon destroys squad and smaller-sized elements, and fixes elements larger than a squad. The platoon maintains sufficient fighting force capable of conducting further combat operations. Reports of enemy locations and contact are forwarded. If not detected by the enemy, the PL initiates a hasty attack. The platoon sustains no casualties from friendly fire. The platoon is prepared to initiate further movement within 25 minutes of contact, and all personnel and equipment are accounted for.

NOTES

Chapter 6
BATTLE DRILLS

Infantry battle drills describe how platoons and squads apply fire and maneuver to commonly encountered situations. They require leaders to make decisions rapidly and to issue brief oral orders quickly.

Section I. INTRODUCTION

This section defines and describes the format for battle drills.

6-1. **DEFINITION.** FM 7-1 defines a battle drill as "*a collective action rapidly executed without applying a deliberate decision-making process.*"

a. Characteristics of a battle drill are—
- They require minimal leader orders to accomplish and are standard throughout the Army.
- Sequential actions are vital to success in combat or critical to preserving life.
- They apply to platoon or smaller units.
- They are trained responses to enemy actions or leader's orders.
- They represent mental steps followed for offensive and defensive actions in training and combat.

b. A unit's ability to accomplish its mission often depends on Soldiers and leaders to execute key actions quickly. All Soldiers and their leaders must know their immediate reaction to enemy contact, as well as follow-up actions. Drills are limited to situations requiring instantaneous response; therefore, Soldiers must execute drills instinctively. This results from continual practice. Drills provide small units with standard procedures essential for building strength and aggressiveness.

(1) They identify key actions that leaders and Soldiers must perform quickly.

(2) They provide for a smooth transition from one activity to another; for example, from movement to offensive action to defensive action.

(3) They provide standardized actions that link Soldier and collective tasks at platoon level and below. (Soldiers perform individual tasks to CTT or SDT standard.)

(4) They require the full understanding of each individual and leader, and continual practice.

6-2. **FORMAT.** Drills include a ***title***, a ***situation*** to cue the unit or leader to start the drill, the ***required actions*** in sequence, and supporting illustrations. Where applicable, drills are cross-referenced with material in other chapters, or other drills, or both. Training standards for battle drills are in the mission training plan (MTP).

Section II. DRILLS

This section provides the battle drills for small units.

BATTLE DRILL 1
REACT TO CONTACT

SITUATION: A squad or platoon receives fires from enemy individual or crew-served weapons.

REQUIRED ACTIONS: See Figure 6-1.

1. Soldiers immediately assume the nearest covered positions.
2. Soldiers return fire immediately on reaching the covered positions.
3. Squad/team leaders locate and engage known or suspected enemy positions with well-aimed fire, and they pass information to the platoon/squad leader.
4. Fire team leaders control the fire of their Soldiers by using standard fire commands (initial and supplemental) containing the following elements:
   - Alert
   - Direction
   - Description of target
   - Range
   - Method of fire (manipulation and rate of fire)
   - Command to commence firing

5. Soldiers maintain contact (visual or oral) with the Soldiers on their left and right.

6. Soldiers maintain contact with their team leaders and indicate the location of enemy positions.

7. The leaders (visual or oral) check the status of their personnel.

8. The squad/team leaders maintain visual contact with the platoon/squad leader.

9. The platoon/squad leader moves up to the squad/fire team in contact and links up with its leader.
   a. The platoon leader brings his RTO, platoon FO, the nearest squad's squad leader, and one machine gun team.
   b. The squad leader of the trail squad moves to the front of his lead fire team.
   c. The PSG also moves forward with the second machine gun team and links up with the platoon leader, ready to assume control of the base-of-fire element.

10. The platoon/squad leader determines whether or not his unit must move out of an EA.

11. The platoon/squad leader determines whether his unit can gain and maintain suppressive fires with the element already in contact (based on the volume and accuracy of enemy fires against the element in contact). The platoon/squad leader assesses the situation. He identifies—
   - The location of the enemy position and obstacles.
   - The size of the enemy force engaging the unit in contact. The number of enemy automatic weapons, the presence of any vehicles, and the employment of indirect fires are indicators of the enemy strength.
   - Vulnerable flanks.
   - Covered and concealed flanking routes to the enemy position.

12. The platoon/squad leader determines the next COA, such as fire and movement, assault, breach, knock out bunker, enter and clear a building or trench.

13. The platoon/squad leader reports the situation to the company commander/platoon leader and begins to maneuver his unit.

14. Platoon leader directs platoon FO to call for and adjust indirect fires (mortars or artillery). Squad leaders relay requests through the platoon leader. The platoon/squad leader in conjunction with the platoon FO maintains accurate battle tracking of all friendly elements to facilitate quick clearance of fires.

15. Leaders relay all commands and signals from the platoon chain of command.

16. The PSG positions the base of fire element to observe and to provide supporting fires.

Note: Once the platoon has executed the React to Contact drill, the platoon leader quickly assesses the situation, for example, enemy size and location. He picks a COA. The platoon leader reports the situation to the company commander.

Figure 6-1. REACT TO CONTACT

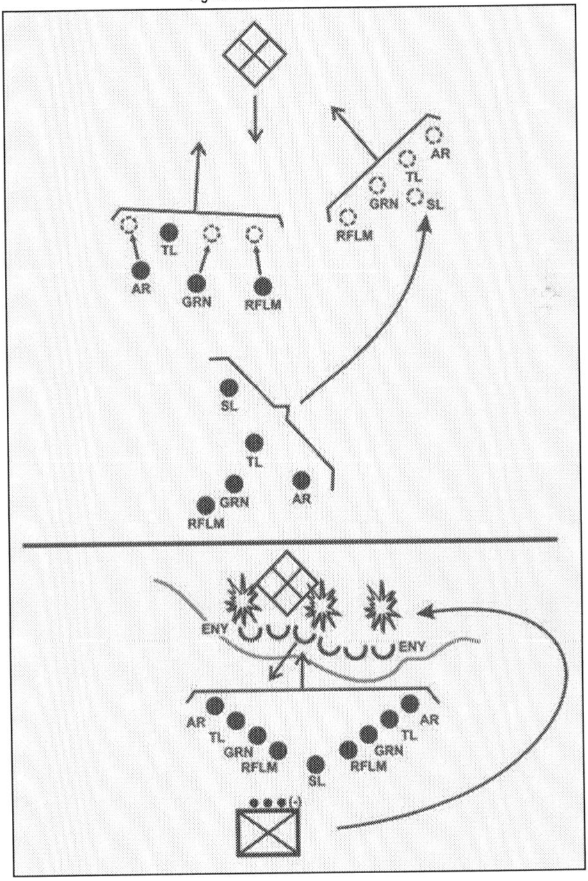

BATTLE DRILL 2
BREAK CONTACT

SITUATION: The squad or platoon is under enemy fire and must break contact.

REQUIRED ACTIONS: See Figure 6-2.

1. The platoon/squad leader directs fire support for the disengagement. He accomplishes this by—
   - Directing one squad/fire team in contact to support the disengagement of the remainder of the unit.
   - Considering the use of indirect fires for breaking contact.
   - Clearing the location, task, purpose, and method of conducting the fire mission with the platoon FO.

2. The platoon/squad leader orders a distance and direction, a terrain feature, or last ORP for the movement of the first squad/fire team. In conjunction with the platoon FO, the leader maintains accurate battle tracking of all friendly elements to facilitate quick clearance of fires.

3. The base of fire squad/team continues to suppress the enemy. The platoon/squad leader directs the platoon FO to execute the fire mission, if needed.

4. The moving squad/team assumes the overwatch position. The squad/team should use M203 grenade launchers, throw fragmentation and concussion grenades, and use smoke grenades to mask movement. The platoon/squad leader directs the platoon FO to execute smoke mission to screen friendly elements movement, if needed.

5. The moving squad/team takes up the designated position and engages the enemy position.

6. The platoon leader directs the base-of-fire element to move to its next location. Based on the terrain and the volume and accuracy of the enemy's fire, the moving fire team or squad may need to use fire and movement techniques.

7. The platoon/squad continues to bound away from the enemy until—
   - It breaks contact. It must continue to suppress the enemy as it does this.
   - It passes through a higher-level SBF position.
   - Its squad/fire teams are in the assigned position to conduct the next mission.

8. The leader should consider changing his unit's direction of movement once contact is broken. This will reduce the ability of the enemy to place effective indirect fires on the unit.

9. If the platoon/squad becomes disrupted, Soldiers stay together and move to the last designated rally point.

10. The platoon/squad leaders account for Soldiers, report, reorganize as necessary, and continue the mission. With the platoon FO, the platoon/squad develops a quick fire plan to support the hasty defense or new route of march.

Figure 6-2. BREAK CONTACT

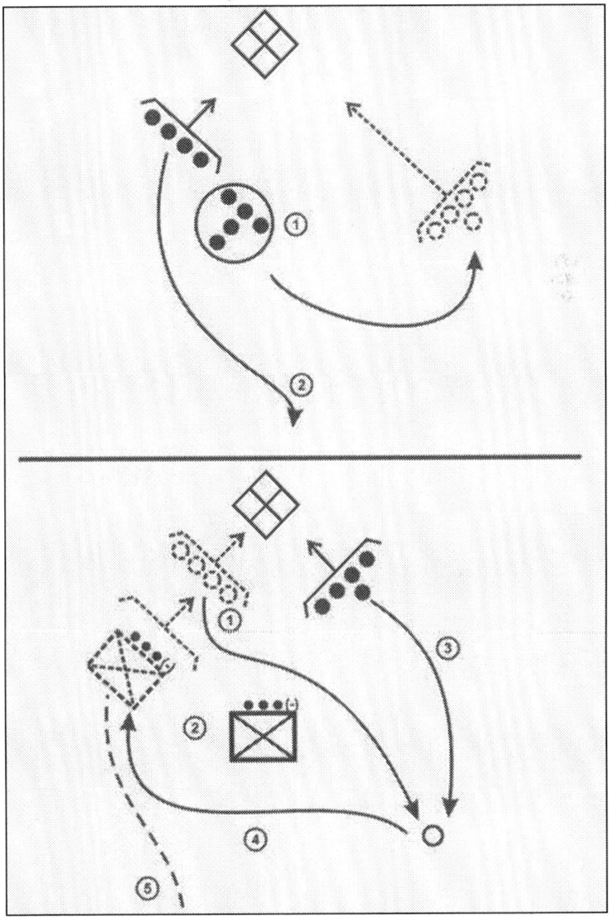

BATTLE DRILL 3
REACT TO AMBUSH

SITUATION: The squad or platoon enters a kill zone and the enemy initiates an ambush with a casualty-producing device and a high volume of fire.

REQUIRED ACTIONS: See Figure 6-3.

1. Near ambush (within hand-grenade range).
   a. Soldiers in the kill zone return fire immediately. How they go about this depends on the terrain. If—
      - Cover is not available, Soldiers immediately, without order or signal, assume prone position and throw concussion or fragmentation and smoke grenades.
      - Cover is available, without order or signal, Soldiers seek the nearest covered position, assume the prone position, and throw fragmentation or concussion and smoke grenades.

   b. Immediately after the explosion of the concussion or fragmentation grenades, Soldiers in the kill zone return fire and assault through the ambush position using fire and movement.
   c. Soldiers not in the kill zone identify the enemy's position. Fire is shifted as the personnel in the kill zone begin to assault.
   d. Soldiers in the kill zone continue the assault to eliminate the ambush or until contact is broken.
   e. The platoon/squad conducts consolidation and reorganization.

2. Far ambush (out of hand-grenade range). Soldiers receiving fire immediately return fire, take up covered positions, and suppress the enemy by—
   - Destroying or suppressing enemy crew-served weapons.
   - Sustaining suppressive fires.

3. Soldiers (squad/teams) not receiving fires move by a covered and concealed route to a vulnerable flank of the enemy position and assault using fire and movement techniques.

4. Soldiers in the kill zone continue suppressive fires and shift fires as the assaulting squad/team fights through the enemy position.

5. The platoon FO calls for and adjusts indirect fires as directed by the platoon leader. On order, he shifts or ceases fires to isolate the enemy position or to attack them with indirect fires as they retreat.

6. The platoon/squad leader reports, reorganizes as necessary, and continues the mission.

Figure 6-3. REACT TO AMBUSH

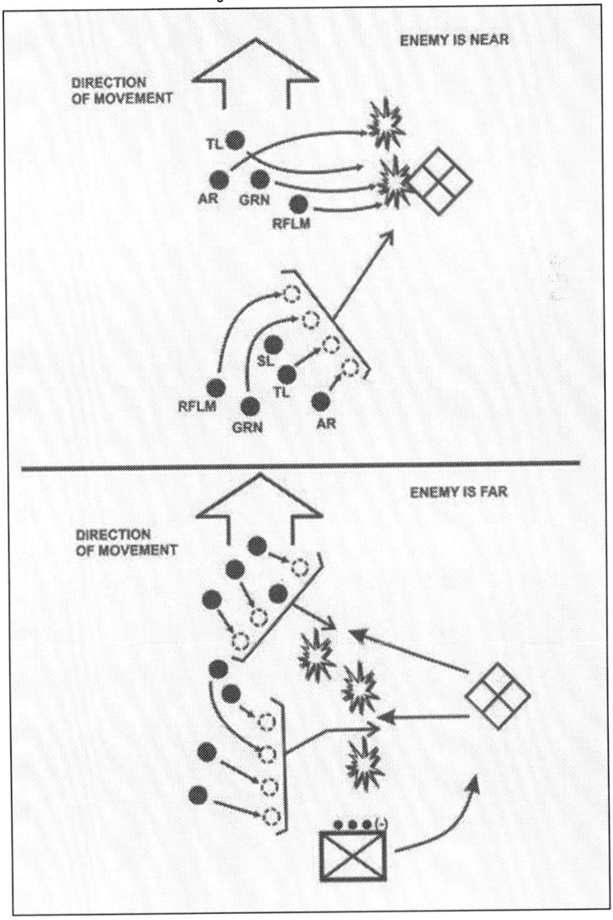

BATTLE DRILL 4
KNOCK OUT BUNKERS

SITUATION: While moving as part of a larger force, the platoon identifies the enemy in bunkers.

REQUIRED ACTIONS: See Figures 6-4 and 6-5.

1. The platoon initiates contact.
   a. The squad in contact establishes a base of fire.
   b. The platoon leader, his RTO, platoon FO, and one machine gun team move forward to link up with the squad leader of the squad in contact.
   c. The PSG moves forward with the second machine gun team and assumes control of the base-of-fire squad.
   d. The base-of-fire squad—
      - Destroys or suppresses enemy crew-served weapons first.
      - Obscures the enemy position with smoke (M203).
      - Sustains suppressive fires at the lowest possible level.
   e. The platoon leader directs platoon FO to call for and adjust indirect fires. The platoon leader in conjunction with the platoon FO maintains accurate battle tracking of all friendly elements to facilitate quick clearance of fires.
2. The platoon leader determines that he can maneuver by identifying—
   - Enemy bunkers, other supporting positions, and any obstacles.
   - Size of the enemy force engaging the platoon. The number of enemy automatic weapons, the presence of any vehicles, and the employment of indirect fires are indicators of enemy strength.
   - A vulnerable flank of at least one bunker.
   - A covered and concealed flanking route to the flank of the bunker.
3. The platoon leader determines which bunker is to be assaulted first and directs one squad (not in contact) to knock it out.
4. If necessary, the PSG repositions a squad, fire team, or machine gun team to isolate the bunker as well as to continue suppressive fires.
5. The assaulting squad, with the platoon leader and his RTO, move along the covered and concealed route and take action to knock out the bunker. The following occurs.
   a. On the platoon leader's signal, the support squad ceases or shifts fires to the opposite side of the bunker from which the squad is assaulting.
   b. At the same time, the platoon FO shifts indirect fires to isolate enemy positions.
6. The assaulting squad leader reports to the platoon leader and reorganizes his squad.
7. The platoon leader—
   - Directs the supporting squad to move up and knock out the next bunker.
   - Directs the assaulting squad to continue and knock out the next bunker.
   - Rotates squads as necessary.
8. The platoon leader reports, reorganizes as necessary, and continues the mission. The company follows up the success of the platoon attack and continues to assault enemy positions.

Figure 6-4. KNOCK OUT A BUNKER (SQUAD)

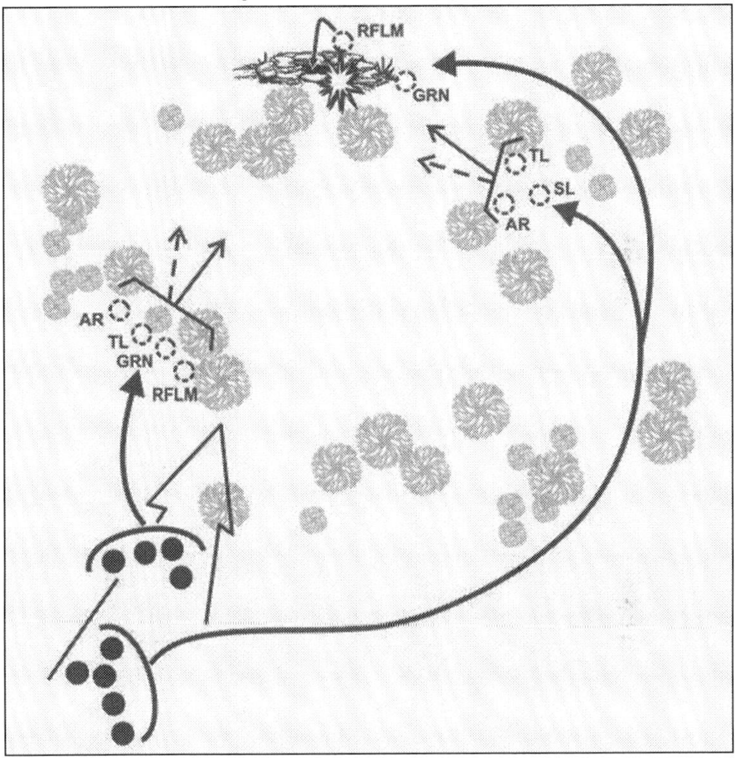

Figure 6-5. KNOCK OUT BUNKERS (PLATOON)

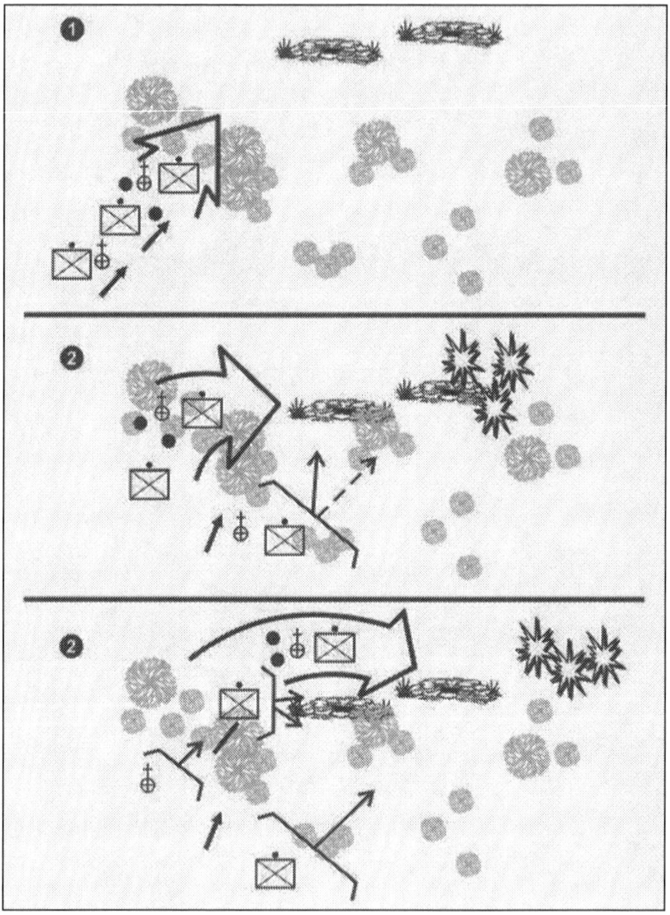

BATTLE DRILL 5
ENTER BUILDING/CLEAR ROOM

SITUATION: Operating as part of a larger force (in daylight or darkness), the squad is tasked to participate in clearing a building. The platoon leader directs the squad to enter the building or to clear a room. An entry point or breach has already been identified or will be created before initiating the entry. For a detailed discussion of urban entry breaching techniques, see FM 3.06-11. Enemy forces and noncombatants may or may not be present in the room and or building to be cleared.

REQUIRED ACTIONS: See Figures 6-6 through 6-14.

1. Platoon and squad leaders must consider the task and purpose they have been given and the method they are to use to achieve the desired results. They must operate IAW the ROE and must be aware of the affects that platoon weapons will have on the type and composition of the buildings.

2. To seize or gain control of a building may not always require committing troops into the structure or closing with the enemy. Before initiating this action and exposing members of the clearing squad to direct enemy contact and risking casualties, the platoon leader should consider/direct employment of all organic, crew-served, and supporting weapon systems onto the objective area to suppress and neutralize the threat, providing the mission, purpose, building composition, and ROE permit.

3. When conducting urban operations, Soldiers must be equipped at all times with a properly-mounted and immediately-useable NVD or light source to illuminate the immediate area.

Note: The following discussion assumes that only the platoon's organic weapons are supporting the infantry squad. Urban situations may require precise application of firepower. This situation is especially true of an urban environment in which the enemy is mixed with noncombatants. Noncombatants may be found in the room, which can restrict the use of fires and reduce the combat power available to a squad leader. His squad may have to operate in no-fire areas. ROE can prohibit the use of certain weapons until a specific hostile action takes place. All Soldiers must be aware of the ROE. Leaders must include the specific use of weapons in their planning for precision operations in urban terrain. Leaders should always consider the use of snipers or designated marksmen to apply precise fires to the objective.

4. Clearing team members must approach the entry point quickly, quietly, and in standard order. The squad leader must ensure he is in a position to control the actions of both fire teams. This approach preserves the element of surprise and allows for quick entry and domination of the room. If a breach is required, the order may be slightly modified based on the breach technique (FM 3-06.11). The members of the fire team are assigned numbers one through four. The rifleman is # 1 and the grenadier is # 3. If one member of the clearing team is armed with the SAW rather than an M16 rifle or carbine, he should be designated # 4. The team leader is normally the # 2 man, because he will have the most immediate decision to make as he enters the room.

5. The entire team enters the room as quickly and smoothly as possible and clears the doorway immediately. If possible, the team moves from a covered or concealed position already in their entry order. Ideally, the team arrives and passes through the entry point without having to stop. If the team must stop to effectively "*stack*" outside the entry point, it must do so only momentarily, and it must provide cover.

6. The door is the focal point of anyone in the room. It is known as the "*fatal funnel*," because it focuses attention at the precise point where the individual team members are the most vulnerable. Moving into the room quickly reduces the chance of anyone being hit by enemy fire directed at the doorway.

7. For this battle to be effectively employed, each member of the team must know his sector of fire and how his sector overlaps and links with the sectors of the other team members. Team members do not move to the points of domination and then engage their targets. Rather, they engage targets as they calmly and quickly move to their designated points. Engagements must not slow movement to their points of domination. Team members may shoot from as short a range as 1 to 2 inches. They engage the most immediate threat first and then the less immediate threats in sector. Immediate threats are personnel who—

- Are armed and prepared to return fire immediately.
- Block movement to the position of domination.
- Are within arm's reach of a clearing team member.
- Are within 3 to 5 feet of the breach point.

8. The squad leader designates the assault team and identifies the location of the entry point for them.

9. The squad leader positions the follow-on assault team to provide overwatch and supporting fires for the initial assault team.

10. Assault team members use available cover and concealment, and move as close to the entry point as possible.

a. If an explosive breach or a ballistic breach is to be performed by a supporting element, the assault team remains in a covered position until the breach is made. They may provide overwatch and fire support for the breaching element if necessary.

b. All team members must signal one another that they are ready before the team moves to the entry point.

c If stealth is a consideration, team members avoid the use of verbal signals, which may alert the enemy and remove the element of surprise.

d. Assault team members must move quickly from the covered position to the entry point, minimizing the time they are exposed to enemy fire.

11. The assault team enters through the entry point or breach. Unless a grenade will be thrown prior to entry, the team should avoid stopping outside the point of entry.

a. If required, the # 2 man throws a grenade of some type (fragmentary, concussion, stun) into the room before entry.

b. The use of grenades should be consistent with the ROE and building structure. The grenade should be cooked off before being thrown, if applicable to the type of grenade used.

c. If stealth is not a factor, the thrower should sound off with a verbal indication ("*Frag out,*" "*Concussion out,*" "*Stun out*") that a grenade of some type is being thrown. If stealth is a factor, only visual signals are given as the grenade is thrown.

> **CAUTION**
>
> If walls and floors are thin, fragments from fragmentation grenades and debris created by concussion grenades can injure Soldiers outside the room. If the structure has been stressed by previous explosive engagements, the use of these grenades could cause it to collapse. Leaders must determine the effectiveness of these types of grenades compared to possibilities of harm to friendly troops.

12. On the signal to go, or after the grenade detonates, the assault team moves through the entry point (Figure 6-6 through Figure 6-9) and quickly takes up positions inside the room that allow it to completely dominate the room and eliminate the threat. Unless restricted or impeded, team members stop movement only after they have cleared the door and reached their designated point of domination. In addition to dominating the room, all team members are responsible for identifying possible loopholes and mouseholes in the ceiling, walls, and floor.

Note: Where enemy forces may be concentrated and the presence of noncombatants is highly unlikely, the assault team can precede their entry by throwing a fragmentation or concussion grenade (structure dependent) into the room, followed by bursts of automatic small-arms fire by the # 1 man as he enters. Carefully consider the ROE and building composition before employing this method.

13. The # 1 and # 2 men are initially concerned with the area directly to their front, then along the wall on either side of the door or entry point. This area is in their path of movement, and it is their primary sector of fire. Their alternate sector of fire is from the wall they are moving toward, back to the opposite far corner.

14. The # 3 and # 4 men start at the center of the wall opposite their point of entry and clear to the left if moving toward the left, or to the right if moving toward the right. They stop short of their respective team member (either the # 1 man or the # 2 man).

15. The team members move toward their points of domination, engaging all threat or hostile targets in sequence in their sector. Team members must exercise fire control and discriminate between hostile and noncombatant room occupants. (The most practical way to do this is to identify whether or not the target has a weapon in his or her hands.) Shooting is done without stopping, using reflexive shooting techniques. Because the Soldiers are moving and shooting at the same time, they must move using the careful hurry. Figure 6-10, shows all four team members at their points of domination for a room with a center door and their overlapping sectors of fire.

16. The first man (rifleman) enters the room and eliminates the immediate threat. He has the option of going left or right, normally moving along the path of least resistance to one of two corners. When using a doorway as the point of entry, the team uses the path of least resistance, which they determine initially based on the way the door opens. If the door opens inward, the first man plans to move away from the hinges. If the door opens outward, he plans to move toward the hinged side. Upon entering, his direction is influenced by the size of the room, the enemy situation, and furniture or other obstacles that hinder or channel his movement.

17. The direction each man moves in should not be preplanned unless the exact room layout is known. Each man should go in a direction opposite the man in front of him (Figure 6-6). Every team member must know the sectors and duties of each position.

18. As the first man goes through the entry point, he can usually see into the far corner of the room. He eliminates any immediate threat and continues to move along the wall if possible and to the first corner, where he assumes a position of domination facing into the room.

Figure 6-6. FIRST MAN ENTERS A ROOM, FOLLOWED BY TEAM LEADER

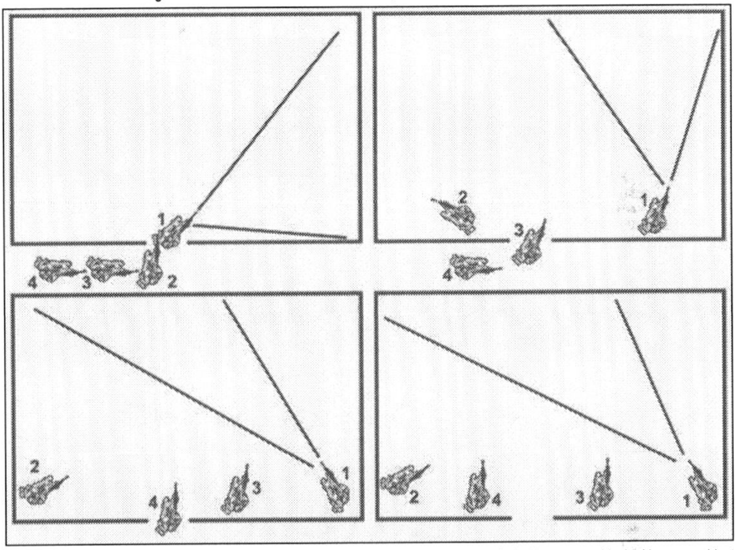

Note: Team members must always stay within 1 meter of the wall. If a team member finds his progress blocked by some object that will force him more than 1 meter from the wall, he must either step over it (if able) or stop where he is and clear the rest of his sector from where he is. If this action creates dead space in the room, the team leader directs which clearing actions to take once other members of the team have reached their points of domination.

19. The second man (normally the team leader), entering almost simultaneously with the first, moves in the opposite direction, following the wall (Figure 6-7). The second man must clear the entry point, clear the immediate threat area, clear his corner, and move to a dominating position on his side of the room. The second man must also immediately determine if he is entering a center door or corner door and act accordingly (Figure 6-7 and Figure 6-10).

Figure 6-7. SECOND MAN (TEAM LEADER) ENTERS A ROOM

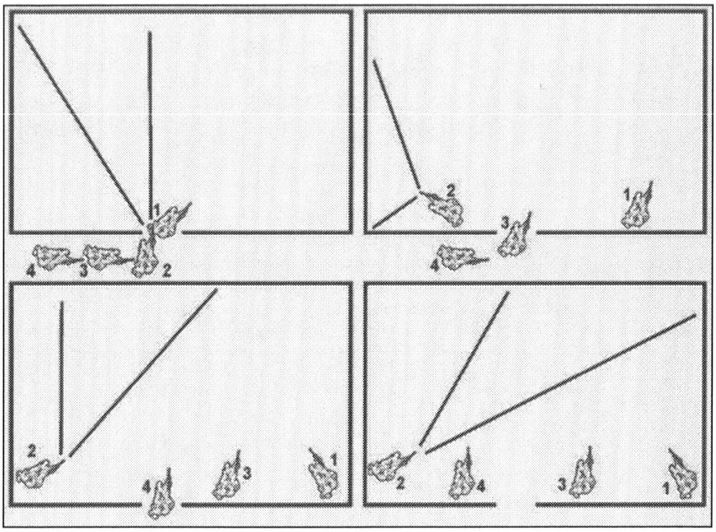

20. The third man (normally the grenadier) simply goes opposite of the second man inside the room, moves at least 1 meter from the entry point, and takes a position that dominates his sector (Figure 6-8).

Figure 6-8. THIRD MAN ENTERS A ROOM

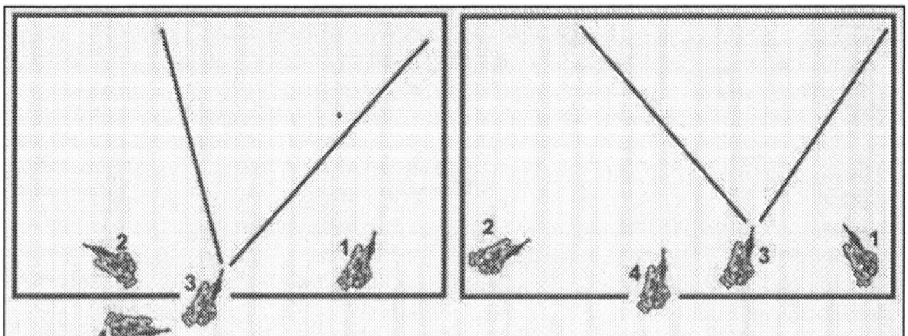

21. The fourth man (normally the SAW gunner) moves opposite of the third man, clears the doorway by at least 1 meter, and moves to a position that dominates his sector (Figure 6-9).

Figure 6-9. FOURTH MAN IN A ROOM

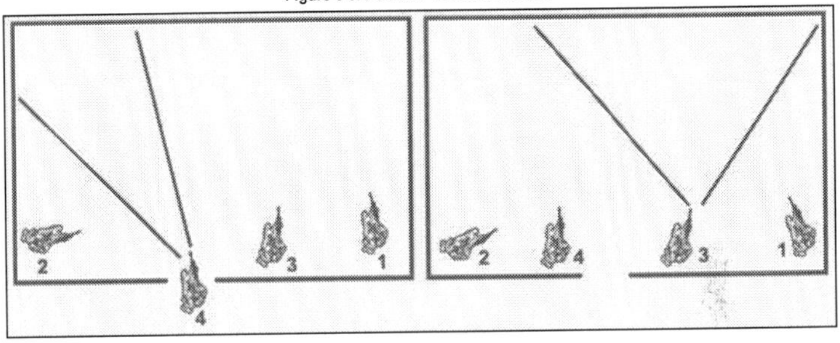

Note: If the path of least resistance takes the first man to the left, then all points of domination are the mirror image of those shown in the diagrams.

22. Points of domination should not be in front of doors or windows so team members are not silhouetted to the outside of the room (Figure 6-10). No movement should mask the fire of any of the other team members.

Figure 6-10. POINTS OF DOMINATION AND SECTORS OF FIRE, CENTER DOOR VERSUS CORNER DOOR

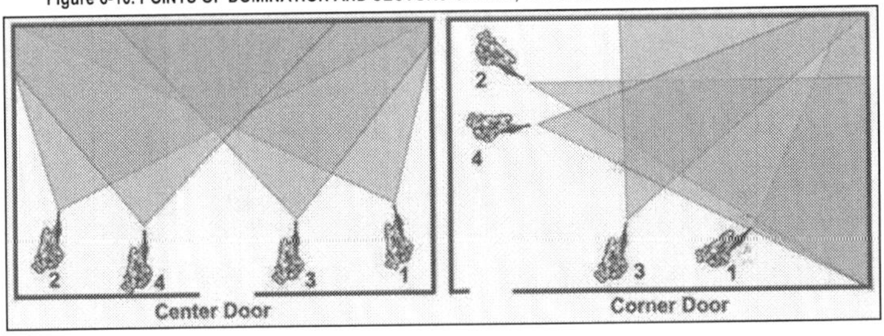

23. On order, any member of the assault team may move deeper into the room overwatched by the other team members. The team leader must control this action.

24. Once the room is cleared, the team leader signals to the squad leader that the room has been cleared. The squad leader marks the room IAW unit SOP (see FM 3.06-11). The squad leader determines whether or not his squad can continue to clear through the building.

25. The squad reorganizes as necessary. Leaders redistribute the ammunition. The squad leader reports to the platoon leader when the room is clear.

Note: If at any point a team member experiences a weapon malfunction in the presence of enemy combatants, he has to make an immediate decision. If the enemy is outside immediate danger distance from the Soldier, then the Soldier should drop to one knee, which indicates that he has experienced a weapon malfunction and prevents fratricide by ensuring that the next Soldier's fields of fire are clear. Once on a knee, the Soldier must remain there until the team leader directs him to stand up. If the kneeling Soldier corrects his weapon malfunction, he can continue to engage targets from his kneeling position. If, however, the enemy Soldier is so close that he presents an immediate threat, and if dropping to a knee would expose the US Soldier to immediate harm, then the US Soldier must try to subdue or disable the enemy Soldier. He can strike the enemy with the muzzle of his weapon or a bayonet to the face, throat, or chest. Or, he could grapple with the enemy soldier and take him to the ground immediately to clear the other team members' fields of fire. Once they have cleared their sectors and eliminated any other threats in the room, the other members of the clearing team assist.

26. Although this battle drill is effective, leaders might have to modify it for the situation. Some reasons and methods for modifying the technique are shown in Table 6-1.

Table 6-1. REASONS FOR AND METHODS OF MODIFYING ENTRY TECHNIQUES

REASON	METHOD
Objective rooms are consistently small	Clear in teams of 2 or 3
Shortage of personnel	Clear in teams of 2 or 3
Enemy poses no immediate threat	2 or 3 men search rooms to ensure no enemy or noncombatants are present
No immediate threat, speed is critical	1 to 3 men visually search each room

27. When full four-man teams are not available for room clearing, two- and three-man teams can be used. If the # 1 or # 2 man discovers that the room is very small, he can yell, "*Short room*" or "*Short,*" which tells the # 3 or # 4 man (whoever following the # 1 or # 2 man) to stay outside the room. Figures 6-11 and 6-12 show the points of domination and sectors of fire for a two-man clearing team. Figures 6-13 and 6-14 show the actions for a three-man team.

DANGER

RICOCHETS POSE A HAZARD. ALL SOLDIERS MUST BE AWARE OF THE TYPE OF WALL CONSTRUCTION OF THE ROOM BEING CLEARED. THE WALLS OF AN ENCLOSED ROOM PRESENT MANY RIGHT ANGLES. COMBINED WITH HARD SURFACES SUCH AS CONCRETE, A BULLET MAY CONTINUE TO RICOCHET AROUND A ROOM UNTIL ITS ENERGY IS SPENT. AFTER HITTING THREAT PERSONNEL, BALL AMMUNITION MAY PASS THROUGH THE BODY AND RICOCHET. BODY ARMOR AND THE KEVLAR HELMET PROVIDE SOME PROTECTION FROM THIS HAZARD.

Figure 6-11. POINTS OF DOMINATION AND SECTORS OF FIRE (TWO-MAN TEAM, CENTER DOOR)

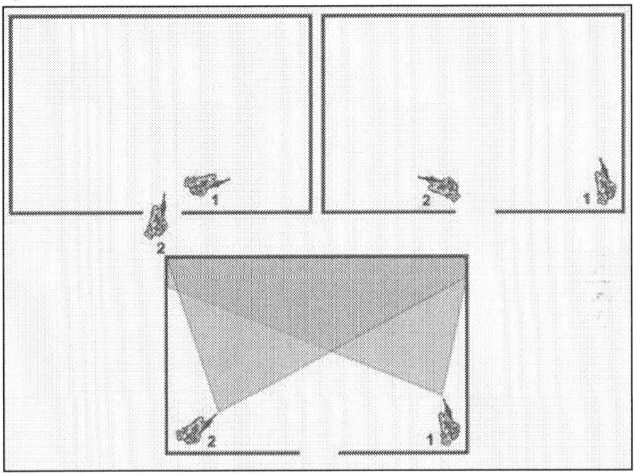

Figure 6-12. POINTS OF DOMINATION AND SECTORS OF FIRE (TWO-MAN TEAM, CORNER DOOR)

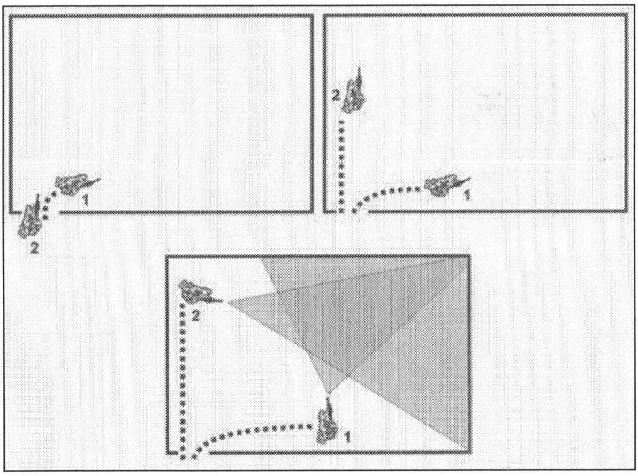

Figure 6-13. POINTS OF DOMINATION AND SECTORS OF FIRE (THREE-MAN TEAM, CENTER DOOR)

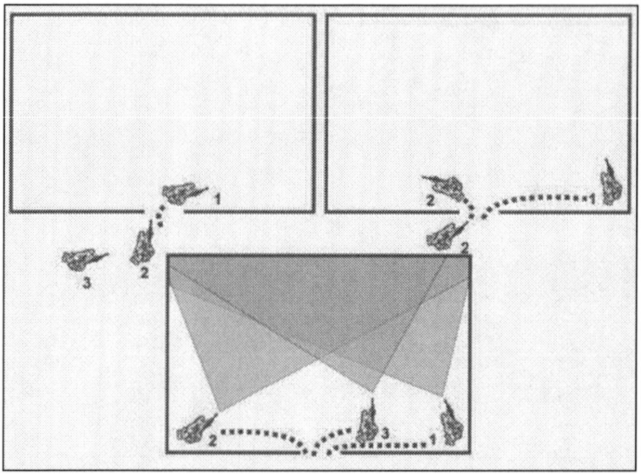

Figure 6-14. POINTS OF DOMINATION AND SECTORS OF FIRE (THREE-MAN TEAM, CORNER DOOR)

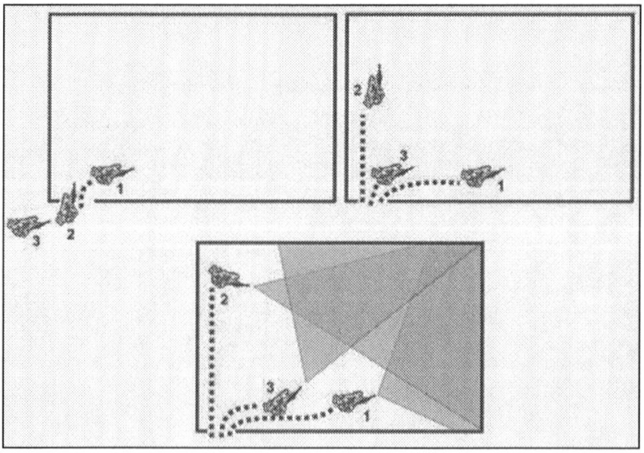

28. While moving through a building Soldiers may encounter the following architectural building features.

a. **Multiple Team/Multiple Rooms**. Figure 6-15 shows this.

STEP 1: First team enters and clears Room 1.

STEP 2: Squad leader determines the direction from which the second clearing team must enter Room 1, based on the location of the Room 2 entry point.

STEP 3: First team collapses inward to allow the second team to move into the room.

STEP 4: Second team 'stacks left' and prepares to enter Room 2.

Figure 6-15. MULTIPLE TEAMS / MULTIPLE ROOMS

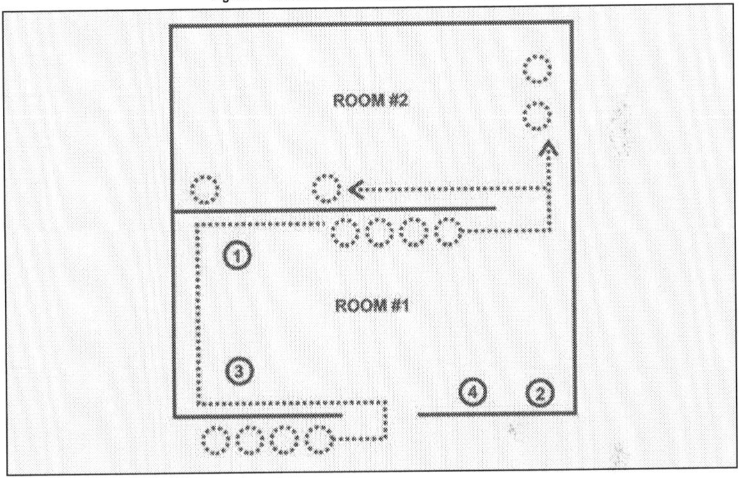

b. **Open Stairwell**. This is a gap between flights of stairs that allows a person to look up and down between the flights. Figure 6-16 shows an open stairwell.

STEP 1: The # 1 man pulls security on the highest point he can see / engage.

STEP 2: The # 2 man moves up the stairs on the inside with the # 3 man to a point that he can see / engage the next landing, where he turns around and continues to move up to the next landing.

STEP 3: The # 3 man moves up the stairs with the # 2 man on the outside and engages the threat on the immediate landing.

STEP 4: The # 4 man moves up the stairs with the # 1 man, on the squeeze, the # 2 man turns around to engage the next landing.

STEP 5: The flow continues with the # 2 man picking up the sector of the # 1 man had. The # 3 man picks up where the # 2 man was. The # 4 man picks up where the # 3 man was. The # 1 man picks up where the # 4 man was.

Note: Most stairwells require a second team.

Figure 6-16. OPEN STAIRWELL

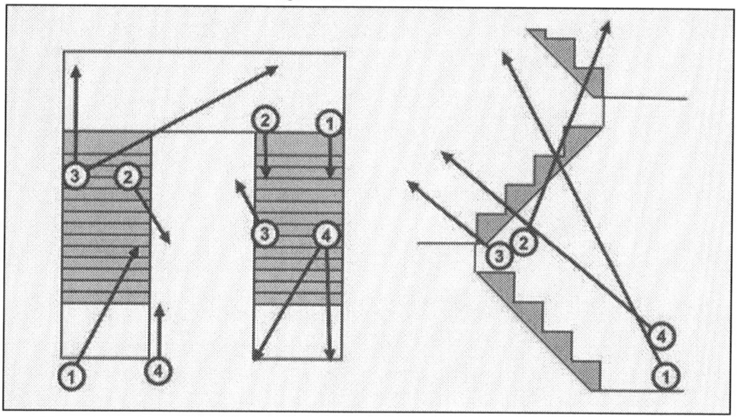

c. **Closed Stairwell**. This is any stairwell with walls separating the flights of stairs. Figure 6-17 shows a closed stairwell.

STEP 1: The # 1 man checks high to ensure there is no opening on the landing or between the stairs.

STEP 2: The # 2 man pulls long security to the next bend or landing.

STEP 3: The # 1 or # 3 man moves up the steps with the # 2 man. As they approach the corner, the # 2 man signals his presence by tapping the # 1 man on the shoulder.

STEP 4: Keying off the # 1 man's movement, the other men simultaneously break around the corner.

STEP 5: If no fire is received, the # 2 man moves to the opposite wall. Both men continue to move up until they reach their objective.

STEP 6: The # 3 and # 4 men will continue to move 3 to 4 steps behind.

Note: Avoid getting locked into a security position such as inside a stairwell. Also, avoid spreading yourselves thin or getting separated by more than one floor of stairs.

Figure 6-17. CLOSED STAIRWELL

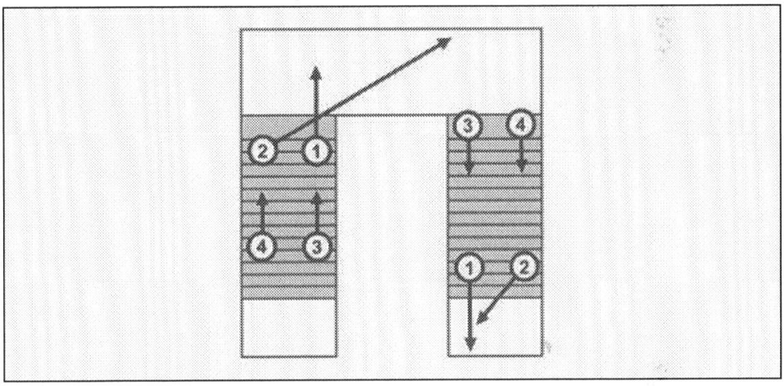

d. **Hallway Movement**. Clearing team(s) move down the hallway using the frontal security (cross cover) technique (Figure 6-18).

Figure 6-18. HALLWAY MOVEMENT

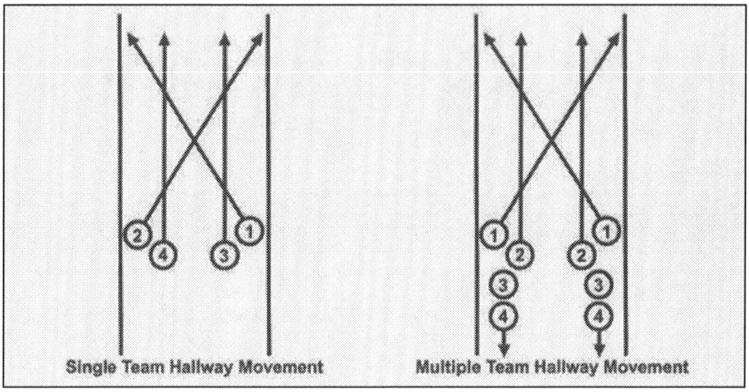

e. **T-Shaped Intersection**. See Figure 6-19. This technique can incorporate the dynamic corner clear (Figure 6-20).

STEP 1: Each # 1 man goes to a knee covering his sector.

STEP 2: On a predetermined signal, each two-man team will break the corner, picking up their sectors of fire.

Figure 6-19. T-SHAPED INTERSECTION

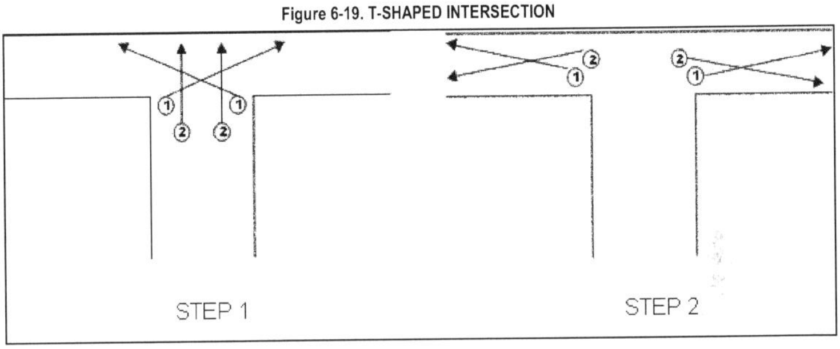

f. **Dynamic Corner**. (See Figure 6-20.)

STEP 1: The #1 and #2 man, as they approach the corner, have to clear. Do not slow down.

STEP 2: The #2 man will tap the #1 man on the shoulder about 2-3 meters away from the corner letting the #1 man know the #2 man is with him.

STEP 3: Keying off of the #1 man's movement, they both break the corner simultaneously.

STEP 4: The #1 man goes low to a knee, the #2 man stays high.

STEP 5: If the Rangers are not receiving fire the #2 man rabbits, or moves, to the far side.

STEP 6: The #1 and #2 man take up sectors of fire.

STEP 7: The #3 and #4 man take long security in the direction of movement.

Figure 6-20. DYNAMIC CORNER

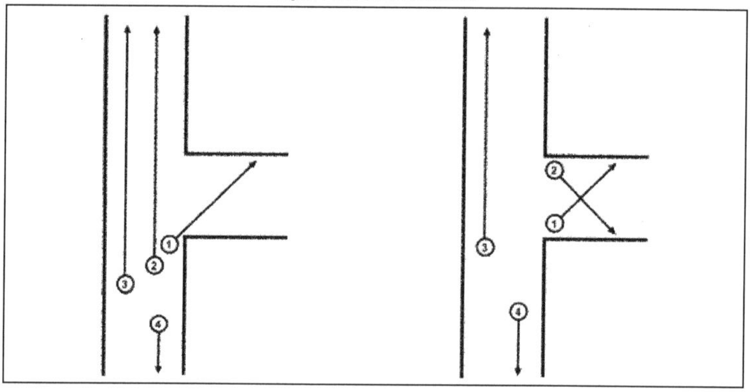

g. **Three-Way Intersection**. (See Figure 6-21.)

STEP 1: The #1 through #4 man will use one of the corner clearing techniques to clear the intersection.

STEP 2: When the intersection is secure, the #5 and #6 man move through the intersection and enter either more hallway or a room. If they are entering more hallway, the #1 through #4 man provide cross coverage as the #5 and #6 man move through the intersection into more hallway. If they are moving into a room, the #5 and #6 man move to one side of the door and signal the #1 and #3 or #2 and #4 man to enter the stack depending on which side of the door they stack on. The remaining members of the team, not in the stack, will continue to provide security down the hallway.

STEP 3: If the #5 and #6 man move into more hallway the #1 through #4 man will enter the stack and proceed down the hallway with the #3 and #4 man providing rear security.

Figure 6-21. THREE WAY INTERSECTION

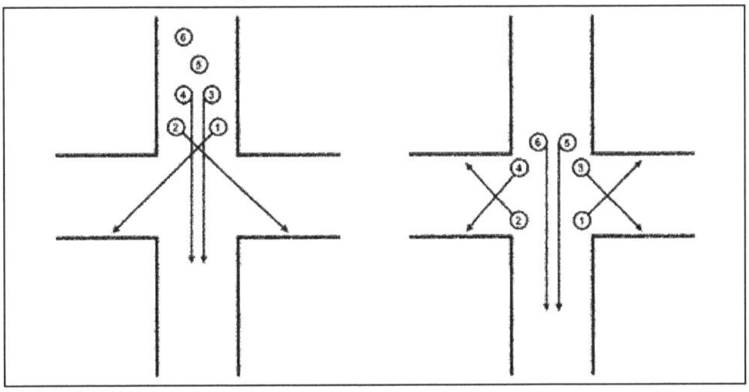

BATTLE DRILL 6
ENTER/CLEAR A TRENCH

SITUATION: The platoon is attacking as part of a larger force and identifies the enemy in a trench line. The platoon deploys and establishes a base of fire. The platoon leader determines that he has sufficient combat power to maneuver and assault the trench line.

REQUIRED ACTIONS: See Figure 6-22 and Figure 6-23.

1. The platoon leader directs one squad to enter the trench and secure a foothold.
2. The platoon leader designates the entry point of the trench line and the direction of movement once the platoon begins clearing.
3. The PSG positions Soldiers and machine guns to suppress the trench and isolate the entry point.
4. Platoon leader directs platoon FO to initiate fire mission, if necessary, in support of assault. They maintain accurate battle tracking of all friendly elements to facilitate quick clearance of fires. The platoon FO ceases or shifts fires to isolate the OBJ as assault team advances.
5. The assaulting squad executes actions to enter the trench and establish a foothold. The squad leader signals to the platoon leader that the foothold is secure, and the follow on elements can move into the trench. The squad leader remains at the entry point and marks it. The platoon follows the success of the seizure of the foothold with the remainder of the platoon as part of the platoon actions to clear a trench line.
6. The platoon leader moves into the trench with the assaulting squad.

7 The platoon leader directs one of the base-of-fire squads to move into the trench and begin clearing it in the direction of movement from the foothold.

8. The base-of-fire element repositions as necessary to continue suppressive fires.

9. The assaulting squad passes the squad that has secured the foothold and executes actions to take the lead and clear the trench.

a. The squad leader designates a lead fire team and a trail fire team.

b. The lead fire team and the squad leader move to the forward most secure corner or intersection. The squad leader tells the team securing that corner or intersection that his squad is ready to continue clearing the trench. The trail fire team follows, maintaining visual contact with the last Soldier of the lead team.

Notes: 1. The fire support element must be able to identify the location of the lead fire team in the trench at all times.

2. Throughout this battle drill, the team leader positions himself at the rear of the fire team to have direct control (physically, if necessary) of his Soldiers. Other Soldiers in the fire team rotate the lead to change magazines and prepare grenades. Rotating the lead provides constant suppressive fires down the trench and maintains the momentum of the attack as the squad clears the trench.

c. The lead fire team passes the element securing the foothold. The following then occurs.

(1) The lead Soldier of the fire team moves abreast of the Soldier securing the corner or intersection, taps him, and announces, "*Taking the lead*."

(2) The Soldier securing the corner or intersection acknowledges that he is handing over the lead by shouting, "*Okay!*" He allows the fire team to pass him.

d. The lead fire team starts clearing in the direction of movement. They arrive at a corner or intersection. The following then occurs.

(1) Allowing for cook-off (2 seconds maximum) and shouting, "*Frag out*," the second Soldier prepares and throws a grenade around the corner.

(2) Upon detonation of the grenade, the lead Soldier moves around the corner firing three-round bursts and advancing as he fires. The entire fire team follows him to the next corner or intersection.

e. The squad leader—

- Follows behind the team.
- Ensures that the trailing fire team moves up and is ready to pass the lead at his direction.
- Rotates fire teams as necessary to keep his Soldiers fresh and to maintain the momentum of the attack.
- Requests indirect fires, if required, through the platoon leader. The squad leader also directs the employment of the M203 to provide immediate suppression against positions along the trench line.
- Ensures fire teams maintain sufficient interval to prevent them from being engaged by the same enemy fires.

f. At each corner or intersection, the lead fire team performs the same actions previously described.

g. If the lead Soldier finds that he is nearly out of ammunition before reaching a corner or intersection, he announces, "*Ammo*," and then--

(1) The lead Soldier stops and moves against one side of the trench, ready to let the rest of the team pass. He continues to aim his weapon down the trench in the direction of movement.

(2) The next Soldier ensures that he has a full magazine, moves abreast of the lead Soldier, taps him, and announces, "*Taking the lead*."

(3) The lead Soldier acknowledges that he is handing over the lead by shouting, "*Okay*." Positions rotate and the squad continues forward.

h. The trailing fire team secures intersections and marks the route within the trench as the squad moves forward. The trailing fire team leader ensures that follow-on squads relieve his buddy teams to maintain security.

i. The squad leader reports the progress of the clearing operation. The base-of-fire element must be able to identify the location of the lead fire team in the trench at all times.

10. The platoon leader rotates squads to keep the Soldiers fresh and to maintain the momentum of the assault.

11. The PSG calls forward ammunition resupply and organizes teams to move it forward into the trench.

12. The base-of-fire element ensures that all friendly forces move into the trench only through the designated entry point to avoid fratricide.

13. The platoon leader reports to the company commander that the trench line is secured, or he is no longer able to continue clearing. If trench line is secured, then platoon leader directs platoon FO to develop a fire plan to support the defense of the platoon position.

> **DANGER**
>
> **FIRE TEAMS: MAINTAIN SUFFICIENT INTERVALS TO AVOID ENGAGEMENT BY THE SAME ENEMY FIRES.**

Figure 6-22. ENTER A TRENCH (SQUAD)

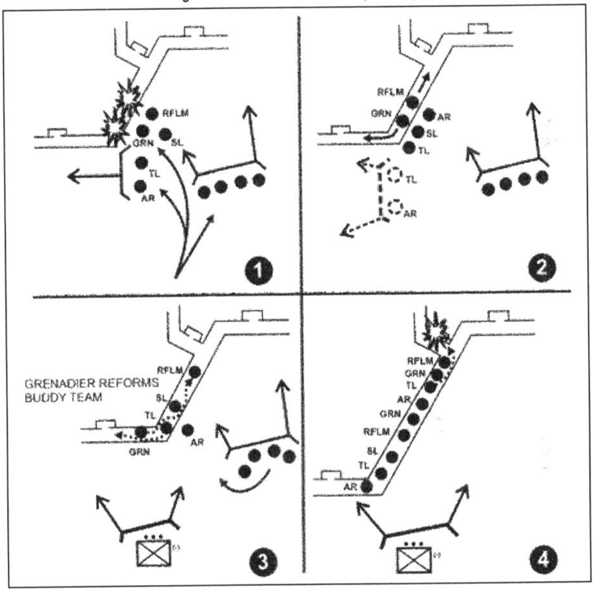

Figure 6-23. CLEAR A TRENCH LINE (PLATOON)

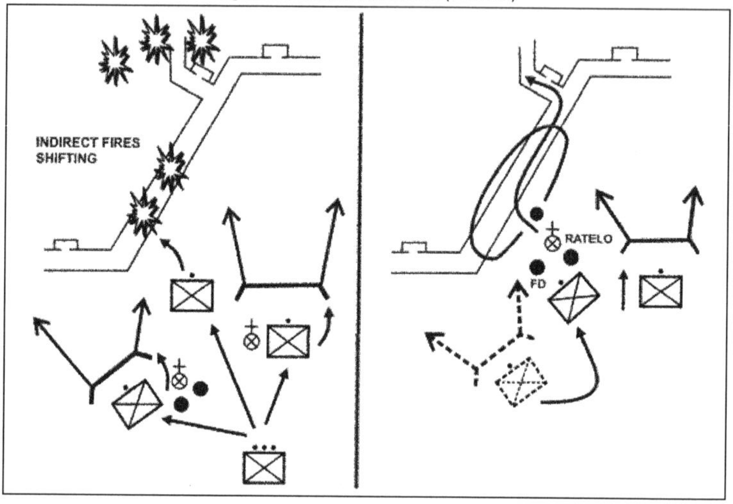

BATTLE DRILL 7
CONDUCT INITIAL BREACH OF A MINED WIRE OBSTACLE (PLATOON)

SITUATION: The platoon is operating as part of a larger force. The lead squad identifies a wire obstacle reinforced with mines that cannot be bypassed and enemy positions on the far side of the obstacle.

REQUIRED ACTIONS: See Figure 6-24 and Figure 6-25.

1. The squad in contact reacts to contact.
2. The platoon gains suppressive fires. The following then occurs.
   a. The squad in contact establishes a base-of-fire position.
   b. The platoon leader, his RTO, platoon FO, and the squad leader of the next squad with one machine gun team move forward to link up with the squad leader of the squad in contact.
3. The platoon leader determines that he can maneuver by identifying—
   - The obstacle and enemy positions.
   - The size of the enemy force engaging the squad, for example, the number of enemy automatic weapons, the presence of any vehicles, and the employment of indirect fires are indicators of enemy strength.
   - A breach point.
   - A covered and concealed route to the breach point.
4. The platoon leader directs the squad in contact to support the movement of another squad to the breach point. He indicates the base-of-fire position and the route to it, the enemy position to be suppressed, and the breach point and route the rest of the platoon will take to it. The platoon leader also clears the location, task, purpose, and method of conducting the fire mission with the platoon FO.

5. On the platoon leader's signal, the base-of-fire squad—
    - Destroys or suppresses enemy weapons that are firing effectively against the platoon.
    - Obscures the enemy position with smoke.
    - Continues suppressive fires at the lowest possible level.

6. The platoon leader designates one squad as the breach squad and the remaining squad as the assault squad once the breach has been made. The assault squad may add its fires to the base-of-fire squad. Normally, it follows the covered and concealed route of the breach squad and assaults through immediately after the breach is made.

7. The base-of-fire squad moves to the breach point and establishes a base of fire.

8. The PSG moves forward to the base-of-fire squad with the second machine gun and assumes control of the squad.

9. The platoon leader leads the breach and assault squads along the covered and concealed route.

10. The platoon FO calls for and adjusts indirect fires as directed by the platoon leader. The platoon leader in conjunction with the platoon FO maintains accurate battle tracking of all friendly elements to facilitate quick clearance of fires.

11. The breach squad executes actions to breach the obstacle.

    a. The squad leader directs one fire team to support the movement of the other fire team to the breach point.

    b. The squad leader identifies the breach point.

    c. The base-of-fire element continues to provide suppressive fires and isolates the breach point.

    d. The breaching fire team, with the squad leader, moves to the breach point using the covered and concealed route. From there, the following takes place:

        (1) The squad leader and breaching fire team leader employ smoke grenades to obscure the breach point. The platoon base-of-fire element shifts direct fires away from the breach point and continues to suppress key enemy positions. The platoon FO ceases indirect fires or shifts them beyond the obstacle.

        (2) The breaching fire team leader positions himself and the automatic rifleman on one flank of the breach point to provide close-in security.

        (3) The grenadier and rifleman of the breaching fire team probe for mines and cut the wire obstacle, marking their path as they proceed. If available, a bangalore is preferred.

        (4) Once the obstacle has been breached, the breaching fire team leader and the automatic rifleman move to the far side of the obstacle and take up covered and concealed positions with the rifleman and grenadier. The team leader signals to the squad leader when they are in position and ready to support.

        (5) The squad leader signals the base-of-fire team leader to move his fire team up and through the breach. He then moves through the obstacle and joins the breaching fire team, leaving the grenadier (or anti-armor specialist) and rifleman of the supporting fire team on the near side of the breach to guide the rest of the platoon through.

        (6) Using the same covered and concealed route as the breaching fire team, the base-of-fire team moves through the breach and takes up covered and concealed positions on the far side.

12. The squad leader reports the situation to the platoon leader and posts guides at the breach point.

13. The platoon leader leads the assault squad through the breach in the obstacle and positions them beyond the breach to support the movement of the remainder of the platoon or assaults the enemy position covering the obstacle. Platoon leader directs the platoon FO to shift indirect fires off the enemy position.

14. The breaching squad continues to widen the breach to allow vehicles to pass through.

15. The platoon leader reports the situation to the company commander, and then he directs his base-of-fire squad to move up and through the obstacle. The platoon leader appoints Soldiers to guide the company through the breach point.

16. The PSG brings the remaining elements forward and through the breach on the platoon leader's command.

17. The company follows up the success of the platoon as it conducts the breach and continues the assault against the enemy positions.

Figure 6-24. CONDUCT INITIAL BREACH OF A MINED WIRE OBSTACLE (SQUAD)

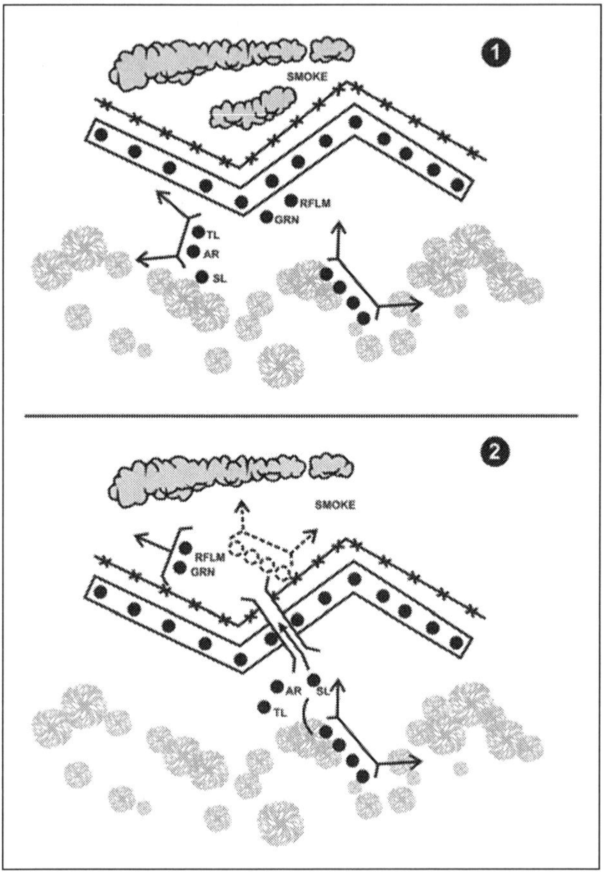

Figure 6-25. CONDUCT INITIAL BREACH OF A MINED WIRE OBSTACLE (PLATOON)

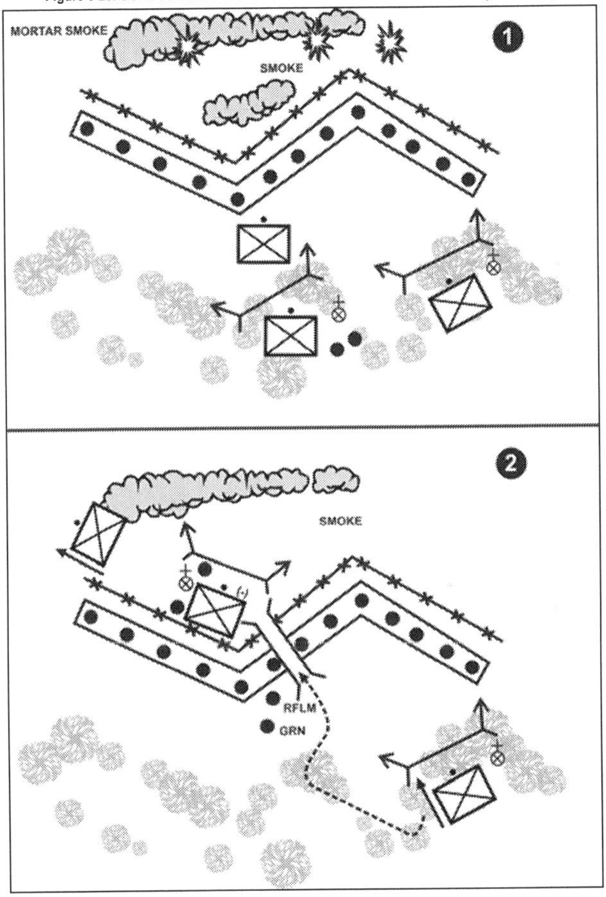

BATTLE DRILL 8
REACT TO INDIRECT FIRE

SITUATION: A platoon or squad while moving or at the halt (not dug in) receives indirect fire.

REQUIRED ACTIONS: NA

Any Soldier announces, "*Incoming.*" Then, if the platoon/squad is moving, the following takes place:

1. The platoon/squad leader gives direction and distance for the platoon/squad to move to a rally point by ordering direction and distance, for example, "*Four o'clock, three hundred meters.*"

2. Platoon/squad members move rapidly along the direction and distance to the rally point.

3. At the rally point, the leader immediately accounts for personnel and equipment, and forms the platoon/squad for a move to an alternate position.

4. The senior leader submits a SITREP to higher headquarters.

5. If the platoon/squad is halted (not dug in) or is preparing to move because they hear incoming artillery, the following takes place:

   a. The platoon/squad leader gives direction and distance for the platoon/squad to move to a rally point by ordering direction and distance, for example, "*Four o'clock, three hundred meters.*"

   b. Platoon/squad members secure all mission-essential equipment and ammunition and move rapidly along the direction and distance to the rally point.

   c. At the rally point, the leader immediately accounts for personnel and equipment, and forms the platoon/squad for a move to an alternate position.

   d. The senior leader submits a SITREP to higher headquarters.

Note: If platoon/squad members are in defensive (dug in) positions, then members will remain in those positions if appropriate. Senior leader submits SITREP.

NOTES

Chapter 7
COMMUNICATIONS

The basic requirement of combat communications is to provide rapid, reliable, and secure interchange of information. Communications are a vital aspect in successful mission accomplishment. The information in this chapter helps the Ranger squad/platoon maintain effective communications and correct any radio antenna problems.

Section I. INTRODUCTION

This section discusses military frequency modulated radios and automated net control devices (ANCDs).

7-1. **MILITARY FREQUENCY MODULATED RADIOS**. Military frequency modulated (FM) radios are composed of a receiver and a transmitter. Together, the receiver-transmitter has many capabilities and features that enable you to perform your mission more effectively. The radio can operate in Single Channel mode or frequency hopping mode. The radio has about 2,320 SC Channels and includes voice and digital communication. The operating voltage for the manpack radio is 13.5 volts from the primary battery. The range of the manpack radio is 5-10 KM on Hi power. This range is based upon line of sight and is derived from averages achieved under normal atmospheric and weather conditions. Ranges depend upon location, sighting, weather, and surrounding noise level, among other factors. the automated net control device (ANCD) is an essential tool. It replaces the signal operating instructions booklet. The ANCD has many capabilities. The main purpose of the ANCD for your mission is to obtain signal operating instructions electronically. The following discusses assembly, operation, and troubleshooting for the AN/PRC-119; provides step-by-step instructions to obtain SOI from an ANCD; and explains how to repair broken antennas:

a. **Manpack Radio Assembly**. To assemble a manpack radio you must first check and install a battery.
- (1) Inspect the battery box for dirt or damage.
- (2) Stand RT on front panel guards.
- (3) Check battery life condition (you will be using the rechargeable BB-390 batteries).
- (4) Place battery in box.
- (5) Close battery cover, and secure using latches.
- (6) Return radio to upright position.
- (7) If used battery was installed, enter the battery life condition into radio by performing the following:
  - (a) Set **FCTN** to **LD**.
  - (b) Press **BAT**, then **CLR**.
  - (c) Enter number recorded on side of battery.
  - (d) Press **STO**.
  - (e) Set **FCTN** to **SQ ON**.

b. **Antenna**.
- (1) Inspect whip antenna connector on antenna and on radio for damage.
- (2) Screw whip antenna into base.
- (3) I land tighten.
- (4) Carefully mate antenna base with RT ANT connector.
- (5) Hand tighten.
- (6) Position antenna as needed by bending goose neck.

Note: Keep antenna straight, if possible. If the antenna is bent to a horizontal position, it may be necessary to turn the radio in order to receive and transmit messages.

c. **Handset**.
- (1) Inspect the handset for damage.
- (2) Push handset on **AUD/DATA** and twist clockwise to lock in place.

d. **Field Pack**.
- (1) Place radio transmitter (RT) in field pack with antenna on the left shoulder.
- (2) Fold top flap of field over RT and secure flap to field pack using straps and buckles.

e. **Presets**.
- (1) Set **CHAN** to **1**.
- (2) Set **MODE** to **SC**.
- (3) Set **RF PWR** to **HI**.

(4) Set **VOL** to mid range.
(5) Set **DIM** full clockwise.
(6) Set **FCTN** to **LD**.
(7) Set **DATA RATE** to **OFF**.

f. **Single Channel Loading Frequencies**.
(1) Obtain Ranger SOI.
(2) Set **FCTN** to **LD**.
(3) Set **MODE** to **SC**.
(4) Set **CHAN** to **MAN**, **Cue**, or desired channel (**1** to **6**) where frequency is to be stored.
(5) Press **FREQ** (display will show "**00000**," or the frequency the RT is currently set to).
(6) Press **CLR** (display will show five lines).
(7) Enter the number of the new frequency.
(8) If you make a mistake with a number, press **CLR**.
(9) Press **STO** (display will blink).
(10) Set **FCTN** to **SQ ON**.

g. **Clearing of Frequencies**.
(1) Set **MODE** to **SC**.
(2) Set **CHAN** to **MAN**, **Cue** or the channel whose frequency is to be cleared.
(3) Press **FREQ**.
(4) Press **CLR**.
(5) Press **Load**; then press **STO**.
(6) Set **FCTN** to **SQ ON**.

h. **Scanning of Multiple Frequencies**.
(1) Load all desired frequencies using "Single Channel Loading Frequencies" instructions (see 7-2h).
(2) Set **CHAN** to **CUE**.
(3) Set **SC** to **FH**.
(4) Set **FCTN** to **SQ ON**.
(5) Press **STO** (display will say **SCAN**).
(6) Press **8**. You will now be able to scan more than one frequency.

7-2. **AUTOMATED NET CONTROL DEVICE**. The RTO retrieves all necessary COMSEC and information from the ANCD. To retrieve the SOI from an ANCD:

a. Press the **ON** button on the ANCD keypad.
b. Press the letter lock button to unlock the keys on the ANCD.
c. Press **Main Menu** key.
d. Once you press the main menu key, this is what should show on the screen.

Appl Date Time Setup
Util Bit.

e. Using the arrow keys, scroll over to **Appl** (as shown above) and press the enter button on the keypad.

f. Once you have entered into the **Appl** (applications) menu using the arrow keys scroll over to **SOI** and press the enter button on the key pad. For example--

RDS SOI Radio.
(Appl) *<--(Indicates that you are in the applications menu).*

SOI menu will look like this:
Qref Group Net Sufx Pyro
Tmpd Set C/S Find Memo

g. Using the arrow keys scroll over to **Set** and press the enter button on the keypad.
h. Press enter on choose once in the **Set** menu.

EXAMPLE:
SELECT:
Choose Send Receive.

i. Once in the **Choose** menu press the enter key on set **1-5** or **6-10**.

j. After you choose your selection out of the **Set** menu, the ANCD will automatically return back to the **SOI** menu.

k. Using the keypad scroll over to **Tmpd** and enter the number out of the set you need for that day. After this operation, the ANCD will automatically return back to the **SOI** menu.

l. Using the arrow keys on the keypad scroll over to **Net** and press the enter button on the keypad. The Net menu should look like the following example:

EXAMPLE:
T05 <--(time period) 1 <--(platoon) / A <--(company) /A3Q <--(prefix).
C 60.000.

Note: When in **Net** menu, this is where you will obtain your frequency and the first part of your call sign.

m. After you obtain the information you need out of **Net**, press the **Abort** button on the keypad. This will return you to the **SOI** menu.

n. Using the keypad, scroll over to **Sufx** and press the **Enter** button.

o. Once you are in the **Sufx** menu, use the up and down arrow keys on the ANCD to locate your two-digit designator. These two digits will go at the end of your prefix; your prefix plus the two digits equal your call sign.

p. Once you are done, press the **OFF** button on the ANCD to end operation.

Section II. ANTENNAS

This section discusses repair techniques, construction and adjustment, field-expedient antennas, antenna length and orientation, and improvement of marginal communications.

7-3. **REPAIR**. Antennas are sometimes broken or damaged, causing either a communications failure or poor communications. If a spare antenna is available, the damaged antenna is replaced. When there is no spare, the Ranger squad/platoon may have to construct an emergency antenna. The following paragraphs contain suggestions for repairing antennas and antenna supports and the construction and adjustment of emergency antennas.

DANGER

SERIOUS INJURY OR DEATH CAN RESULT FROM CONTACT WITH THE RADIATING ANTENNA OF A MEDIUM-POWER OR HIGH-POWER TRANSMITTER. TURN THE TRANSMITTER OFF WHILE ADJUSTING THE ANTENNA.

a. **Whip Antennas**. When a whip antenna is broken into two sections, the part of the antenna that is broken off can be connected to the part attached to the base by joining the sections.

(1) Use the method shown in A, Figure 7-1 when both parts of the broken whip are available and usable.

(2) Use the method in B, Figure 7-1, when the part of the whip that was broken off is lost, or when the whip is so badly damaged that it cannot be used.

(3) To restore the antenna to its original length, a piece of wire is added that is nearly the same length as the missing part of the whip. The pole support is then lashed securely to both sections of the antenna. The two antenna sections are cleaned thoroughly to ensure good contact before connecting them to the pole support. If possible, the connections are soldered.

b. **Wire Antennas**. Emergency repair of a wire antenna may involve the repair or replacement of the wire used as the antenna or transmission line; or the repair or replacement of the assembly used to support the antenna.

(1) When one or more wires of an antenna are broken, the antenna can be repaired by reconnecting the broken wires. To do this, lower the antenna to the ground, clean the ends of the wires, and twist the wires together. Whenever possible, solder the connection.

(2) If the antenna is damaged beyond repair, construct a new one. Make sure that the lengths of the wires on the substitute antenna match the length of the original.

(3) Antenna supports may also require repair or replacement. A substitute item may be used in place of a damaged support and, if properly insulated, can be of any material of adequate strength. If the radiating element is not properly insulated, field antennas may be shorted to ground and be ineffective. Many commonly found items can be used as field-expedient insulators.

(4) The best of these items are plastic or glass to include plastic spoons, buttons, bottle necks, and plastic bags. Though less effective than plastic or glass but still better than no insulator at all are wood and rope. The radiating element--the actual antenna wire-should touch only the antenna terminal and should be physically separated from all other objects, other than the supporting insulator. (Figure 7-2 shows how to make emergency insulators.)

Figure 7-1. USE OF BROKEN ANTENNA

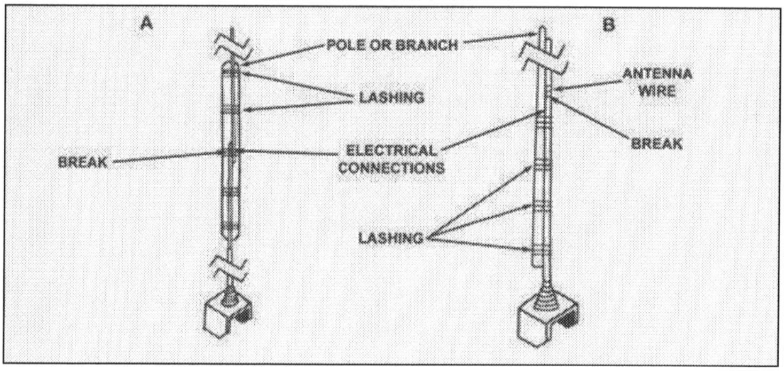

Figure 7-2. METHODS OF CONSTRUCTING EMERGENCY INSULATORS

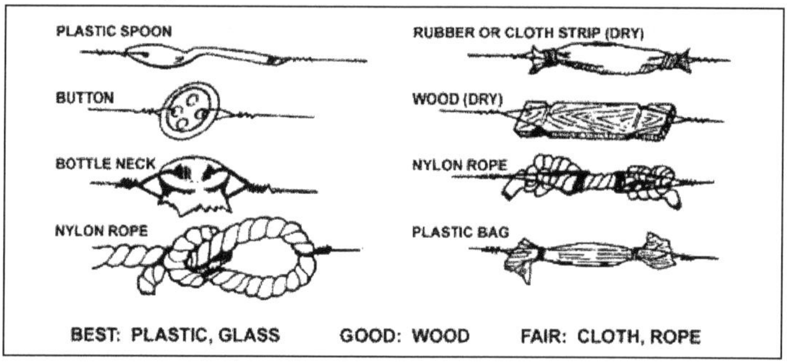

7-4. **CONSTRUCTION AND ADJUSTMENT**. Ranger squad/platoons may use the following methods to construct and adjust antennas:

a. **Construction**. The best kinds of wire for antennas are copper and aluminum. In an emergency, however, Rangers use any type of wire that is available.

(1) The exact length of most antennas is critical. The emergency antenna should be the same length as the antenna it replaces.

(2) Antennas supported by trees can usually survive heavy wind storms if the trunk of a tree or a strong branch is used as a support. To keep the antenna taut and to prevent it from breaking or stretching as the trees sway, the Ranger attaches a spring or old inner tube to one end of the antenna. Another technique is to pass a rope through a pulley or eyehook. The rope is attached to the end of the antenna and loaded with a heavyweight to keep the antenna tightly drawn.

(3) Guidelines used to hold antenna supports are made of rope or wire. To ensure the guidelines will not affect the operation of the antenna, the Ranger cuts the wire into several short lengths and connects the pieces with insulators.

b. **Adjustment**. An improvised antenna may change the performance of a radio set. The following methods can be used to determine if the antenna is operating properly:

(1) A distant station may be used to test the antenna. If the signal received from this station is strong, the antenna is operating satisfactorily. If the signal is weak, the Ranger adjusts the height and length of the antenna and the transmission line to receive the strongest signal at a given setting on the volume control of the receiver. This is the best method of tuning an antenna when transmission is dangerous or forbidden.

(2) In some radio sets, the Ranger uses the transmitter to adjust the antenna. First, he sets the controls of the transmitter to normal; then, he tunes the system by adjusting the antenna height, the antenna length, and the transmission line length to obtain the best transmission output.

7-5. **FIELD-EXPEDIENT OMNIDIRECTIONAL ANTENNAS**. Vertical antennas are omnidirectional. The omnidirectional antenna transmits and receives equally well in all directions.

a. **Vertical Antennas**. Most tactical antennas are vertical. For example, the man-pack portable radio uses a vertical whip and so do the vehicular radios in tactical vehicles. A vertical antenna can be made by using a metal pipe or rod of the correct length, held erect by means of guidelines. The lower end of the antenna should be insulated from the ground by placing it on a large block of wood or other insulating material. A vertical antenna may also be a wire supported by a tree or a wooden pole (Figure 7-3).

Figure 7-3. VERTICAL ANTENNA MADE FROM A WIRE SUPPORTED BY A TREE OR WOODEN POLE

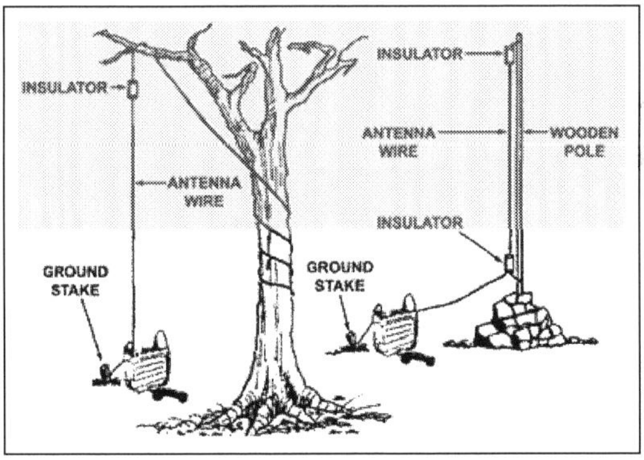

b. **Modification of Connection on Shorter Vertical Antennas**. For short vertical antennas, the pole may be used without guidelines (if properly supported at the base). If the length of the vertical mast is not long enough to support the wire upright, it may be necessary to modify the connection at the top of the antenna (Figure 7-4; see FM 24-18).

Figure 7-4. MODIFIED CONNECTION AT TOP OF ANTENNA

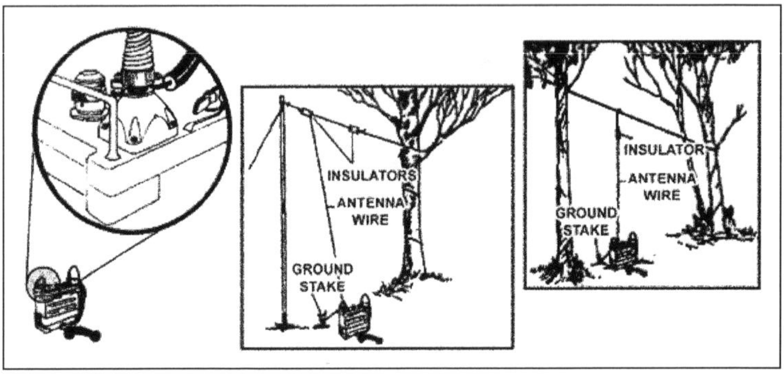

c. **End-Fed, Half-Wave Antenna**. An emergency, end-fed half-wave antenna (Figure 7-5) can be constructed from available materials such as field wire, rope, and wooden insulators. The electrical length of this antenna is measured from the antenna terminal on the radio set to the far end of the antenna. The best performance can be obtained by constructing the antenna longer than necessary and then shortening it, as required, until the best results are obtained. The ground terminal of the radio set should be connected to a good earth ground for this antenna to function efficiently.

Figure 7-5. END-FED, HALF-WAVE ANTENNA

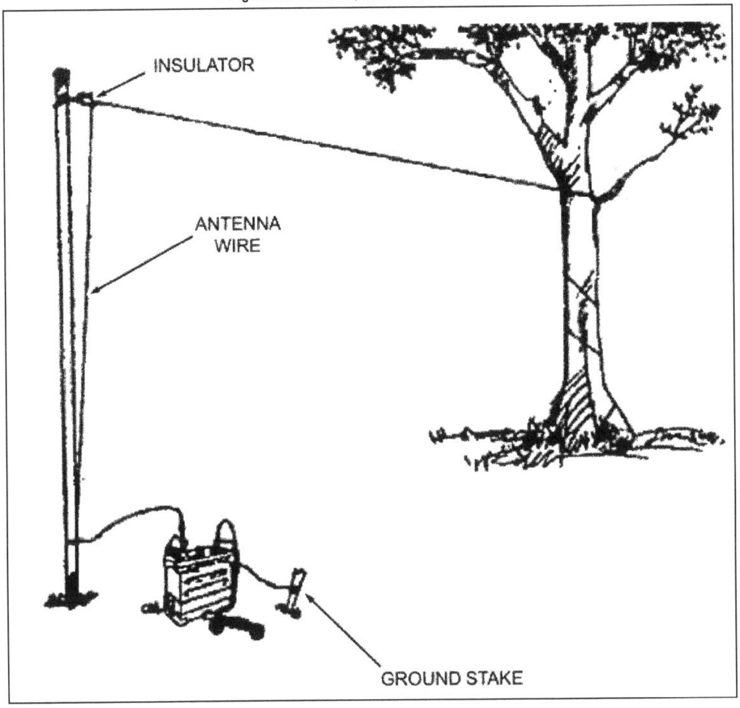

d. **Center-Fed, Doublet Antenna**. The center-fed doublet is a half-wave antenna consisting of two quarter wavelength sections on each side of the center (Figure 7-6). Doublet antennas are directional broadside to their length, which makes the vertical doublet antenna omnidirectional. This is because the radiation pattern is doughnut-shaped and bidirectional.

Figure 7-6. CENTER-FED, DOUBLET ANTENNA

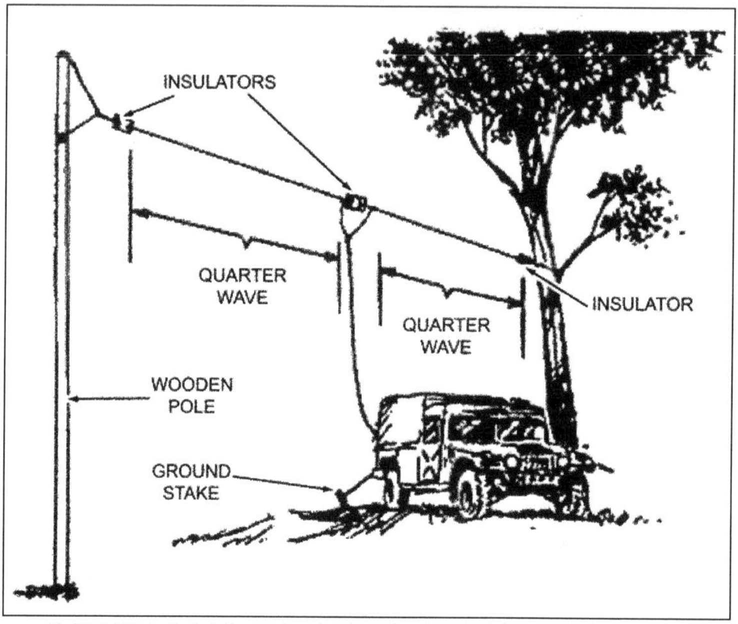

(1) Compute the length of a half-wave antenna by using the formula previously discussed. Cut the wires as close as possible to the correct length; this is very important.

(2) Uses transmission line for conducting electrical energy from one point to another and for transferring the output of a transmitter to an antenna. Although it is possible to connect an antenna directly to a transmitter, the antenna is usually located some distance away.

(3) Support center-fed half-wave FM antennas entirely with pieces of wood. (A horizontal antenna of this type is shown in A, Figure 7-7 and a vertical antenna in B, Figure 7-7.) Rotate these antennas to any position to obtain the best performance.

(a) If the antenna is erected vertically, bring out the transmission line horizontally from the antenna for a distance equal to at least one-half of the antenna's length before it is dropped down to the radio set.

Figure 7-7. CENTER-FED ANTENNAS SUPPORTED WITH WOOD

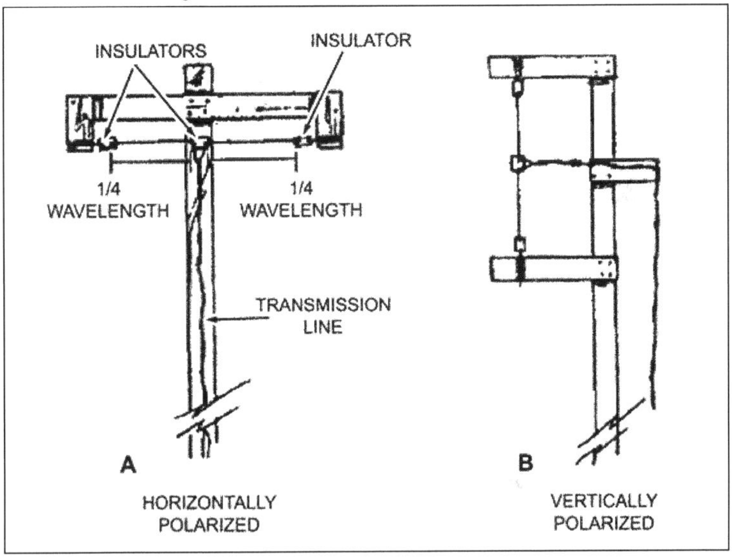

(b) The half-wave antenna is used with FM radios (Figure 7-8). It is effective in heavily wooded areas to increase the range of portable radios. Connect the top guidelines to a limb or pass it over the limb and connect it to the tree trunk or a stake.

Figure 7-8. HALF-WAVE ANTENNA USED WITH FM RADIO

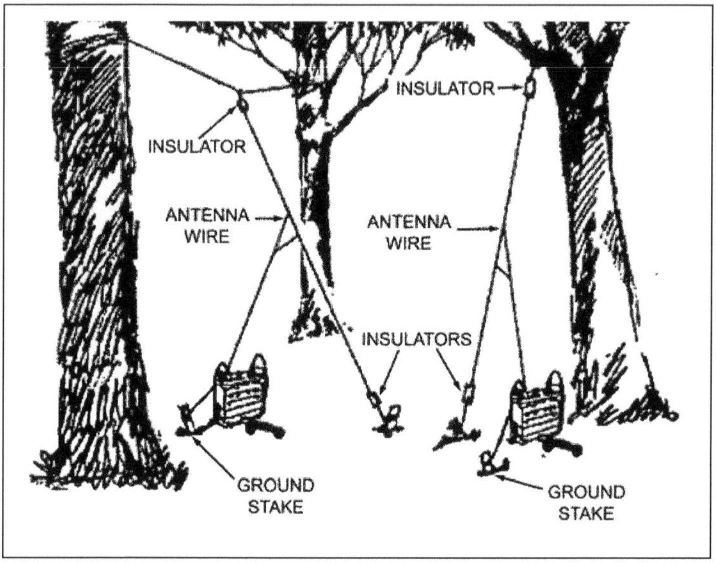

c. **Expedient 292-Type Antenna**. The expedient 292-type antenna was developed for use in the jungle and, if used properly, can increase the team's ability to communicate. In its entirety, the antenna is bulky, heavy, and not acceptable for sniper team operations. The team can, however, carry only the mast head and antenna sections, mounting these on wood poles or hanging them from trees; or, the team can make a complete expedient 292-type antenna (Figure 7-9), using WD-1, wire, and other readily available material. The team can also use almost any plastic, glass, or rubber objects for insulators. Dry wood is acceptable when nothing else is available. (See Figure 7-2 for types of insulators that may be used.) The following describes how to make this antenna:

(1) Use the quick-reference table (Table 7-2) to determine the length of the elements (one radiating and three ground planes) for the frequency that will be used. Cut these elements (A, Figure 7-9) from WD-1 field wire (or similar wire). Cut spacing sticks (B, Figure 7-9) the same length. Place the ends of the sticks together to form a triangle and tie the ends with wire, tape, or rope. Attach an insulator to each corner. Attach a ground-plane wire to each insulator. Bring the other ends of the ground-plane wires together, attach them to an insulator (C, Figure 7-9), and tie securely. Strip about 3 inches of insulation from each wire and twist them together (D, Figure 7-9).

(2) Tie one end of the radiating element wire to the other side of insulator C and the other end to another insulator (D, Figure 7-9). Strip about 3 inches of insulation from the radiating element at insulator C.

(3) Cut enough WD-1 field wire to reach from the proposed location of the antenna to the radio set. Keep this line as short as possible, because excess length reduces the efficiency of the system. Tie a knot at each end to identify it as the *hot* lead. Remove insulation from the *hot* wire and tie it to the radiating element wire at insulator C. Remove insulation from the other wire and attach it to the bare ground-plane element wires at insulator C. Tape all connections and do not allow the radiating element wire to touch the ground-plane wires.

(4) Attach a rope to the insulator on the free end of the radiating element and toss the rope over the branches of a tree. Pull the antenna as high as possible, keeping the lead-in routed down through the triangle. Secure the rope to hold the antenna in place.

(5) At the radio set, remove about 1 inch of insulation from the *hot* lead and about 3 inches of insulation from the other wire. Attach the *hot* line to the antenna terminal (doublet connector, if so labeled). Attach the other wire to the metal case-the handle, for example. Be sure both connections are tight or secure.

(6) Set up correct frequency, turn on the set, and proceed with communications.

Figure 7-9. COMPLETE EXPEDIENT 292-TYPE ANTENNA

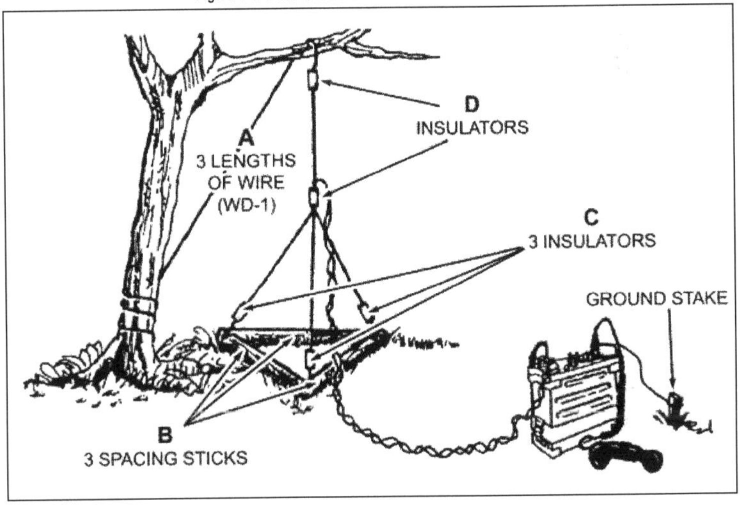

Table 7-2. QUICK REFERENCE TABLE

OPERATING FREQUENCY IN MHz	ELEMENT LENGTH (RADIATING ELEMENTS AND GROUND-PLANE ELEMENTS)	
30	2.38 m	(7 ft 10 in)
32	2.23 m	(7 ft 4 in)
34	2.1 m	(6 ft 11 in)
36	1.98 m	(6 ft 6 in)
38	1.87 m	(6 ft 2 in)
40	1.78 m	(5 ft 10 in)
43	1.66 m	(5 ft 5 in)
46	1.55 m	(5 ft 1 in)
49	1.46 m	(4 ft 9 in)
52	1.37 m	(4 ft 6 in)
55	1.3 m	(4 ft 3 in)
58	1.23 m	(4 ft 0 in)
61	1.17 m	(3 ft 10 in)
64	1.12 m	(3 ft 8 in)
68	1.05 m	(3 ft 5 in)
72	0.99 m	(3 ft 3 in)
76	0.94 m	(3 ft 1 in)

7-6. **FIELD-EXPEDIENT DIRECTIONAL ANTENNAS**. The vertical half-rhombic antenna (Figure 7-10) and the long-wire antenna (Figure 7-11) are two field-expedient directional antennas. These antennas consist of a single wire, preferably two or more wavelengths long, supported on poles at a height of 3 to 7 meters (10 to 20 feet) above the ground. The antennas will, however, operate satisfactorily as low as 1 meter (about 3 feet) above the ground--the radiation pattern is directional. The antennas are used mainly for either transmitting or receiving high-frequency signals.

Figure 7-10. VERTICAL HALF-RHOMBIC ANTENNA

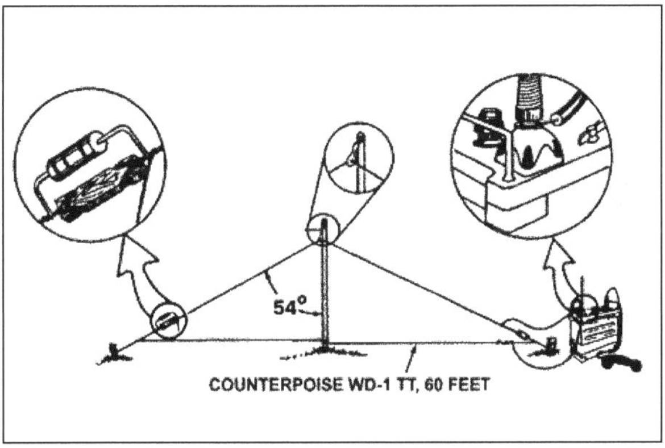

Figure 7-11. LONG-WIRE ANTENNA

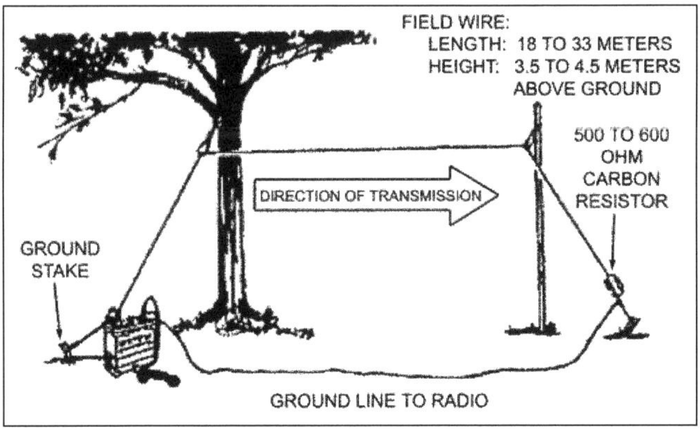

a. The vee antenna (Figure 7-12) is another field-expedient directional antenna. It consists of two wires forming a V with the open area of the "V" pointing toward the desired direction of transmission or reception. To make construction easier, the legs should slope downward from the apex of the "V." This is called a sloping vee antenna (Figure 7-13). The angle between the legs varies with the length of the legs to achieve maximum performance. (Use Table 7-1 to determine the angle and the length of the legs.)

Figure 7-12. VEE ANTENNA

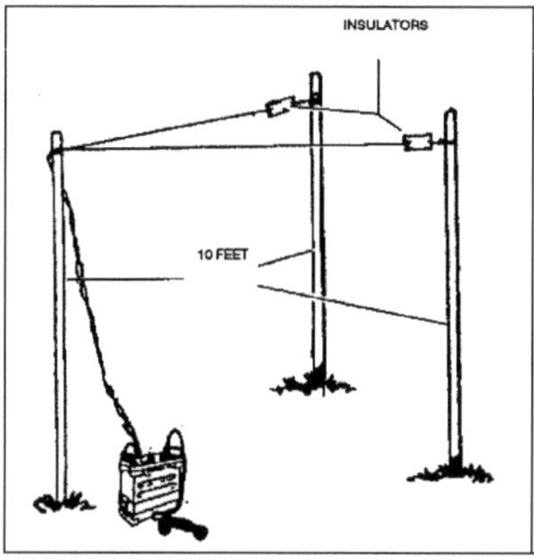

Figure 7-13. SLOPING VEE ANTENNA

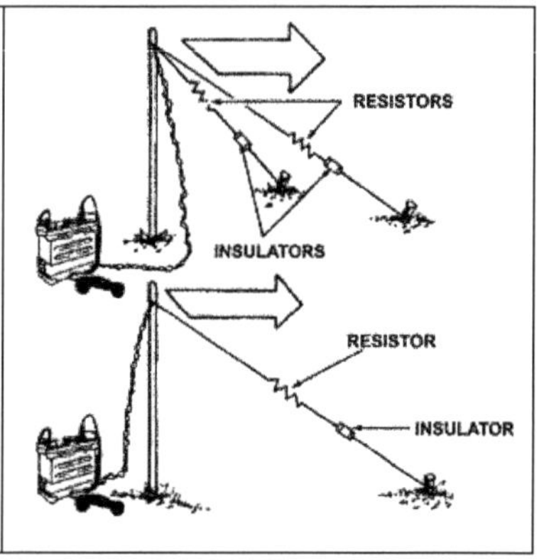

b. When the antenna is used with more than one frequency or wavelength, use an apex angle that is midway between the extreme angles determined by the chart. To make the antenna radiate in only one direction, add noninductive terminating resistors from the end of each leg (not at the apex) to ground. (See TM 11-666.)

7-7. **ANTENNA LENGTH**. The length of an antenna must be considered in two ways: both a physical and an electrical length. These two lengths are never the same. The reduced velocity of the wave on the antenna and a capacitive effect (known as end *effect*) make the antenna seem longer electrically than it is physically. The contributing factors are the ratio of the diameter of the antenna to its length and the capacitive effect of terminal equipment, such as insulators and clamps, used to support the antenna.

a. To calculate the physical length of an antenna, use a correction of 0.95 for frequencies between 3.0 and 50.0 MHz The figures given below are for a half-wave antenna.

b. Use the following formula to calculate the length of a long-wire antenna (one wavelength or longer) for harmonic operation:

EXAMPLE:

If N = the number of half-wavelengths in the total length of the antenna = 3,
and if the frequency in MHz = 7, then--

$$\text{Length (meters)} = \frac{150(N-0.05)}{\text{Frequency in MHz}} = \frac{150(3-.05)}{7} =$$

$$\frac{150 \times 2.95}{7} = \frac{442.50 = 63.2 \text{ meters}}{7}$$

7-8. **ANTENNA ORIENTATION**. If the azimuth of the radio path is not provided, the azimuth should be determined by the best available means. The accuracy required in determining the azimuth of the path depends on the radiation pattern of the directional antenna. In transportable operation, the rhombic and V antennas may have such a narrow beam as to require great accuracy in azimuth determination. The antenna should be erected for the correct azimuth. Great accuracy is not required in erecting broad-beam antennas. Unless a line of known azimuth is available at the site, the direction of the path is best determined by a magnetic compass.

7-9. **IMPROVEMENT OF MARGINAL COMMUNICATIONS**. Under certain situations, it may not be feasible to orient directional antennas to the correct azimuth of the desired radio path. As a result, marginal communications may suffer. To improve marginal communications, the following procedure can be used:

a. Check, tighten, and tape cable couplings and connections.
b. Return all transmitters and receivers in the circuit.
c. Ensure antennas are adjusted for the proper operating frequency.
d. Change the heights of antennas.
e. Move the antenna a short distance away and in different locations from its original location.

NOTES

Chapter 8
ARMY AVIATION

Army aviation and infantry units can be fully integrated with other members of the combined arms team to form powerful and flexible air assault task forces that can project combat power throughout the entire depth, width, and breadth of the modern battlefield with little regard for terrain barriers. These combat operations are deliberate, precisely planned, and vigorously executed. They are assigned to strike the enemy when and where he is most vulnerable (See page 2-28, *Air Movement Annex* and page 2-37, *Army Aviation Coordination Checklist*).

8-1. **REVERSE PLANNING SEQUENCE**. Successful air assault execution is based on a careful analysis of METT-TC and detailed, precise reverse planning. Five basic plans that comprise the reverse planning sequence are developed for each air assault operation. The battalion is the lowest level that has sufficient personnel to plan, coordinate, and control air assault operations. When company size or lower operations are conducted, the bulk of the planning takes place at battalion or higher headquarters. They are:

a. **Ground Tactical Plan**. The foundation of a successful air assault operation is the commander's ground tactical plan. All additional plans must support this plan. The plan specifies actions in the objective area to ultimately accomplish the mission and address subsequent operations.

b. **Landing Plan**. The landing plan must support the ground tactical plan. This plan outlines a sequence of events that allows elements to move into the area of operations, and ensures that units arrive at designated locations at prescribed times prepared to execute the ground tactical plan.

c. **Air Movement Plan**. The air movement plan is based on the ground tactical and landing plans. It specifies the schedule and provides instructions for air movement of troops, equipment, and supplies from PZs to LZs.

d. **Loading Plan**. The loading plan is based on the air movement plan. It ensures that troops, equipment, and supplies are loaded on the correct aircraft. Unit integrity is maintained when aircraft loads are planned. Cross-loading may be necessary to ensure survivability of command and control assets, and that the mix of weapons arriving at the LZ is ready to fight.

e. **Staging Plan**. The staging plan is based on the loading plan and prescribes the arrival time of ground units (troops, equipment and supplies) at the PZ in the order of movement

8-2. **SELECTION AND MARKING OF PICKUP AND LANDING ZONES**

a. Small unit leaders should consider the following when selecting a PZ/LZ:

(1) ***Size***. Minimal circular landing point separation from other aircraft and obstacles is needed:

- OH-58D – 25 meters.
- UH-1, AH-1 – 35 meters.
- UH-60, AH-64 – 50 meters.
- Cargo helicopters – 80 meters.

(2) ***Surface Conditions***. Avoid potential hazards such as sand, blowing dust, snow, tree stumps, or large rocks.

(3) ***Ground Slope***.

- 0% - 6 % -- land upslope.
- 7% - 15% -- land sideslope.
- over 15% -- no touchdown (aircraft may hover).

(4) ***Obstacles***. An obstacle clearance ratio of 10 to 1 is used in planning approach and departure of the PZ and LZ, for example, a 10-foot tall tree requires 100 feet of horizontal distance for approach or departure. Obstacles will be marked with a red chemlight at night, or red panels in daytime. Avoid using markings if the enemy would see them.

(5) ***Approach/Departure***. Approach and departure are made into the wind and along the long axis of the PZ/LZ.

(6) ***Loads***. The greater the load, the larger the PZ/LZ must be to accommodate the insertion or extraction.

b. **Marking of PZs and LZs**.

(1) ***Day***. A ground guide will mark the PZ or LZ for the lead aircraft by holding an M16/M4 rifle over his head, by displaying a folded VS-17 panel chest high, or by other coordinated and identifiable means.

(2) ***Night***. The code letter "Y" (inverted "Y") is used to mark the landing point of the lead aircraft at night (Figure 8-1). Chemical lights or "beanbag" lights are used to maintain light discipline. A swinging chemlight may also be used to mark the landing point.

Figure 8-1. INVERTED "Y"

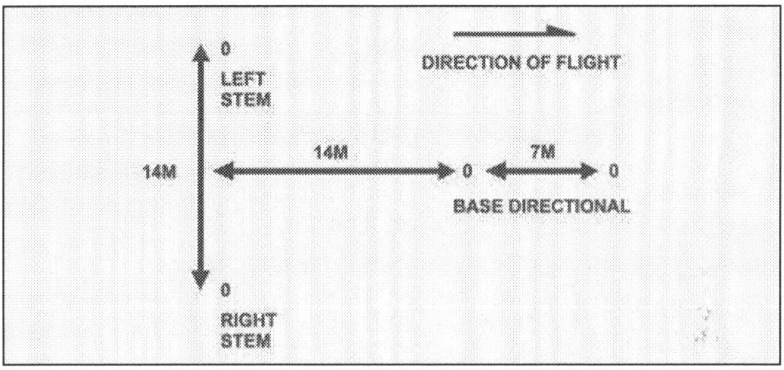

8-3. **AIR ASSAULT FORMATIONS**. Aircraft supporting an operation may use any of the following PZ/LZ configurations which are prescribed by the air assault task force (AATF) commander working with the air mission commander (AMC):

a. ***Heavy Left or Right***. Requires a relatively long, wide landing area; presents difficulty in pre-positioning loads; restricts suppressive fire by inboard gunners; provides firepower to front and flank (Figure 8-2).

Figure 8-2. HEAVY LEFT/HEAVY RIGHT

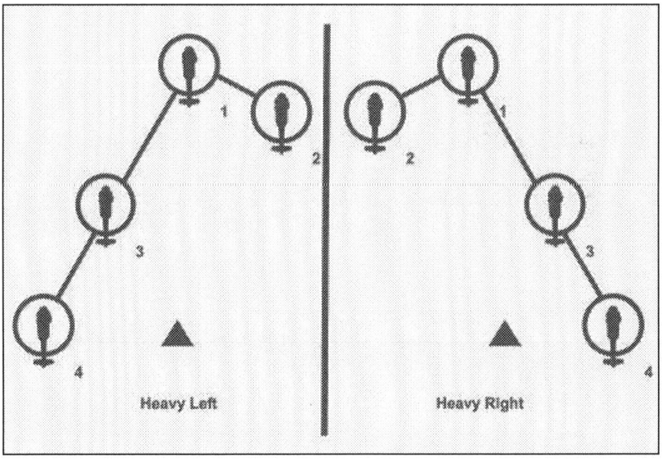

b. **Diamond**. Allows rapid deployment for all-round security; requires small landing area; presents some difficulty in pre-positioning loads; restricts suppressive fire of inboard gunners (Figure 8-3).

Figure 8-3. DIAMOND

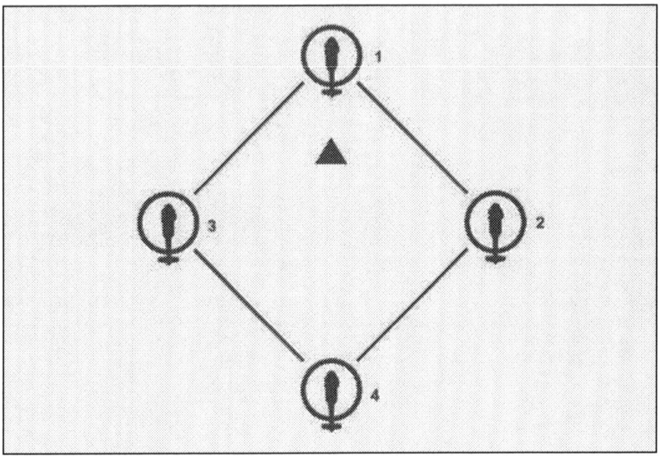

c. **Vee**. Requires a relatively small landing area; allows rapid deployment of forces to the front; restricts suppressive fire of inboard gunners; presents some difficulty in pre-positioning loads (Figure 8-4).

Figure 8-4. VEE

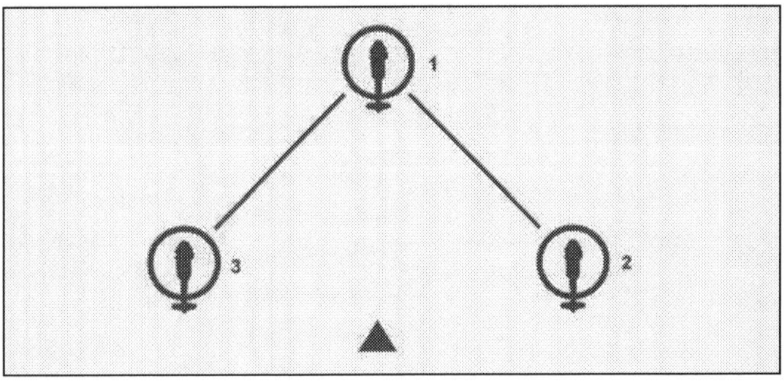

d. **Echelon Left or Right**. Requires a relatively long, wide landing area; presents some difficulty in pre-positioning loads; allows rapid deployment of forces to the flank; allows unrestricted suppressive fire by gunners (Figure 8-5).

Figure 8-5. ECHELON LEFT/ECHELON RIGHT

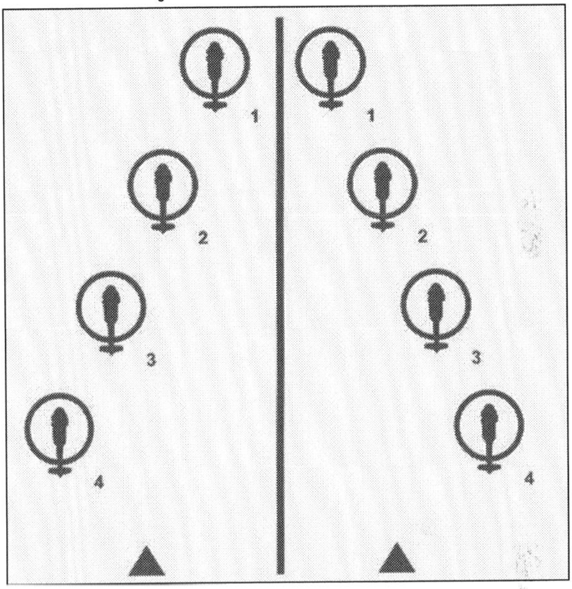

e. **Trail**. Requires a relatively small landing area; allows rapid deployment of forces to the flank; simplifies pre-positioning loads; allows unrestricted suppressive fire by gunners (Figure 8-6).

Figure 8-6. TRAIL

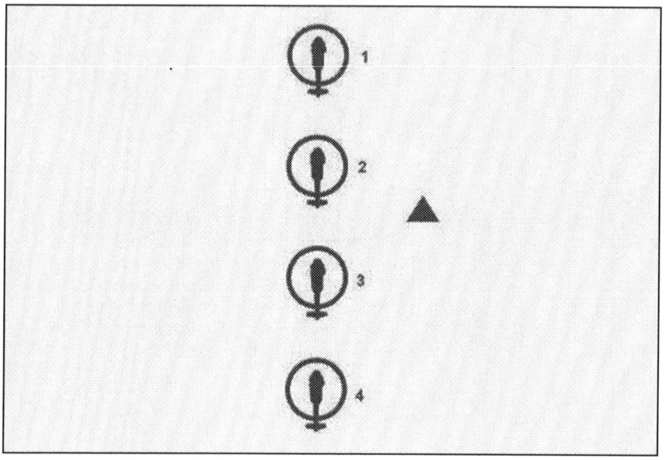

f. **Staggered Trail Left or Right**. Requires a relatively long, wide landing area; simplifies pre-positioning loads; allows rapid deployment for all round security; gunners' suppressive fire restricted somewhat (Figure 8-7).

Figure 8-7. STAGGERED TRAIL LEFT/STAGGERED TRAIL RIGHT

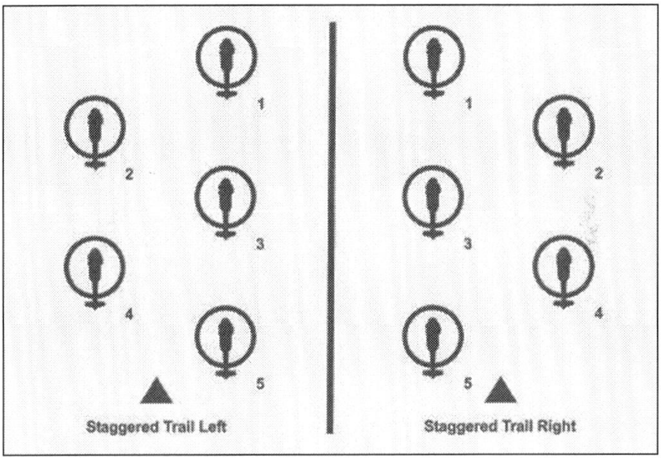

8-4. **PICKUP ZONE OPERATIONS**. Prior to arrival of aircraft, the PZ is secured, PZ control party is positioned, and the troops and equipment are positioned in platoon/squad assembly areas. In occupying a patrol/squad assembly area, the patrol/squad leader does the following. Figure 8-8 shows an example of a large, one-sided PZ. Figures 8-8 through 8-12 show loading/unloading procedures and techniques:

- Maintain all-around security of the assembly area.
- Maintain communications.
- Organize personnel and equipment into chalks and loads.
- Conduct safety briefing and equipment check of troops.

Figure 8-8. LARGE, ONE SIDED PZ

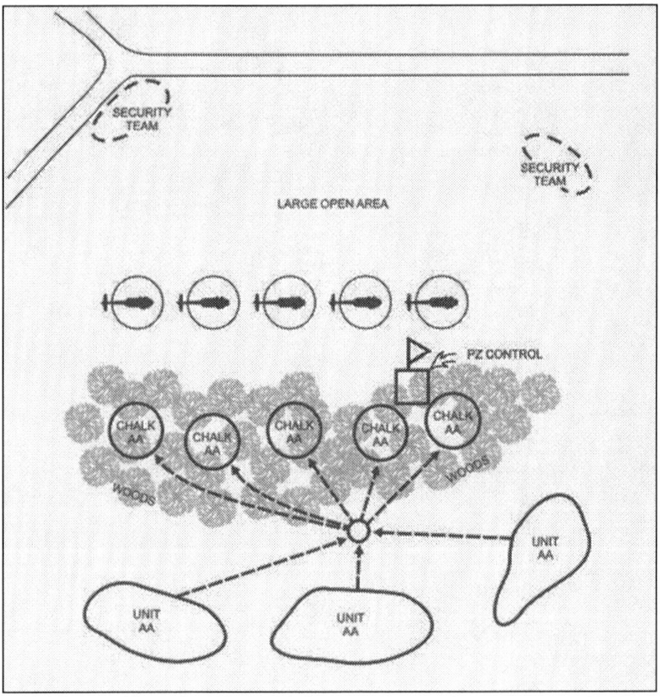

Figure 8-9. UH-60 LOADING SEQUENCE

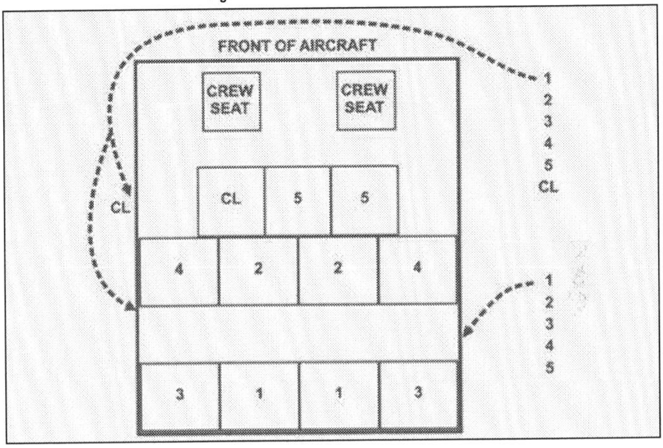

Figure 8-10. UNLOADING SEQUENCE

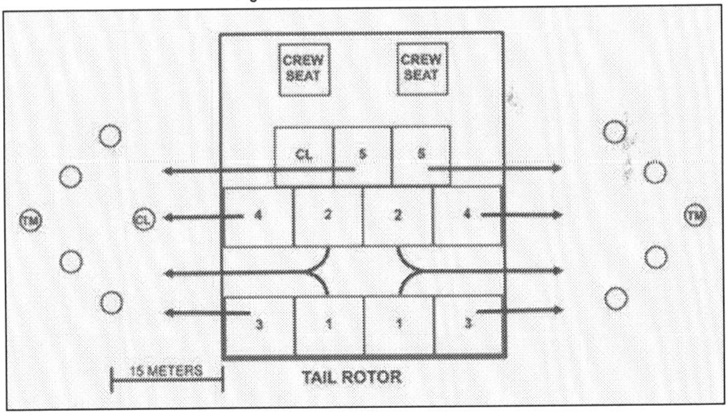

Figure 8-11. TACTICAL LOADING SEQUENCE

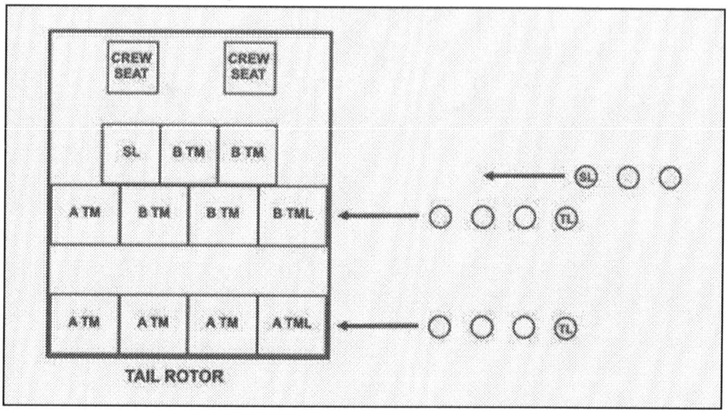

Figure 8-12. TACTICAL UNLOADING USING DOOR NEAREST COVER, CONCEALMENT

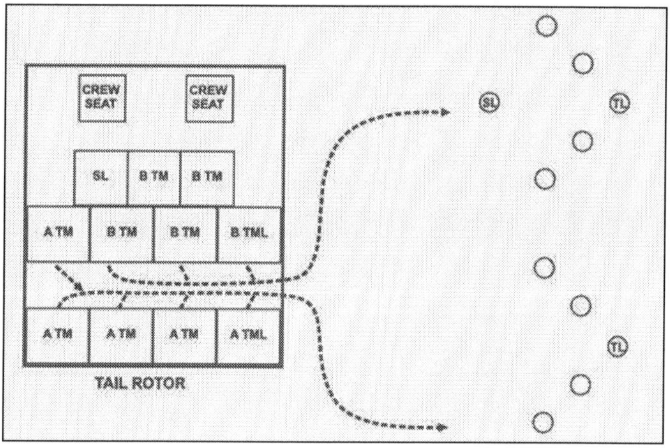

8-5. **SAFETY**. Safety is the primary concern of all leaders when operating in/around aircraft. The inclusion of aircraft into Ranger operations carries an inherent high-risk factor. The following guidelines are to be considered.

a. Approach the aircraft from 90 degrees to 45 degrees off the nose.

b. Weapons with blank firing adapters are pointed muzzles up.

c. Weapons loaded with live ammunition are muzzles down.

d. Wear the ballistic helmet.

e. When possible, conduct an air crew safety brief with all personnel. At a minimum, cover loading/off-loading, emergencies, and egress procedures.

f. Leaders need to carry a manifest and turn a copy into higher.

8-6. **CLOSE COMBAT ATTACK AVIATION**. Close combat attack is defined as a hasty or deliberate attack in support of units engaged in close combat. During CCA, armed helicopters engage enemy units with direct fire that impacts near friendly forces. Targets may range from a few hundred meters to a few thousand meters. Close combat attack is coordinated and directed by a team, platoon or company level ground unit soldiers using standardized CCA procedures in unit SOP's.

a. **Aircraft Capabilities and Limitations**.

(1) AH-64D (Apache):

- Air speed (knots): 164 max, 120 Cruise combat radius: 200 KM
- 30 mm chain gun: 1200 rds, 3,500 m range
- 2.75" rockets: 19 per pod (76), 3-5 km range
- Hellfire Missile: 8 per side (16), 5-8 km range

(2) Capabilities.

- MOBILITY - AO can cover corps or division area
- SPEED - 100-120 knots day/80-100 Knots night
- VERSATILITY - scout-weapons teams vs. pure attack
- LETHALITY - Attack bn can engage 288 targets
- VIDEO RECONNAISSANCE - Provide near real time intelligence

(3) ***Limitations***.

- Threat ID with FLIR
- Low ceilings (Clouds) <500' AGL degrades hellfire capability
- Combat service support consume large amounts of class III, V, and IX

(4) **OH-58D (Kiowa)**.

- Air speed (knots): 125 max, 100 cruise combat radius: 120 KM
- .50 cal MG: 500 rds, 2,000 m range
- 2.75" rockets: 7 per pod (14), 3-5 km range
- Hellfire: 2 per side (4), 5-8 km range
- Stinger: 2 per side (4), 4 km range

(5) ***Capabilities***.

- Mast-mounted sight
- Ability to designate targets while remaining masked
- Thermal imaging system (day/night)
- Laser designator / Aim1 laser
- Video image crosslink (VIXL)
- Moving map display
- Video

(6) ***Limitations***.

- Power limited
- Infrared crossover
- Battlefield obscurants
- Low ceilings - Hellfire
- Remote designation constraints
- Instrument meteorological conditions

c. **CCA Call-for-Fire Format**.

(1) IP/BP/ABF or friendly location_________
- Grid/Lat Long

(2) Known point
- Terrain feature

(3) HDG to TGT ________ (magnetic)
- Specify from IP/BP/ABF or friendly location

(4) DST to TGT _________ (m) (Specify from IP/BP/ABF or friendly location.)

(5) TGT elevation_________ (feet, mean sea level)

(6) TGT Description _________________

(7) TGT location _____________
- Grid
- Known point
- Terrain Feature

(8) Type TGT Mark __________ (day/night)

(9) location of friendly __________
- Omit if previously given
- Grid/Lat Long
- Known point
- Terrain feature

(10) Egress direction___________
- Cardinal direction not to over fly threats

d. **Marking Method Chart**. Table 8-1 shows this chart.

Table 8-1. MARKING METHOD CHART

Method	D	N	NVG	NVS	Friendly	Target
Smoke	X				X	X
Smoke IR	X		X		X	X
Signal Mirror	X				X	
IR Laser			X		X	X
Tracer	X	X	X			X
Glint Tape				X	X	
CID	X				X	
Strobe		X	X		X	
IR Strobe			X		X	
IR Panel				X	X	
Chem Light		X	X		X	
AN/PAQ-4			X			X
VS-17 Panel	X				X	
Spot Light		X	X		X	
MRE Heater				X	X	
AN/PEQ-2			X			X
Hydra 70 Illum	X	X	X	X		X
Ref FM 3-04.111 Appendix Q for more detail						

e. **Risk-Estimate Distances**. REDs for CCA munitions are shown in Table 8-2.

Table 8-2. RISK ESTIMATE DISTANCES FOR CCA MUNITIONS

WEAPON	*DANGER CLOSE*	*MAX EFF RANGE)*
50 cal	150 meters	2,000 meters
30 mm	150 meters	3,500 meters
2.75 rocket	175 meters	8,000 meters
Hellfire Missile	500 meters	8,000 meters
Danger Close = Risk-Estimate Distances for 0.1% (1/10th of a percent) probability of incapacitation (PI) per FM 3-09.32.		

Chapter 9
WATERBORNE OPERATIONS

This chapter discusses rope bridges, poncho rafts, and watercraft.

9-1. **ROPE BRIDGE TEAM.** The Ranger patrol seldom has ready-made bridges, so they must know how to employ field-expedient stream crossing techniques.

a. **Personnel**.

(1) # 1 man................Lead safety swimmer and far side lifeguard.

(2) # 2 man................Rope puller, swims water obstacle pulling 150-foot rope, ties off rope on far side anchor point.

(3) # 3 man................Near side lifeguard is the last man to cross water obstacle.

(4) # 4 man................Bridge team commander (BTC; most knowledgeable person on team).

(5) # 5/6 men............Rope tighteners.

b. **One-Rope Bridge (Wet Crossing).** Special equipment follows:

- Two snaplinks per piece of heavy equipment.
- Two snaplinks for every 120 feet of rope.
- One 14-foot utility rope per person (Swimmer safety line).
- Two snaplinks per person.
- One waterproof bag per RTO.
- Two B-7 life preservers.
- Three flotation work vests.
- Two 150-foot nylon ropes.

c. **Planning**. A stream crossing annex is prepared in conjunction with the unit's operation order. Special organization is accomplished at this time. For a platoon size patrol, a squad is normally given the task of providing the bridge team. The squad leader designates the most technically proficient soldier in the squad as the bridge team commander (BTC).

d. **Rehearsals and Inspections**.

(1) Rehearse the entire stream crossing, emphasizing--

- Security and actions on enemy contact.
- Actual construction of the rope bridge within the 8-minute time standard on dry land.
- Individual preparation.
- Order of crossing.
- All signals and control measures.
- Reorganization.

(2) Conduct rehearsals as realistically as possible.

(3) Ensure personnel are proficient in the mechanics of a stream crossing operation.

(4) Inspect for equipment completeness, correct rigging and preparation, personnel knowledge and understanding of the operation.

(5) Actions of the # 4 man (BTC) during the preparation phase.

(a) Rehearse the bridge team.

(b) Account for all equipment in the bridge kit.

(c) Ensure 150-foot rope is coiled.

e. **Execution Phase**.

(1) Steps for the establishment and conduct of bridge stream crossing:

(a) Leader halts short of the river, local security is established, and a recon is conducted of the area for the presence of the enemy and for crossing site suitability/ necessity. He directs the BTC to construct the bridge.

(b) Security is established up and downstream while unit leader briefs BTC on anchor points. The unit leader counts individuals across.

(c) The bridge team begins to establish the rope bridge while unit members begin individual preparation.

(d) The BTC is responsible for construction of one-rope bridge and selection of the near side anchor point as well as the far side anchor point if visibility permits. He will tie a swimmer's safety line to anchor himself into the bridge. Tying an end of the line bowline around the waist secured with an overhand knot, and on the free running end, an end of line bowline with an overhand knot. A snaplink will be placed on the loop portion of the end of line bowline, which will extend no further than arms length away from the soldier (standard waterborne uniform). This will ensure that the soldier is never more than one arms length away from the rope bridge should he lose his grip.

(e) Noise and light discipline are enforced and security is maintained.

(f) Individual soldiers put a snaplink in their end of the line bowline and the front sight assembly of every M-4/M16 or M203. M240 gunners put a snaplink through the front sight assembly and rear swivel of their M240 MG. RTOs (and others with heavy rucksacks) place an additional snaplink on their rucksack frame, top center.

(2) The responsibilities of the bridge team while constructing the rope bridge are as follows:

(a) # 1 man (Lead safety swimmer and far side lifeguard). He grounds his rucksack (with snaplink through top of frame) to the rear of the near side anchor point. He wears equipment in the following order (body out). He carries a safety line to assume duties of far side lifeguard:

-- Waterborne uniform (top button buttoned, pants unbloused)
-- B7 life preserver or engineer work vest
-- LBV
-- Weapon (across the back)

(b) # 1 man enters the water upstream from the # 2 man and stays an arms length away from the # 2 man on the upstream side. He identifies the far side anchor point upon exiting the water and once the # 2 man has exited the water moves to his far side lifeguard position downstream of the rope bridge with knotted safety line on wrist, LBV LBV/weapon grounded, and work vest held in throwing hand. He continues to wear the B7.

(c) # 2 man (rope puller) in waterborne uniform (same as # 1 man) wears his equipment in the following order. He grounds his rucksack (with snaplink through top of frame) to the rear of the near side anchor point. His duties are to swim across the water obstacle pulling the rope. He ties off the rope on the anchor point identified by the # 1 man with a round turn and two half hitches with a quick release. The direction of the round turn is the same direction as the flow of water (current) to facilitate exit off the rope bridge:

- Work vest.
- LBV.
- Weapon (across the back).
- Swimmer safety line.

(d) # 3 man (near side lifeguard) in the same waterborne uniform as the far side lifeguard. The # 3 man positions himself on the downstream side of the bridge before the # 1/2 men enter the water. He (# 1 man) grounds his rucksack (with snaplink through top of frame) on rear of near-side anchor point. After the PSG crosses and verifies the headcount, the # 1 man unties the quick release at the near side anchor point. The # 3 man reties his safety line into an Australian rappel seat, hooks the end of line bowline into his snaplink, and connects his snaplink to the one on the end-of-line bowline . # 3 man is the last pulled across the water obstacle. Before crossing the water obstacle, he dons his equipment in the following order:

- B7.
- Work vest.
- LBV.
- Weapon.

(e) # 4 man (BTC). He is in the standard waterborne uniform with LBV and sling rope tied in safety line (round the waist bowline with end of line bowline no more than one arms length). He is responsible for construction of rope bridge and organization of bridge team. He is also responsible for back feeding the rope and tying end of line bowlines. He designates the near side anchor point, ties the Figure Eight slip of the transport tightening system, and hooking all personnel to the rope bridge. He ensures that the transport tightening knot is on the upstream side of the rope bridge. He ensures that all individuals are in the waterborne uniform, hooked into the rope facing the current with the safety line routed through the trailing shoulder of the individual's LBV and rucksack. He ensures that the weapon is hooked onto the rope. He controls the flow of traffic on the bridge. He is responsible for crossing with the # 1 man's rucksack. He is generally the next to the last man to cross (follows PSG who is keeping head count).

(f) # 5 and # 6 men (rope tighteners) wear the waterborne uniform with LBV and safety line. They tighten the transport tightening knot. They also take the rucksacks of the # 2 and # 3 men across. Once they reach the far side, # 5 and # 6 men pull the last man (# 3 man) across.

(g) The rucksacks of # 1/2/3 men are transported across by # 4/5/6 men. The rucksacks of the # 1/2/3 men are hooked into the rope by the snaplink through the top of the frames and the # 4/5/6 men pull them across. The weapons of # 4/5/6 men are attached between the # 4/5/6 men and the rucksack that they are pulling across the bridge.

(h) BTC rehearses the bridge team during the planning sessions and directs construction and emplacement. The unit leader selects the crossing site which complements the tactical plan.

- # 3 man positions himself downstream of crossing site.
- # 1 man enters water upstream of # 2. He stays one arm's length from the # 2 man and is prepared to render any assistance to the # 2 man. Both swim in conjunction upstream to compensate for the current. BTC feeds rope out of rucksack positioned on the downstream side of near side anchor point.

(i) The # 1 man exits and identifies the far side anchor point (if BTC cannot identify it for the # 2 man). # 2 man exits on the upstream side of the far side anchor point. The rope is now routed to facilitate movement onto and off the bridge.

(j) Radios and heavy equipment are waterproofed and rigged. All individuals don waterborne uniform and tie safety lines. PSG moves to anchor point and maintains accountability through headcount.

(k) The # 2 man signals the BTC that the rope is temporarily attached to the far side anchor point, and the BTC pulls out excess slack and ties the transport tightening system using a Figure Eight slip. The BTC signals the # 2 man to pull the knot twelve to fifteen feet from the near side anchor point. After this is done the # 2 two man ties round turns 18 to 24 inches off the water with the remaining rope and secures the rope, to itself, with a snaplink. The # 2 man signals the BTC and the pulling team (# 4/5/6) tightens the bridge, pulling the transport tightening system as close as possible to the near side anchor point.

(l) The # 1 man moves downstream and assumes his duties as the far side lifeguard. The bridge team commander will tie off the rope with a round turn and two half hitches around the near side anchor point. The BTC will place himself on the upstream side of the bridge (facing downstream). He begins to hook individuals into the rope inspecting them for safety.

Note: Any Ranger identified as a weak swimmer will cross with no other personnel on the rope bridge. The weak swimmer crosses individually to allow the near and far side lifeguards to focus their attention exclusively on the weak swimmer and not be distracted by other personnel crossing the bridge.

(m) The # 2 man moves to the upstream side of the rope bridge, assists personnel off the rope on the far side, and maintains head count. The # 5 and # 6 men cross with the # 1 and # 2 men's rucksacks..

(n) The BTC maintains the flow of traffic ensuring that no more than three men are on the bridge at any one time (one hooking up, one near the center, and one being unhooked). Once the PSG has accounted for everyone on the near side, he withdraws left and right (L/R) security and sends them across. PSG follows security across. The # 3 man hooks the BTC (with # 3's ruck) onto the rope. Once the BTC has crossed, the # 3 man unhooks near side anchor point and the BTC unties far side anchor point. The # 3 man ties an Australian rappel seat with snaplink to the front, hooks onto the snaplink that is in the end of the line bowline on the 120-foot rope, and then signals the # 4/5/6 men to take in slack. The # 3 man extends arms in front of his head, slightly upstream to fend off debris and is pulled across by the # 4/5/6. All individuals (except # 1/2/3 and RTOs) wear rucksacks across. The # 4/5/6 men hook the rucksacks of the # 1/2/3 men onto the bridge by the snaplink. All the men cross facing upstream.

(o) Once the far side headcount, weapons and equipment are verified (between PSG and # 5 man), personnel reorganize, and then they continue the mission.

(p) Personnel with heavy equipment:

M240 All major groups are tied together with 1/4-inch cord. An anchor line bowline runs through the rear swivel, down the left side of gun. Tie a round turn through the trigger guard. Route the cord down the right side and tie off two half hitches around the forearm assembly with a round turn and two half hitches through the front sight posts. The remainder of the working end is tied off with an end of the rope bowline approximately one foot from the front sight post large enough to place leading hand through. The M240 is secured to the bridge by snaplinks on the front sight post and rear swivel. The M240 is pulled across by the trailing arm of the M240 gunner.

AN-PRC/II9s These are waterproofed before crossing a one-rope bridge. Once far side FM commo is established, the near side RTO breaks down and waterproofs his radio, and prepares to cross the bridge. He puts a snaplink in the top center of the rucksack frame (same as for # 1/2/3 men). The BTC will hook the rucksack to the rope.

Note: Using two snaplinks binds the load on the rope. Adjust arm straps all the way out. RTO pulls radio across the rope bridge.

9-2. **PONCHO RAFT.** Normally a poncho raft is constructed to cross rivers and streams when the current is not swift. A poncho raft is especially useful when the unit is still dry and the platoon leader desires to keep the individuals equipment dry.

a. **Equipment Requirements**.

(1) Two serviceable ponchos.

(2) Two weapons (poles can be used in lieu of weapons).

(3) Two rucksacks per team.

(4) 10 feet of utility cord per team.

(5) One sling rope per team.

b. **Conditions**. Poncho rafts are used to cross water obstacles when any or all of the following conditions are found:

(1) The water obstacle is too wide for 150-foot rope.

(2) No sufficient near or far shore anchor points are available to allow rope bridge construction.

(3) Under no circumstances will poncho rafts be used as a means to cross a water obstacle if an unusually swift current is present.

c. **Choosing a Crossing Site**. Before a crossing site is used, a thorough reconnaissance of the immediate area is made. Analyzing the situation using METT-TC, the patrol leader chooses a crossing site that offers as much cover and concealment as possible and has entrance and exit points that are as shallow as possible. For speed of movement, it is best to choose a crossing site that has near and far shore banks that are easily traversed by an individual Ranger.

d. **Execution Phase**. Steps for the construction of a poncho raft:

(1) Pair off the unit/patrol in order to have the necessary equipment.

(2) Tie off the hood of one poncho and lay out on the ground with the hood up.

(3) Weapons are then placed in the center of the poncho, approximately 18 inches apart, muzzle to butt.

(4) Next, rucksacks and LBV are placed between the weapons with the two individuals placing their rucksacks as far apart as possible.

(5) The two will then start to undress (bottom to top), first with their boots, taking the laces completely out for subsequent use as tie downs if necessary).

(6) The boots are then placed over muzzle/butt of weapon toe in.

(7) Members continue to undress, folding each item neatly and placing it on top of their boots.

(8) Once all of the equipment is placed between the two weapons or poles, the poncho is snapped together. The snapped portion of the poncho is then lifted into the air and tightly rolled down to the equipment. Start at the center and work out to the end of the raft creating pigtails at the end. This is accomplished much easier if done by both soldiers together. The pigtailed ends are then folded in toward the center top of the raft and tied off with a single boot lace.

(9) The other poncho is then laid out on the ground with the hood up and the first poncho with equipment is placed in its center. The second poncho is then snapped, rolled and tied in the same manner as the first poncho. The third and fourth boot laces (or utility cord) are then tied around the raft approximately one foot from each end for added security. The poncho raft is now complete.

Note: The patrol leader must analyze the situation using METT-TC and make a decision on the uniform for crossing the water obstacle such as weapons inside the poncho raft or slung across the back, remaining dressed or stripped down with clothes inside raft.

9-3. **WATERCRAFT.** Use of inland and coastal waterways may add flexibility, surprise, and speed to tactical operations. Use of these waterways will also increase the load carrying capacity of normal dismounted units. Watercraft are employed in reconnaissance and assault operations.

a. **Inflation Method**. Inflatable with foot pumps using four separate valves located on the inside of the buoyancy tubes. Each of the four valves are used to section off the assault boat into eight separate airtight compartments. To pump air into the boat, turn all valves into the "orange" or "inflate" section of the valve. Once the assault boat is filled with air, turn all valves onto the "green" or "navigation" section. This will section the assault boat into eight separate compartments.

b. **Characteristics**.

- Overall length: 15 feet, 5 inches.
- Overall width: 6 feet, 3 inches.
- Weight: 265 pounds.
- Maximum payload: 2,756 pounds.
- Crew : 1 coxswain + either 10 paddlers or a 65 HP short-shaft outboard motor.

c. **Preparation**.

(1) ***Rubber Boat***.

- All rubber boats will have a twelve-foot bowline secured to the front starboard D-Ring. This rope will be tied with an anchor-line bowline, and the knot will be covered with 100-MPH tape.
- All rubber boats will have a fifteen foot center line tied to the rear floor D-Ring. The same procedure for securing the bowline will be used for the centerline.
- All rubber boats will be filled to 240 millibars of air, and checked to ensure that all valve caps are tight, and set in the NAVIGATE position.
- All rubber boats will be equipped with one foot pumps. These will be placed in the front pouches of the rubber boat. If no pouches are present, the foot pumps will be placed on the floor.
- All rubber boats will be inspected with the maintenance chart.

(2) ***Personnel and Equipment***.

- All personnel will wear work vest or kapoks (or another suitable positive flotation device).
- LBV is worn over the work vest, unbuckled at the waist.
- Individual weapon is slung across the back, muzzle pointed down and facing toward the inside of the boat.
- Crew served weapons, radios, ammunition and other bulky equipment is lashed securely to the boat to prevent loss if the boat should overturn. Machine guns with hot barrels are cooled prior to being lashed inside the boats.
- Radios and batteries are waterproofed.
- Pointed objects are padded to prevent puncturing the boat.

NOTES

d. **Positions.** Assign each individual a specific boat position (Figure 9-1).

e. **Duties**.

(1) Designate a commander for each boat, (normally coxswain)

(2) Designate a navigator (normally a leader within the platoon) - observer team as necessary.

(3) Crew is positioned as shown in Figure 9-2.

(4) Duties of the Coxswain.

(a) Responsible for control of the boat and actions of the crew.

(b) Supervises the loading, lashing and distribution of equipment.

(c) Maintains the course and speed of the boat.

(d) Gives all commands.

(5) # 2 paddler (long count) is responsible for setting the pace.

(6) # 1 paddler is the observer, stowing and using the bowline unless another observer is assigned.

f. **Embarkation and Debarkation Procedures**.

(1) When launching, the crew will maintain a firm grip on the boat until they are inside it: similarly, when beaching or debarking, they hold on to the boat until it is completely out of the water. Loading and unloading is done using the bow as the entrance and exit point.

(2) Keep a low center of mass when entering and existing the boat to avoid capsizing. Maintain 3 points of contact at all times.

(3) The long count is a method of loading and unloading by which the boat crew embarks or debarks individually over the bow of the boat. It is used at river banks, on loading ramps, and when deep water prohibits the use of the short count method.

(4) The short count is a method of loading or unloading by which the boat crew embarks or debarks in pairs over the sides of boat while the boat is in the water. It is used in shallow water allowing the boat to be quickly carried out of the water. The short count method of organization is primarily used during surf operations.

(5) Beaching the boat is a method of debarking the entire crew at once into shallow water and quickly carrying the boat out of the water.

g. **Commands**. Commands are issued by the coxswain to ensure the boat is transported over land and controlled in the water. All crew members learn and react immediately to all commands issued by the coxswain. T he various commands are as follows:

(1) "Short Count...count off," Crew counts off their position by pairs, such as 1,2,3,4,5 (passenger # 1, # 2, if applicable) coxswain.

(2) "Long Count...count off," Crew counts off the position by individual, such as 1,2,3,4,5,6,7,8,9,10, (Passenger # 1, # 2, if applicable), coxswain.

(3) "Boat Stations," Crew takes positions along side the boat.

(4) "High Carry...Move," (used for long distance move overland).

(a) On the preparatory command of "High Carry," the crew faces the rear of the boat and squats down grasping carrying handles with the inboard hand.

(b) On the command "Move," the crew swivels around, lifting the boat to the shoulders so that the crew is standing and facing to the front with the boat on their inboard shoulders.

(c) Coxswain guides the crew during movement.

(5) "Low Carry...Move," (used for short distance moves overland).

(a) On preparatory command of "Low Carry," the crew faces the front of the boat, bent at the waist, and grasps the carrying handles with their inboard hands.

(b) On the command of "Move," the crew stands up straight raising the boat approximately six to eight inches off the ground.

(c) Coxswain guides the crew during movement.

(6) "Lower the Boat----Move," Crew lowers the boat gently to the ground using the carrying handles.

(7) "Give Way Together," Crew paddles to front with # 2 setting the pace.

(8) "Hold," Entire crew keeps paddles straight downward motionless in the water stopping the boat.

(9) "Left side hold (Right)," Left crew holds, right continues with previous command.

(10) "Back paddle," Entire crew paddles backward propelling the boat to the rear.

Figure 9-1. BOAT POSITIONS

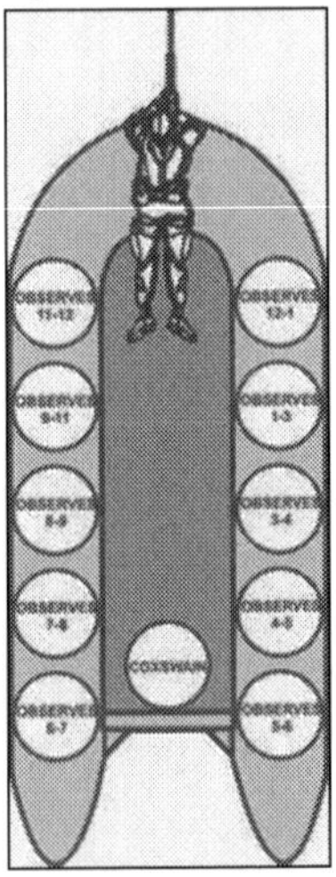

Figure 9-2. CREW POSITIONS, LONG COUNT AND SHORT COUNT

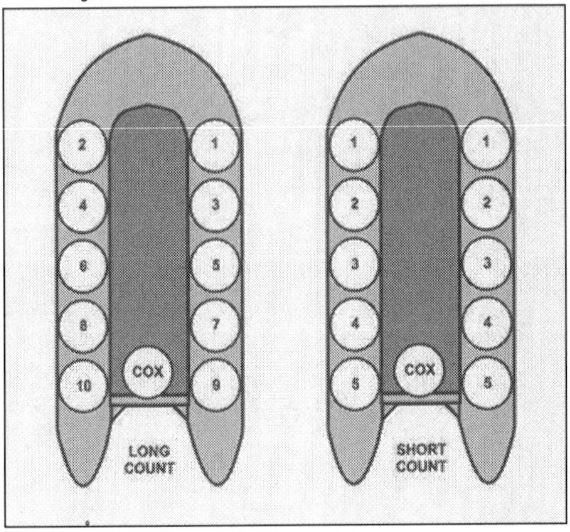

(11) "Back Paddle Left" (Right), Left crew back paddles causing the boat to turn left, right crew continues with previous command.

(12) "Rest Paddles," Crew members place paddles on their laps with blades outboard. This command may be given in pairs such as "# 1s, rest paddles").

h. **Securing of Landing Site.**

(1) If the landing site cannot be secured prior to the waterborne force landing, some form of early warning, such as scout swimmers, is considered. These personnel swim to shore from the assault boats and signal the boats to land. All signals and actions are rehearsed prior to the actual operation.

(2) If the patrol is going into an unsecured landing site, it can provide security by having a security boat land, reconnoiter the landing site and then signal to the remaining boats to land. This is the preferred technique.

(3) The landing site can be secured by force with all the assault boats landing simultaneously in a line formation. While this is the least desirable method of securing a landing site, it is rehearsed in the event the tactical situation requires its use.

(4) Arrival at the debarkation point.

(a) Unit members disembark according to leaders order (Figure 9-3).

(b) Local security is established.

(c) Leaders account for personnel and equipment.

(d) Unit continues movement.

- Soldiers pull security initially with work vest on.
- Coxswains and two men unlash and de-rig rucksacks.
- Soldiers return in buddy teams to secure rucksack and drop off work vest.
- Boats are camouflaged/cached if necessary prior to movement.

Figure 9-3. DEBARKATION

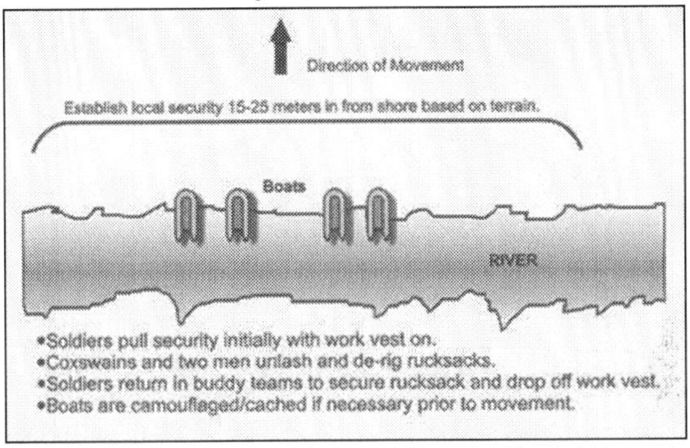

i. **Capsize Drill.** The following commands and procedures are used for capsize drills or to right an over-turned boat,

(1) ***PREPARE TO CAPSIZE***. This command alerts the crew and they raise paddles above their heads, with the blade pointed outward. Before capsizing, the coxswain will conduct a long count.

(2) ***PASS PADDLES***. All paddles are passed back and collected by the # 9/10 men.

(3) ***CAPSIZE THE BOAT***. All personnel slide into the water except the # 3/5/7 men. The # 1 man secures the bowline. The three men in the water grasp the capsize lines (ensuring the lines are routed under the safety lines) and stand on the buoyancy tubes opposite the capsize lines anchor points. The boat is then turned over by the # 3/5/7 men, who lean back and straighten their legs while pulling back on the capsize lines. As the boat lifts off the water, the # 4 man grasps the center carrying handle and rides the boat over. Once the boat is over, the # 4 man helps the # 3/7 men back onto the boat. The # 5 man holds onto the center carrying handle and turns the boat over the same way. The # 5 man rides the boat back over and helps the rest of the crew into the boat. As soon as the boat is capsized, the coxswain commands a long count to ensure that no one sank or was trapped under the boat. Every time the boat is turned over, he conducts another long count.

j. **River Movement.**

(1) ***Characteristics of River***.

(a) Know local conditions prior to embarking on river movement.

(b) A bend is a turn in the river course.

(c) A reach is a straight portion of river between two curves.

(d) A slough is a dead end branch from a river. They are normally quite deep and can be distinguished from the true river by their lack of current

(e) Dead water is a part of the river, due to erosion and changes in the river course that has no current. Dead water is characterized by excessive snags and debris.

(f) An island is usually a pear-shaped mass of land in the main current of the river. Upstream portions of islands usually catch debris and are avoided.

(g) The current in a narrow part of a reach is normally greater than in the wide portion.

(h) The current is greatest on the outside of a curve; sandbars and shallow water are found on the inside of the curve.

(i) Sandbars are located at those points where a tributary feeds into the main body of a river or stream.

(j) Because the # 1 and # 2 men are sitting on the front left and right sides of the boat, they observe for obstacles as the boat moves downriver. If either notices an obstacle on either side of the boat, he notifies the coxswain. The coxswain then adjusts steering to avoid the obstacle.

(2) ***Navigation***. The patrol leader is responsible for navigation. The three acceptable methods of river navigation are--

(a) Checkpoint and general route. These methods are used when the drop site is marked by a well-defined checkpoint and the waterway is not confused by a lot of branches and tributaries. They are best used during daylight hours and for short distances.

(b) Navigator-observer method. This method is the most accurate means of river navigation and is used effectively in all light conditions.

(c) *Equipment Needed.*

- Compass
- GPS
- Photo map (1st choice)
- Topographical map (2nd choice)
- Poncho (for night use)
- Pencil/Grease pencil
- Flashlight (for night use)

(d) *Procedure.* Navigator is positioned in center of boat and does not paddle. During hours of darkness, he uses his flashlight under the poncho to check his map. The observer (or # 1 man) is at the front of the boat.

(a) The navigator keeps his map and compass oriented at all time.

(b) The navigator keeps the observer informed of the configuration of the river by announcing bends, sloughs, reaches, and stream junctions as shown on his map.

(c) The observer compares this information with the bends, sloughs, reaches, and stream junctions he actually sees. When these are confirmed, the navigator confirms the boat's location on his map.

(d) The navigator also keeps the observer informed of the general azimuths of reaches as shown on his map and the observer confirms these with actual compass readings of the river.

(e) The navigator announces only one configuration at a time to the observer and does not announce another until it is confirmed and completed.

(f) A strip-map drawn on clear acetate backed by luminous tape may be used. The drawing is to scale or a schematic. It should show all curves and the azimuth and distance of all reaches. It may also show terrain features, stream junctions, and sloughs.

k. **Formations.** Various boat formations are used (day and night) for control, speed, and security (Figure 9-4). The choice of which is used depends on the tactical situation and the discretion of the patrol leader. He should use hand and arm signals to control his assault boats. The formations are:

- Wedge
- Line
- File
- Echelon
- Vee

Figure 9-4. FORMATIONS

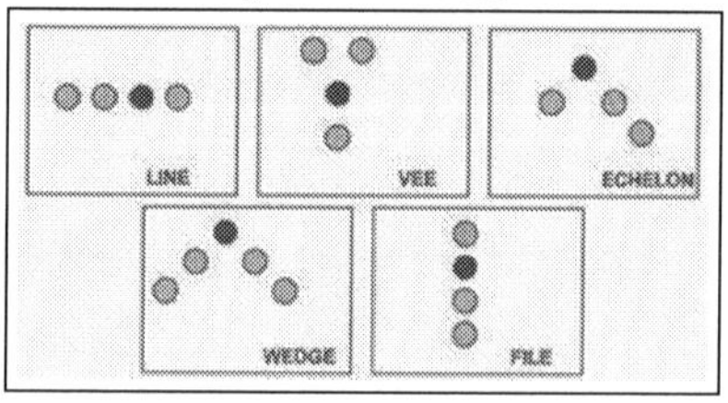

Chapter 10
MILITARY MOUNTAINEERING

In the mountains, commanders face the challenge of maintaining their units' combat effectiveness and efficiency. To meet this challenge, commanders conduct training that provides Rangers with the mountaineering skills necessary to apply combat power in a rugged mountain environment, and they develop leaders capable of applying doctrine to the distinct characteristics of mountain warfare.

10-1. **TRAINING**. Military mountaineering training provides tactical and fundamental mobility skills needed to move units safely and efficiently in mountainous terrain. Rangers should receive extensive training prior to executing combat operations in mountainous environments. Some of the areas to train are as follows:

- Characteristics of the mountain environment.
- Care and use of basic mountaineering equipment.
- Mountain bivouac techniques.
- Mountain communications.
- Mountain travel and walking techniques.
- Mountain navigation, hazard recognition and route selection.
- Rope management and knots.
- Natural and artificial anchors.
- Belay and rappel techniques.
- Installation construction and use such as rope bridges.
- Rock climbing fundamentals.
- Rope bridges and lowering systems.
- Individual movement on snow and ice.
- Mountain stream crossings (to include water survival techniques).

10-2. **DISMOUNTED MOBILITY**. Movement in more technical terrain demands specialized skills and equipment. Before Rangers can move in such terrain, a technical mountaineering team might have to secure the high ground.

10-3. **MOUNTAINEERING EQUIPMENT**. Mountaineering equipment is everything that allows the trained Ranger to accomplish many tasks in the mountains. The importance of this gear to the mountaineer is no less than that of the rifle to the Infantry Soldier.

a. **Ropes and Cords**. Ropes and cords are the most important pieces of mountaineering equipment. They secure climbers and equipment during steep ascents and descents. They are also used to install other ropes and to haul equipment. From WWII until the 1980's, US armed forces mainly used 7/16 *nylon laid rope*, often referred to as *green line* for all mountaineering operations. Since kernmantle ropes were introduced, other, more specialized ropes have in many cases replaced the all-purpose green line. Kernmantle ropes are made much like parachute cord, with a smooth sheath around a braided or woven core. Laid ropes are still used, but should be avoided whenever rope failure could cause injury or loss of equipment.

(1) ***Types***. Kernmantle rope is classified as *dynamic* or *static*:

(a) *Dynamic Ropes*. Ropes used for climbing are classified as *dynamic ropes*. These rope stretch or elongate 8 to 12 percent once subjected to weight or impact. This stretching is critical in reducing the impact force on the climber, anchors, and belayer during a fall by softening the catch. An 11-mm by 150-meter rope is standard for military use. Other sizes are available.

(b) *Static Ropes*. Static ropes are used in situations where rope stretch is undesired, and when the rope is subjected to heavy static weight. Static ropes should never be used while climbing, since even a fall of a few feet could generate enough impact force to injure climber and belayer, and/or cause anchor failure. Static Ropes are usually used when constructing rope bridges, fixed rope installations, vertical haul lines, and so on.

(2) ***Sling Ropes and Cordelettes***. A short section of static rope or static cord is referred to as a sling rope or cordelette. These are a critical piece of personal equipment during mountaineering operations. They are usually 7 to 8mm diameter and up to 21 feet long. The minimum Ranger standard is 8mm by 15 feet.

(3) ***Rope Care***. Ropes used daily should be used no longer that one year. An occasionally used rope can be used generally up to 4 to 5 years if properly cared for.

- Inspect ropes thoroughly before, during, and after use for cuts, frays, abrasion, mildew, and soft or worn spots.
- Never step on a rope or drag it on the ground unnecessarily.
- Avoid running rope over sharp or rough edges (pad if necessary).

- Keep ropes away from oil, acids, and other corrosive substances.
- Avoid running ropes across one another under tension (nylon to nylon contact will damage ropes).
- Do not leave ropes knotted or under tension longer than necessary.
- Clean in cool water, loosely coil, and hang to dry out of direct sunlight. Ultraviolet light rays harm synthetic fibers. When wet, hang rope to drip-dry on a rounded wooden peg, at room temperature (do not apply heat).

(4) ***Webbing and Slings***. Loops of tubular webbing or cord, called slings or runners, are the simplest pieces of equipment and some of the most useful. The uses for these simple pieces are endless, and they are a critical link between the climber, the rope, carabiners, and anchors. Runners are usually made from either 9/16- or 1-inch tubular webbing and are either tied or sewn by a manufacturer.

b. **Carabiners**. The carabiner is one of the most versatile pieces of equipment available in the mountains. This simple piece of gear is the critical connection between the climber, his rope, and the protection attaching him to the mountain. New metal alloys are stronger to hold hard falls; and, they are lighter than steel, which allows the Ranger to carry many at once. Consequently, steel carabiners are used less and less often. Basic carabiners come in several different shapes.

c. **Protection**. This is the generic term used to describe a piece of equipment, natural or artificial, that is used to construct an anchor. Protection, along with a climber, belayer, and climbing rope, forms the lifeline of the climbing team. The rope connects two climbers, and the protection connects them to the rock or ice. The two types of artificial protection are traditional (removable) and fixed (usually permanent) (Figures 10-1 and 10-2).

Figure 10-1. EXAMPLES OF TRADITIONAL (REMOVABLE) PROTECTION USED ON ROCK

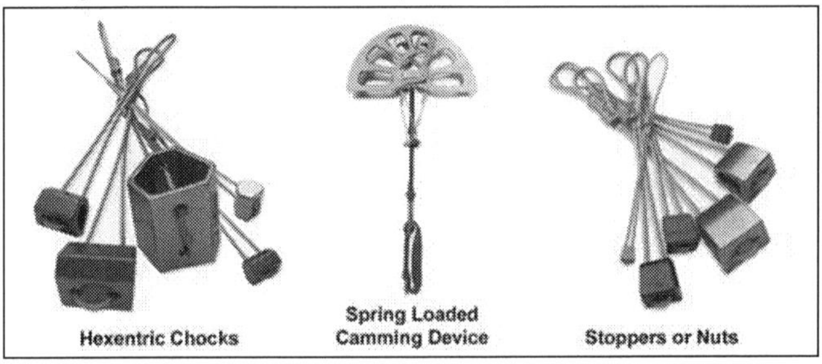

Figure 10-2. EXAMPLES OF FIXED (PERMANENT OR SEMI) PROTECTION USED ON ROCK

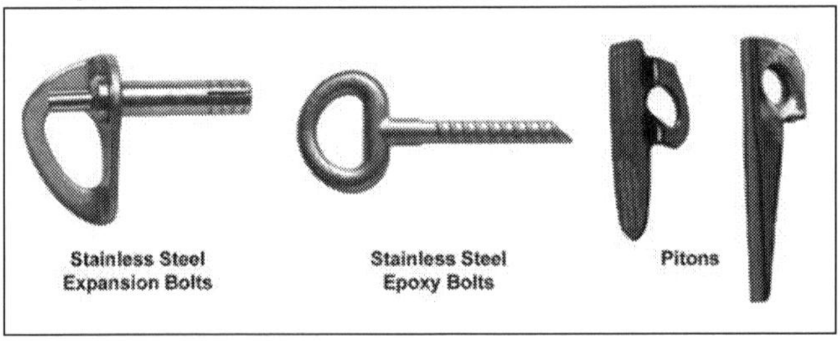

10-4. **ANCHORS**. Anchors form the base for all installations and roped mountaineering techniques. Anchors must be strong enough to support the entire weight of the load or of the impact placed upon them. Several pieces of artificial or natural protection may be incorporated together to make one multipoint anchor. Anchors are classified as artificial or natural:

a. **Artificial Anchors**. Artificial anchors are made from man-made materials. The most common ones use traditional or fixed protection (Figure 10-3).

Figure 10-3. CONSTRUCTION OF THREE-POINT, PRE-EQUALIZED ANCHOR WITH FIXED ARTIFICIAL PROTECTION

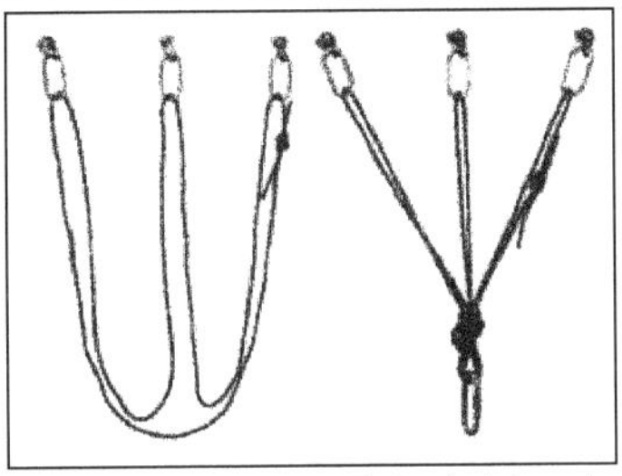

b. **Natural Anchors**. Natural anchors are usually very strong and often simple to construct with little equipment. Trees, shrubs, and boulders are the most common anchors. All natural anchors require is a method of attaching a rope. Regardless of the type of natural anchor used, it must support the weight of the load.

(1) ***Trees***. These are probably the most widely used of all anchors. In rocky terrain, trees usually have a very shallow root system. Check this by pushing or tugging on the tree to see how well it is rooted. Anchor as low as possible to keep the pull on the base of the tree, where it is nearer its own anchor and therefore stronger. Use padding on soft, sap-producing trees to keep sap off ropes and slings.

(2) ***Rock Projections and Boulders***. You can use these, but they must be heavy enough, and they must have a stable enough base to support the load.

(3) ***Bushes and Shrubs***. If no other suitable anchor is available, route a rope around the bases of several bushes. As with trees, place the anchoring rope as low as possible. Make sure all vegetation is healthy and well rooted to the ground.

(4) ***Tensionless Anchor***. Use this to anchor rope on high-load installations such as bridging. The wraps of the rope around the anchor (Figure 10-4) absorb the tension, keeping it off the knot and carabiner. Tie this anchor with at least four wraps around the anchor. A smooth anchor, such as a pipe or rail, might require several more wraps. Wrap the rope from top to bottom. Place a fixed loop into the end of the rope, and attach it loosely back onto the rope with a carabiner.

Figure 10-4. TENSIONLESS NATURAL ANCHOR

10-5. **KNOTS**.

a. **Square Knot**. This joins two ropes of equal diameter (Figure 10-5). It has two interlocking bights; the running ends exit on the same side of the standing portion of the rope. Each tail is secured with an overhand knot on the standing end. When you dress the knot, leave at least a 4-inch tail on the working end.

Figure 10-5. SQUARE KNOT

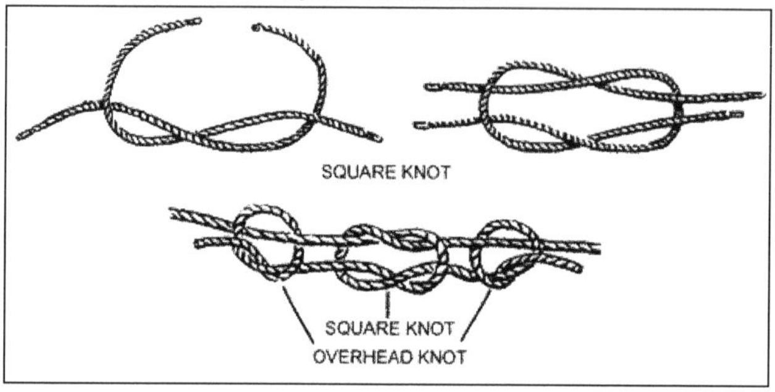

b. **Round-Turn with Two Half Hitches**. This is a constant tension anchor knot (Figure 10-6). The rope forms a complete turn around the anchor point (thus the name round turn), with both ropes parallel and touching, but not crossing. Both half hitches are tightly dressed against the round turn, with the locking bar on top. When you dress the knot, leave at least a 4-inch tail on the working end.

Figure 10-6. ROUND-TURN WITH TWO HALF-HITCHES

c. **End-of-the-Rope Clove Hitch**. The end-of-the-rope clove hitch (Figure 10-7) is an intermediate anchor knot that requires constant tension. Make two turns around the anchor. A locking bar runs diagonally from one side to the other. Leave no more than one rope width between turns of rope. Locking bar is opposite direction of pull. When you dress the knot, leave at least a 4-inch tail on the working end.

Figure 10-7. END-OF-THE-ROPE CLOVE HITCH

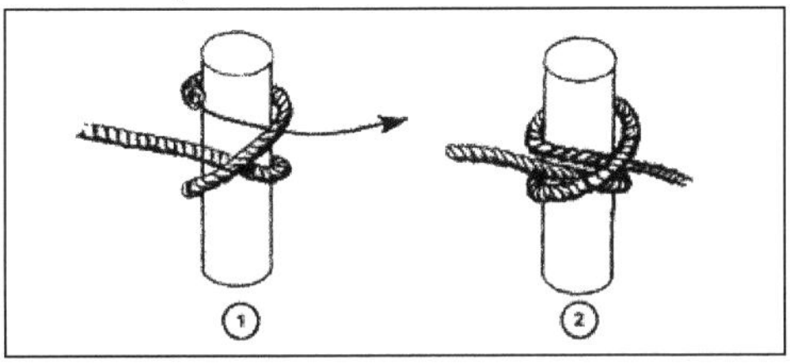

d. **Middle of the Rope Clove Hitch**. This knot (Figure 10-8) secures the middle of a rope to an anchor. The knot forms two turns around the anchor. A locking bar runs diagonally from one side to the other. Leave no more than one rope width between turns. Ensure the locking bar is opposite the direction of pull.

Figure 10-8. MIDDLE-OF-THE-ROPE CLOVE HITCH

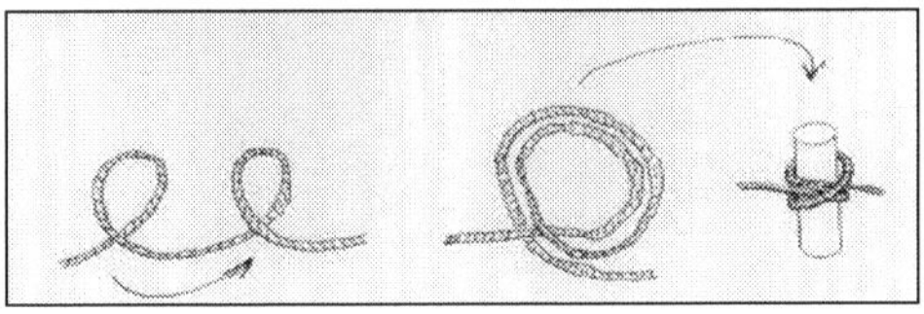

e. **Rappel Seat**. The rappel seat (Figure 10-9) is a rope harness used in rappelling and climbing. It can be tied for use with the left or right hand. Leg straps do not cross, and are centered on buttocks and tight. Leg straps form locking half hitches on rope around waist. A square knot is tied on the right hip and finished with two overhand knots. Tails must be even, no longer than 6 inches. A carabiner is inserted around all ropes, with the gate opening up and away. The carabiner must not contact the square or overhand knot. The rappel seat must be tight enough to prevent a fist from fitting between the rappeller's body and the harness.

Figure 10-9. RAPPEL SEAT

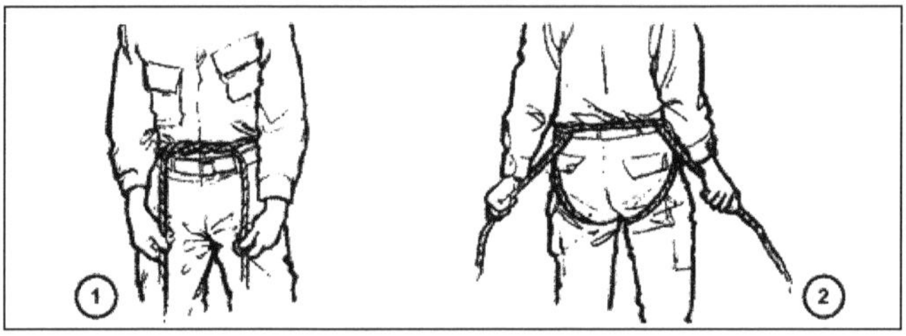

Figure 10-9. RAPPEL SEAT (CONTINUED)

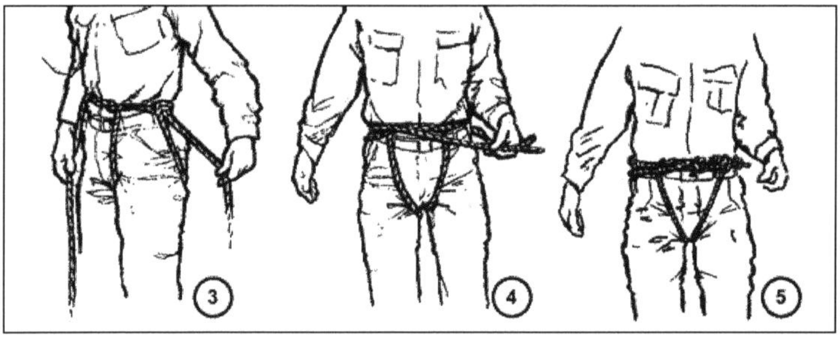

f. **Double Figure Eight**. Use a Figure Eight loop knot (Figure 10-10) to form a fixed loop in the end of the rope. It can be tied at the end of the rope or anywhere along the length of the rope. Figure Eight loop knots are formed by two ropes parallel to each other in the shape of a Figure Eight, no twists are in the Figure Eight. Fixed loops are large enough to insert a carabiner. When you dress the knot, leave at least a 4-inch tail on the working end.

Figure 10-10. DOUBLE FIGURE EIGHT-LOOP KNOT

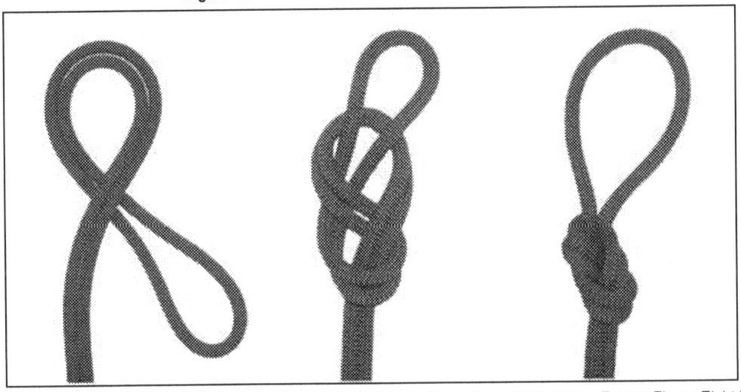

g. **Rerouted Figure Eight Knot**. This anchor knot also attaches a climber to a climbing rope. Form a Figure Eight in the rope, and pass the working end around an anchor. Reroute the end back through to form a double Figure Eight (Figure 10-11). Tie the knot with no twists. When you dress the knot, leave at least a 4-inch tail on the working end.

Figure 10-11. REROUTED FIGURE EIGHT KNOT

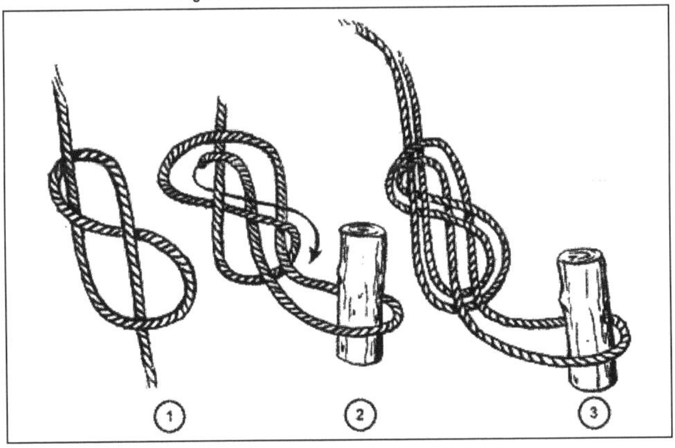

h. **Figure Eight Slip Knot**. The Figure Eight slip knot is used to form an adjustable bight in the middle of a rope. Knot is in the shape of a Figure Eight. Both ropes of the bight pass through the same loop of the Figure Eight. The bight is adjustable by means of a sliding section (Figure 10-12).

Figure 10-12. FIGURE EIGHT SLIP KNOT

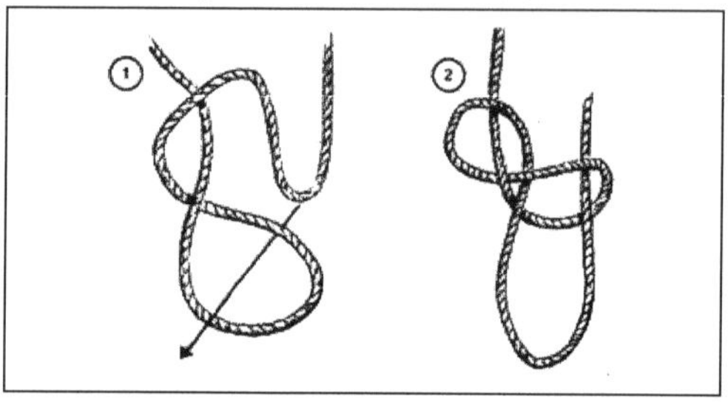

i. **End-of-the-Rope Prusik**. The End-of-the-Rope Prusik (Figure 10-13) attaches a movable rope to a fixed rope. The knot has two round turns, with a locking bar perpendicular to the standing end of the rope. Tie a bowline within 6 inches of the locking bar. When you dress the knot, leave at least a 4-inch tail on the working end.

Figure 10-13. END-OF-THE-ROPE PRUSIK

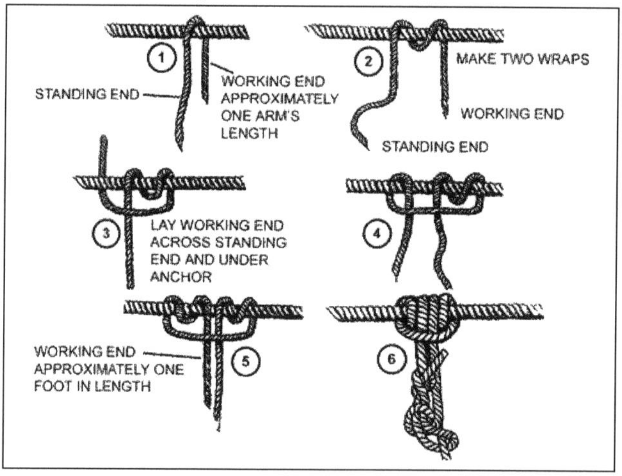

j. **Middle-of-the-Rope Prusik**. The middle-of-the-rope Prusik (Figure 10-14) attaches a movable rope to a fixed rope, anywhere along the length of the fixed rope. Make two round turns with a locking bar perpendicular to the standing end. Ensure the wraps do not cross and that the overhand knot is within 6 inches from the horizontal locking bar. Ensure the knot does not move freely on fixed rope.

Figure 10-14. MIDDLE-OF-THE-ROPE PRUSIK

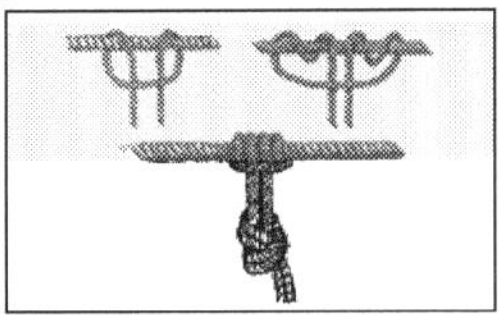

k. **Bowline on a Coil**. This knot (Figure 10-15) secures a climber to the end of a climbing rope. Make at least four parallel wraps around the body, between the hipbone and lower set of ribs. Keep all coils free of clothing, touch and are tight enough to ensure that a fist cannot be inserted between the wraps and the body. Three distinct coils show through the bight of the bowline. The rope coming off the bottom of the coils is on the right side, forward of the hip and forms the bight and the overhand knot. The rope coming off the top of the coils is on the left side, forward of the hips and forms the third and final coil showing through the bight of the bowline. Bowline is centered on the gig line and secured with an overhand knot with a minimum of 4 inch tail remaining after the knot is dressed.

Figure 10-15. BOWLINE ON A COIL

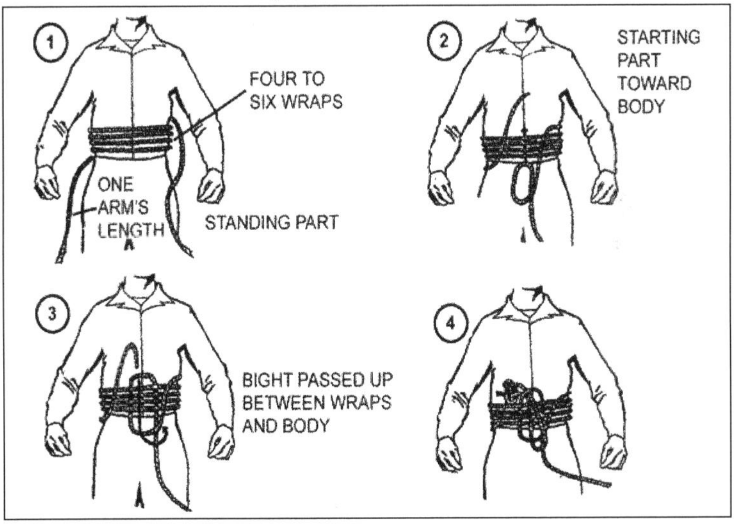

10-6. **BELAYS**. Belaying is any action taken to stop a climber's fall or to control the rate a load descends. The belayer also helps manage the rope for a climber by controlling the amount of rope that is taken out or in, or by controlling a rappeller's rate of descent. The belayer must be stably anchored to prevent him from being pulled out of his position, and losing control of the rope. The two types of belays are body and mechanical belays.

a. **Body Belay**. This belay (Figure 10-16) uses the belayer's body to apply friction by routing the rope around the belayer's body. Caution must be used when belaying from the body since the entire weight of the load may be placed on the belayer's body.

Figure 10-16. BODY BELAY

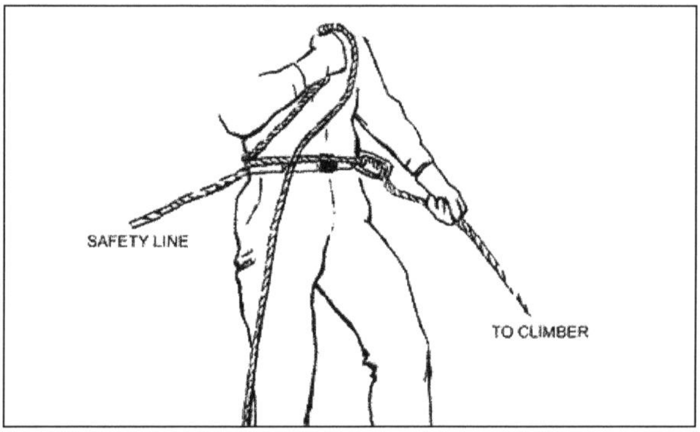

b. **Mechanical Belay**. This belay (Figure 10-17) uses mechanical devices to help the belayer control the rope, as in rappelling. A variety of devices are used to construct a mechanical belay for mountaineering.

Figure 10-17. MECHANICAL BELAY

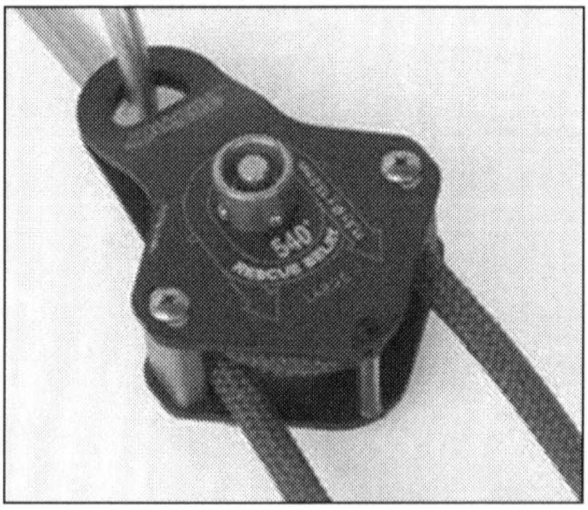

(1) ***Munter Hitch***. One of the most often used belays, the munter hitch (Figure 10-18), requires very little equipment. The rope is routed through a locking, pear-shaped carabiner, then back on itself. The belayer controls the rate of descent by manipulating the working end back on itself with the brake hand.

Figure 10-18. MUNTER HITCH

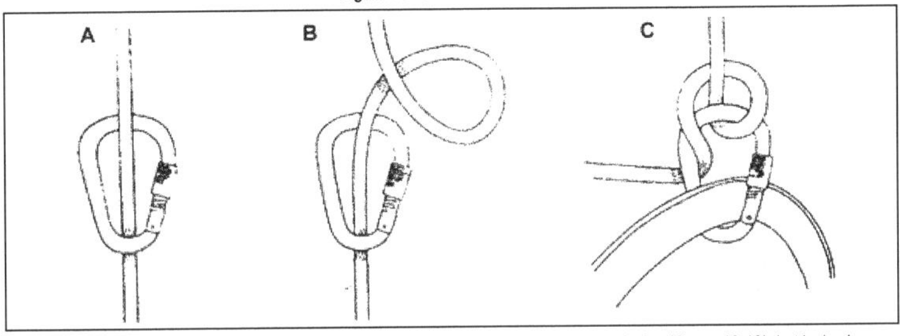

(2) ***Air Traffic Controller***. The ATC (air traffic controller) is a mechanical belay device (Figure 10-19) that locks down on itself when tension is applied in opposite directions. The belayer need apply very little force with his brake hand to control descent or arrest a fall.

Figure 10-19. AIR TRAFFIC CONTROLLER

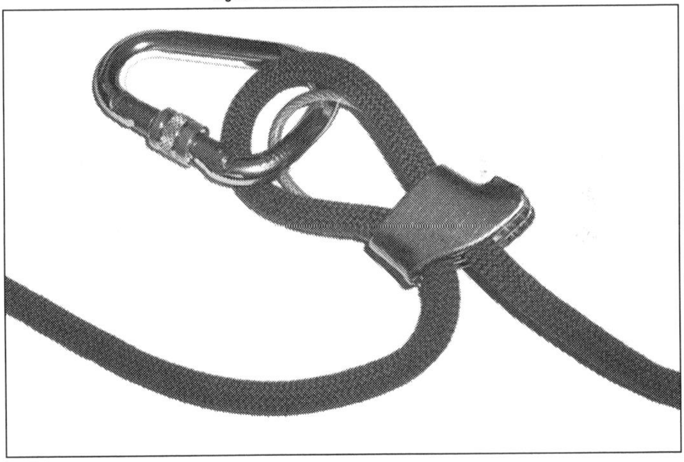

10-7. **CLIMBING COMMANDS**. Table 10-1 shows the sequence of commands used by climber and belayer.

Table 10-1. SEQUENCE OF CLIMBING COMMANDS

SEQUENCE	*COMMAND*	*GIVEN BY*	*MEANING*
1.	BELAY ON, CLIMB	Belayer	Alerts climber that belay is on and climber may climb.
2.	CLIMBING	Climber	Alerts belayer that climber is climbing.
3.	UP-ROPE	Climber	Belayer, remove excess slack in the rope.
4.	BRAKE	Climber	Alerts belayer to immediately apply brake.
5.	FALLING	Climber	Alerts belayer that climber is falling and that the belayer should immediately apply the brake and prepare to arrest the fall.
6.	TENSION	Climber	Alerts belayer to remove all slack from climbing rope until rope is tight, then apply brake and hold position.
7.	SLACK	Climber	Alerts belayer to pull slack into the climbing rope (Belayer may have to assist)
8.	ROCK	Anyone	Alerts everyone that an object is about to fall near them. Belayer immediately applies brake.
9.	POINT	Climber	Alerts belayer that the direction of pull on the climbing rope has changed in the event of a fall.
10.	STAND-BY	Climber or Belayer	Alerts the other to hold position, stand by, I am not ready.
11.	DO YOU HAVE ME?	Climber	Alerts belayer to prepare for a fall or to prepare to lower the climber.
12.	I HAVE YOU	Belay	Alerts climber that the brake is on and the belayer is prepared for the climber to fall, or to lower him.
13.	OFF-BELAY	Climber	Alerts belayer that climber is safetied in or it is safe to come off belay.
14.	3-METERS	Belayer	Alerts climber to the amount of rope between climber and belayer (May be given in feet or meters)
15.	BELAY-OFF	Belayer	Alerts climber that belayer is off belay

10-8. **ROPE INSTALLATIONS.** Rope installations may be constructed by teams to assist units in negotiating natural and manmade obstacles. installation teams consist of a squad sized element, with 2 to 4 trained mountaineers. Installation teams deploy early and prepare the AO for safe, rapid movement by constructing various types of mountaineering installations. Following construction of an installation, the squad, or part of it, remains on site to secure and monitor the system, assist with the control of forces across it, and make adjustments or repairs during its use. After passage of the unit, the installation team may then disassemble the system and deploy to another area as needed.

a. **Fixed Rope Installations**. A fixed rope is anchored in place to help Rangers move over difficult terrain. Its simplest form is a rope tied off on the top of steep terrain. As terrain becomes steeper or more difficult, fixed rope systems may require intermediate anchors along the route. Planning considerations are as follows:

- Does the installation allow you to bypass the obstacle?
- (Tactical considerations) Can obstacle be secured from construction through negotiation, to disassembly?
- Is it in a safe and suitable location? Is it easy to negotiate? Does it avoid obstacles?
- Are natural and artificial anchors available?
- Is the area safe from falling rock and ice?

b. **Vertical Hauling Line**. This installation (Figure 10-20) is used to haul men and equipment up vertical or near vertical slopes. It is often used in conjunction with the fixed rope.

Figure 10-20. VERTICAL HAULING LINE

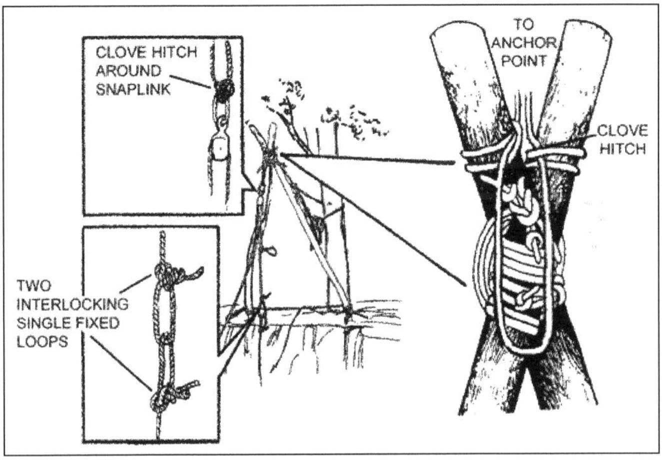

(1) ***Planning Considerations***.

- Does the installation allow you to bypass the obstacle?
- (Tactical considerations) Can you secure the installation from construction through negotiation, to disassembly?
- Does it have good loading and off-loading platforms? Are the platforms natural and easily accessible? Do they provide a safe working area?
- Does it allow sufficient clearance for load? Is there enough space between the slope and the apex of the A-frame to allow easy on-loading and off-loading of troops and equipment?
- Does it have an A-Frame for artificial height?
- Does it allow you to haul line in order to move personnel and equipment up and down slope?
- Does the A frame have a pulley or locking carabiner to ease friction on hauling line?
- Does it have a knotted hand line to help Rangers up the installation?
- Does it allow for Rangers top and bottom to monitor safe operation?

(2) ***Equipment***.

- Three 120-foot (50m) static ropes.
- Three 15-foot sling ropes for constructing A-frame.
- Two A-frame poles, 7 to 9 feet long, 4 to 6 inches in diameter (load dependent).
- Nine carabiners.
- One pulley with steel locking carabiner.

c. **Bridges**. Rope bridges are employed in mountainous terrain to bridge linear obstacles such as streams or rivers where the force of flowing water may be too great or temperatures are too cold to conduct a wet crossing.

(1) ***Construction***. The rope bridge is constructed using a static ropes. The max span that can be bridged is 50 percent of the ropes entire length for a dry crossing, 75 percent for a wet crossing. The ropes are anchored with an anchor knot on the far side

of the obstacle, and are tied off at the near end with a transport tightening system. Rope Bridge Planning Considerations are as follows:

- Does the installation allow you to bypass the obstacle?
- (Tactical considerations) Can you secure the installation from construction through negotiation, to disassembly?
- Is it in the most suitable location such as a bend in the river? Is it easily secured?
- Does it have near and far-side anchors?
- Does it have good loading and off-loading platforms?

(2) ***Equipment (1-Rope Bridge)***.

- One sling rope per-man.
- One steel locking carabineer.
- Two steel ovals.
- Two 120-foot static ropes.

(3) ***Construction Steps***. The first Ranger swims the rope to the far side and ties a tensionless anchor (Fig 10-4) between knee and chest level, with at least 6 to 8 wraps. The BTC ties a transport tightening system (Fig 10-21) to the near side anchor point. Then, he ties a Figure Eight slipknot and incorporates a locking half hitch around the adjustable bight. Insert 2 steel oval carabineers into the bight so the gates are opposite and opposed. The rope is then routed around the near side anchor point at waist level and dropped into the steel oval carabineers.

- A three-Ranger pulling team moves forward from the PLT. No more than three men should tighten the rope. Using more can overtighten the rope, bringing it near failure.
- Once the rope bridge is tight enough, the bridge team secures the transport-tightening system using two half hitches, without losing more than 4 inches of tension.
- Personnel cross using the commando crawl, rappel seat, or monkey crawl method (Figures 10-22 thru 24).

Figure 10-21. TRANSPORT TIGHTENING SYSTEM

Figure 10-22. COMMANDO CRAWL METHOD

Figure 10-23. RAPPEL SEAT (TYROLEAN TRAVERSE) METHOD

Figure 10-24. MONKEY CRAWL METHOD

(4) ***Bridge Recovery***. Once all but two Rangers have crossed the rope bridge, the bridge team commander (BTC) chooses either the wet or dry method to dismantle it. If he chooses the dry method, he should he should first anchor his tightening system with the transport knot.

- The BTC back-stacks all of the slack coming out of the transport knot, then ties a fixed loop and places a carabiner into the fixed loop.
- The next to last Ranger to cross attaches the carabiner to his rappel seat or harness, and then moves across the bridge using the Tyrolean traverse method.
- The BTC then removes all knots from the system. The far side remains anchored. The rope should now only pass around the near side anchor.
- A three-man pull team, assembled on the far side, takes the end brought across by the next to last Ranger and pulls the rope tight again and holds it.
- The BTC then attaches himself to the rope bridge and moves across.
- Once across, the BTC breaks down the far side anchor, removes the knots, and then pulls the rope across. If the BTC chooses a wet crossing, any method can be used to anchor the tightening system.
- All personnel cross except the BTC or the strongest swimmer.
- The BTC then removes all knots from the system.
- The BTC ties a fixed loop, inserts a carabiner, and attaches it to his rappel seat or harness. He then manages the rope as the slack is pulled to the far side.
- The BTC then moves across the obstacle while being belayed from the far side.

d. **Suspension Traverse**. The suspension traverse is used to move personnel and equipment over rivers, ravines, chasms, and up or down a vertical obstacle. By combining the transport tightening system used during the Rope bridge, an A-Frame used for the Vertical haul Line (Figure 10-25), and belay techniques device, Units can construct a suspension traverse (Figures 10-26 and 10-27). Installation of a suspension traverse can be time-consuming and equipment-intensive. All personnel must be well trained and rehearsed in the procedures.

(1) ***Construction***. The suspension traverse must be constructed using a static ropes. The max span that can be bridged is generally 75 percent length of the shortest rope. Planning considerations are the same as rope bridge and vertical haul line combined.

(2) ***Equipment***.

- Three static installation ropes.
- Seven sling Ropes.
- Nine carabiners.
- One heavy duty double pulley.
- One locking carabiner.
- One canvas pad.

Figure 10-25. ANCHORING THE TRAVERSE ROPE TO THE A-FRAME

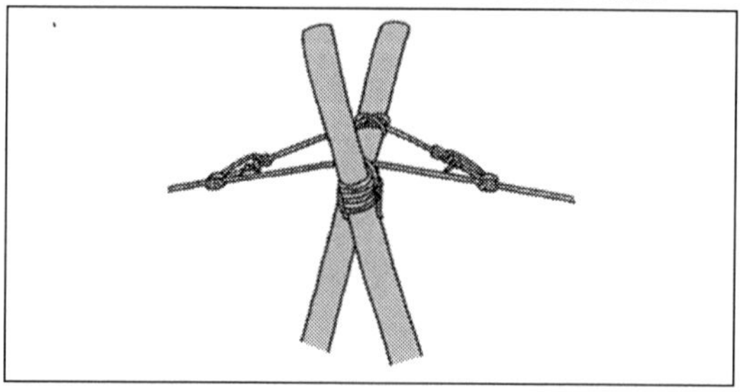

Figure 10-26. CARRYING ROPE FOR USE ON A SUSPENSION TRAVERSE

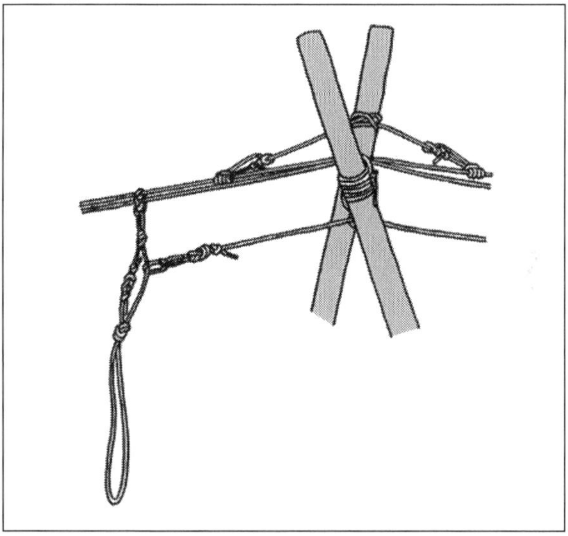

Figure 10-27. SUSPENSION TRAVERSE

10-9. **RAPPELLING**. Rappelling is a quick method of descent but can be extremely dangerous. These dangers include anchor failure, equipment failure, and individual error. Anchors in a mountainous environment should be selected carefully. Great care must be taken to load the anchor slowly and to ensure that no excessive stress is placed on the anchor. The best way to ensure this is to prohibit bounding rappels, and to use only walk-down rappels.

a. **Hasty and Body Rappels**. Hasty (Figure 10-28) and body rappels (Figure 10-29) are easier and faster than other methods, but should only be used on moderate pitches--never on vertical or overhanging terrain. Wear gloves to prevent rope burns.

Figure 10-28. Hasty Rappel

Figure 10-29. Body Rappel

b. **Seat-Hip Rappel**. The seat-hip rappel uses a mechanical rappel device that is inserted in a sling rope seat and is fastened to the rappeller to provide the necessary friction to remain in control. This method provides a faster and more controlled descent than any other method (Figure 10-30). There are several mechanical rappel devices (Figure 10-31 and 10-32) that can be used for the seat-hip rappel.

Figure 10-30. FIGURE EIGHT DESCENDER

Figure 10-30. FIGURE EIGHT DESCENDER (CONTINUED)

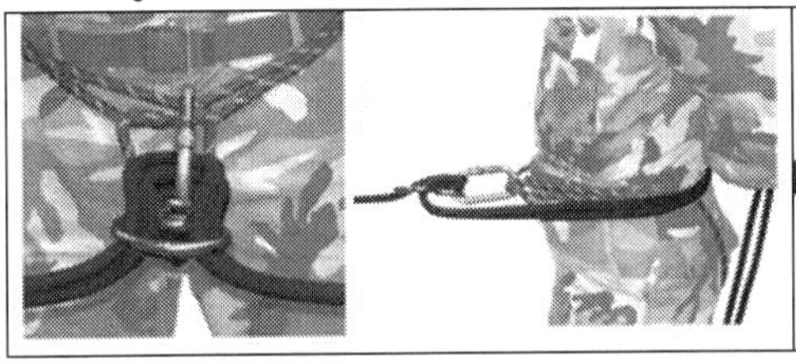

Figure 10-31. CARABINER WRAP

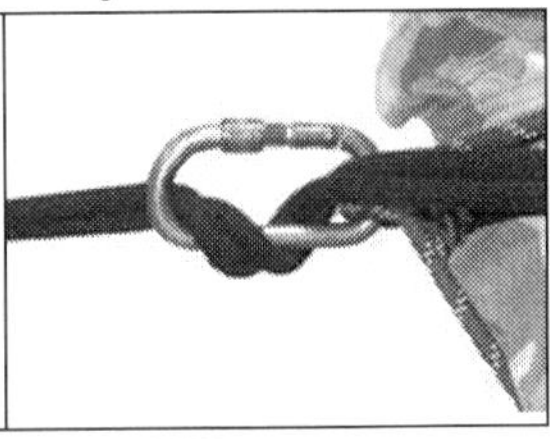

Figure 10-32. SEAT HIP RAPPEL WITH CARABINER WRAP DESCENDER

c. **Site Selection**. The selection of the rappel point depends on factors such as mission, cover, route, anchor points, and edge composition (loose or jagged rocks). There must be good anchors (primary and secondary). The anchor point should be above the rappeller's departure point. Suitable loading and off-loading platforms should be available.

- Each rappel point has primary and secondary anchors.
- Rappel point has equal tension between all anchor points.
- Double rope is used when possible.
- Ropes must reach the off loading platform.
- Site has suitable on and off loading platforms.
- Personnel working near the edge are tied in.
- Select a smooth route free of loose rock and debris.

Chapter 11
EVASION AND SURVIVAL

This chapter will cover basic material dealing with evading the enemy and survival techniques. Field expedient methods of acquiring food/ water, land navigation, constructing shelters, and making fires will be presented.

11-1. **EVASION**. When you become isolated or separated in a hostile area, either as an individual or as a group, your evasion and survival skills will determine whether or not you return to friendly lines.

a. When unable to continue the mission or unable to rejoin your unit, leave the immediate area, and move to your last rally point.

b. Observe activity in the area and form a plan.

c. Traveling alone offers the least possibility of detection, but traveling in groups of two to three is more desirable.

d. Plan a primary and alternate route. Consider distance, cover, food, and water. The easiest and shortest route may not be the best.

e. Food and water are daily requirements. You can do without food for several days; water, however, is essential.

f. Move at night. Use the daylight to observe, plan, and rest in a hide position.

g. Linkup only during daylight hours. Place friendly lines under observation.

h. Attempt to identify the unit you approach, note their movements and routine.

i. After carefully considering your approach route, make voice contact with the unit as soon as possible.

11-2. **SURVIVAL**.
With training, equipment, and the ***will to survive***, you can overcome *any* obstacle you may face. ***You will survive.*** Understand the emotional states associated with survival. "Knowing thyself" is extremely important in a survival situation. It bears directly on how well you cope with serious stresses, anxiety, pain, injury, illness; cold, heat, thirst, hunger, fatigue, sleep deprivation, boredom, loneliness and isolation.

a. **Memory Aid**. You can overcome and reduce the shock of being isolated behind enemy lines if you keep the key word S-U-R-V-I-V-A-L foremost in your mind. Its letters can help guide you in your actions.

(1) S - Size up the situation, the surroundings, your physical condition, and your equipment.

(2) U - Undue haste makes waste; don't be too eager to move. Plan your moves.

(3) R - Remember where you are relative to friendly and enemy units and controlled areas; water sources (most important in the desert); and good cover and concealment. This information will help you make intelligent decisions.

(4) V - Vanquish fear and panic.

(5) I – Improvise/Imagine. You can improve your situation. Learn to adapt what is available for different uses. Use your imagination.

(6) V - Value living. Remember your goal - to get out alive. Remain stubborn. Refuse to give in to problems and obstacles. This will give you the mental and physical strength to endure.

(7) A - Act like the natives; watch their daily routines and determine when, where, and how they get their food and water.

(8) L - Live by your wits. Learn basic skills.

b. **Navigation**. In a survival situation, you might find yourself without a compass. The ability to determine direction can enable you to navigate back to your unit or to a friendly sanctuary. In sunlight, there are two simple ways to determine direction: the shadow-tip and the watch.

(1) ***Shadow Tip***. Use the sun to find approximate true North. Use this in light bright enough to cast shadows. Find a fairly straight stick about 3 feet long, and follow the diagram below (Figure 11-l).

Figure 11-1. SHADOW-TIP METHOD

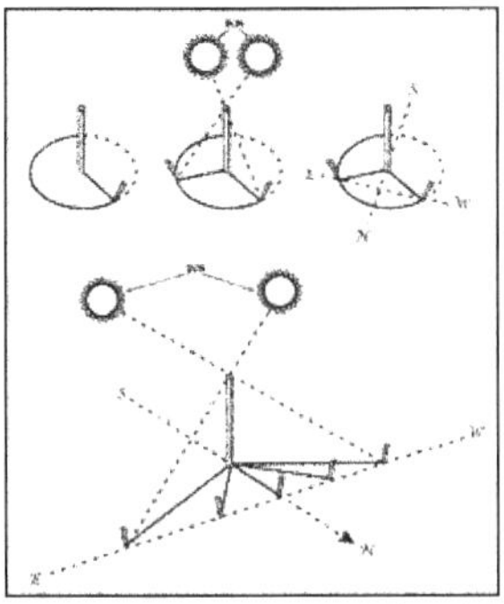

(2) ***Watch Method***. You can also determine direction using a watch (Figure 11-2). The steps you take will depend on whether you are in the northern Temperate Zone or in the southern Temperate Zone. The northern temperate zone is located between 23.4 north and 26.6 north. The southern Temperate Zone is located between 23.4 south and 66.6 south.

Figure 11-2. WATCH METHOD

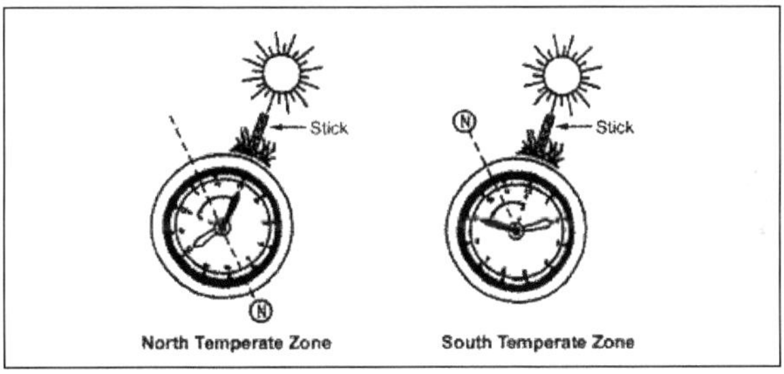

c. **Northern Temperate Zone**. Procedures in the northern temperate zone using a conventional watch are as follows:

(1) Place a small stick in the ground so that it casts a definite shadow.

(2) Place your watch on the ground so that the hour hand points toward and along the shadow of the stick.

(3) Find the point on the watch midway between the hour hand and 12 o'clock and draw an imaginary line from that point through and beyond the center of the watch. This imaginary line is a north-south line. You can then tell the other directions.

Note: If your watch is set on daylight savings time, then use the midway point between the hour hand and 1 o'clock to draw your imaginary line.

d. **Southern Temperate Zone**. Procedures in the southern temperate zone using a conventional watch are as follows:

(1) Place a small stick in the ground so that it casts a definite shadow.

(2) Place your watch on the ground so that 2 o'clock points to and along the shadow.

(3) Find the midway point between the hour and 12 o'clock and draw an imaginary line from the point through and beyond the center of the watch. This is a north-south line.

e. A hasty shortcut using a conventional watch is simply to point the hour hand at the sun in the northern temperate zone (or point the 12 at the sun in the southern temperate zone) and then follow the last step of the watch method above to find your directions. This shortcut, of course, is not as accurate as the regular method but quicker. Your situation will dictate which method to use.

11-3. **WATER**. Water is one of your most urgently needed resources in a survival situation. You can't live long without it, especially in hot areas where you lose so much through sweating. Even in cold areas, you need a minimum of 2 quarts of water a day to maintain efficiency. More than three-fourths of your body is composed of fluids. Your body loses fluid as a result of heat, cold, stress, and exertion. The fluid your body loses must be replaced for you to function effectively. So, one of your first objectives is to obtain an adequate supply of water.

a. **Purification**. Purify all water before drinking. Either--

(1) Boil it for at least one minute (plus 1 more minute for each additional 1,000 feet above sea level) or for a maximum of 10 minutes anywhere.

(2) Use water purification tablets.

(3) Add eight drops of 2-1/2 percent iodine solution to a quart (canteen full) of water. Let it stand for 10 minutes before drinking.

(4) Collect rain water directly in clean containers or on plants. This is generally safe to drink without purifying. Never drink urine or sea water; the salt content is too high. Use old, bluish sea ice, but newer, grayer ice may be salty. Glacier ice is safe to melt and drink.

b. **Desert Environment**. In a desert environment, water has a huge effect on Rangers. If a unit fails to plan properly for water, and resupply is unavailable, then they can run out of water. In the desert, look for four signs of water: animal trails, vegetation, birds, and civilization. Adequate water is critical in a hot environment if a unit is to survive and maintain the physical condition necessary to accomplish the mission. Unit leaders must enforce water discipline and plan for water resupply. The leader can use the following planning considerations for water resupply:

(1) Units average water consumption.

(2) Drop sites.

(3) Aviation support.

(4) DZ and LZ parties.

(5) Caches.

(6) Targets of opportunity (enemy).

c. **Survival Water Still**. Dig a belowground still (Figure 11-3). Select a site where you believe the soil will contain moisture (such as a dry stream bed or a spot where rain water has collected). The soil should be easy to dig, and be in sunlight most of the day:

(1) Dig a bowl-shaped hole about 3 feet across and 2 feet deep.

(2) Dig a sump in center of the hole. The depth and the perimeter of the sump will depend on the size of the container that you have to set inside of it. The bottom of the sump should allow the container to stand upright.

(3) Anchor the tubing to the bottom of the container by forming a loose overhand knot in the tubing.

(4) Place the container upright in the sump.

(5) Extend the unanchored end of the tubing up, over, and beyond the lip of the hole.

(6) Place plastic sheeting over the hole covering the edge with soil to hold it in place.

(7) Place a rock in the center of the plastic.

(8) Allow the plastic to lower into the hole until it is about 15 inches below ground level. The plastic now forms an inverted cone with the rock at its apex. Make sure that the apex of the cone is directly over your container. Also, make sure the plastic cone does not touch the sides of the hole, because the earth will absorb the condensed water.

(9) Put more soil on the edges of the plastic to hold it securely in place and to prevent loss of moisture.

(10) Plug the tube when not being used so that moisture will not evaporate.

d. You can drink water without disturbing the still by using the tube as a straw. You may want to use plants in the hole as a moisture source. If so, when you dig the hole you should dig out additional soil from the sides of the hole to form a slope on which to place the plants. Then proceed as above.

Figure 11-3. SURVIVAL WATER STILL

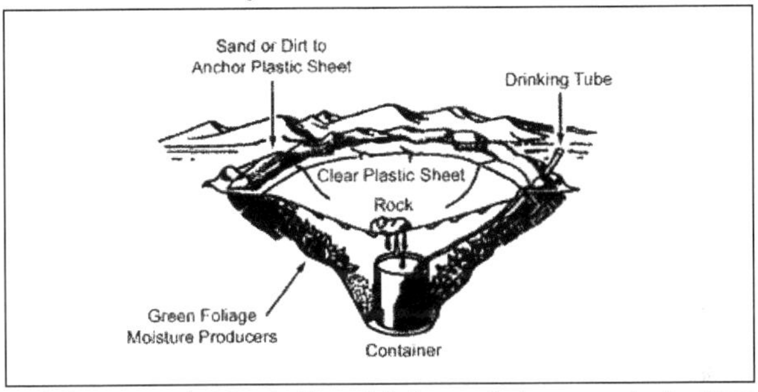

11-4. **PLANT FOOD**. There are many plants throughout the world. Tasting or swallowing even a small portion of some can cause severe discomfort, extreme internal disorders, or death. Therefore, if you have the slightest doubt as to the edibility of a plant, apply the universal edibility test described below before eating any part of it.

a. **Universal Edibility Test**. Before testing a plant for edibility, make sure there are a sufficient number of plants to make testing worth your time and effort. You need more than 24 hours to apply the edibility test outlined below:

(1) Test only one part of a potential food plant at a time.

(2) Break the plant into its basic components, leaves, stems, roots, buds, and flowers.

(3) Smell the food for strong or acid odors. Keep in mind that smell alone does not indicate a plant is edible.

(4) Do not eat for 8 hours before starting the test.

(5) During the 8 hours you are abstaining from eating, test for contact poisoning by placing a piece of the plant you are testing on the inside of your elbow or wrist. Usually 15 minutes is enough time to allow for reaction.

(6) During the test period, take nothing by mouth except purified water and the plant part being tested.

(7) Select a small portion and prepare it the way you plan to eat it.

(8) Before putting the prepared plant part in your mouth, touch a small portion (a pinch) to the outer surface of the lip to test for burning or itching.

(9) If after 3 minutes there is no reaction on your lip, place the plant part on your tongue, holding there for 15 minutes.

(10) If there is no reaction, thoroughly chew a pinch and hold it in your mouth for 15 minutes. *Do not swallow.*

(11) If no burning, itching, numbing, stinging, or other irritation occurs during the 15 minutes, swallow the food.

(12) Wait 8 hours. If any ill effects occur during this period, induce vomiting and drink a lot of water.

(13) If no ill effects occur, eat 1/2 cup of the same plant part prepared the same way. Wait another 8 hours. If no ill effects occur, the plant part as prepared is safe for eating.

b. **Poisonous Plants**. ***Do not*** eat unknown plants that have the below characteristics:

(1) Have a milky sap or a sap that turns black when exposed to air.

(2) Are mushroom like.

(3) Resemble onion or garlic.

(4) Resemble parsley, parsnip, or dill.

(5) Have carrot-like leaves, roots, or tubers.

11-5. **ANIMAL FOOD**. Animal food contains the most food value per pound. Anything that creeps, crawls, swims, or flies is a possible source of food, however you must first catch, kill and butcher it before this is possible. There are numerous methods for catching fish and animals in a survival situation. You can catch fish by using a net across a small stream, (Figure 11-4) or by making fish traps and baskets. Improvise fish hooks and spears as indicated in Figure 11-5, and use them for conventional fishing, spearing and digging.

Figure 11-4. SETTING A GILL NET IN THE STREAM

Figure 11-5. SPEARS AND FISH HOOKS

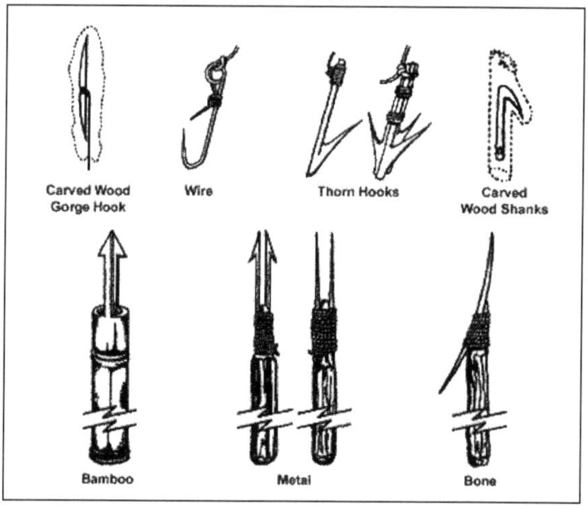

11-6. **TRAPS AND SNARES**. For an unarmed survivor or evader, or when the sound of a rifle shot could be a problem, trapping or snaring wild game is a good alternative. Several well-placed traps have the potential to catch much more game than a Ranger with a rifle is likely to shoot.

a. To be effective with any type of trap or snare, you must---
   - (1) Know the species of animal you intend to catch.
   - (2) Know how to construct a proper trap.
   - (3) Avoid alarming the prey with signs of your presence.

b. There are no catchall traps you can set for all animals. You must determine what species are in a given area and set your traps specifically with those animals in mind. Look for the following:

(1) Runs and trails.
(2) Tracks.
(3) Droppings.
(4) Chewed or rubbed vegetation.
(5) Nesting or roosting sites.
(6) Feeding and watering areas.

c. Position your traps and snares where there is proof that animals pass through. You must determine if it is a "run" or a "trail." A trail will show signs of use by several species and will be rather distinct. A run is usually smaller and less distinct and will only contain signs of one species. You may construct a perfect snare, but it will not catch anything if haphazardly placed in the woods. Animals have bedding areas, waterholes, and feeding areas with trails leading from one to another. You must place snares and traps around these areas to be effective.

d. An evader in a hostile environment must conceal traps and snares. It is equally important, however, to avoid making a disturbance that will alarm the animal and cause it to avoid the trap. Therefore, if you must dig, remove all fresh dirt from the area. Most animals will instinctively avoid a pitfall-type trap. Prepare the various parts of a trap or snare away from the site, carry them in, and set them up. Such actions make it easier to avoid disturbing the local vegetation, thereby alerting the prey. Do not use freshly cut, live vegetation to construct a trap or snare. Freshly cut vegetation will "bleed" sap that has an odor the prey will be able to smell. It is an alarm signal to the animal.

e. You must remove or mask the human scent on and around the trap you set. Although birds do not have a developed sense of smell, nearly all mammals depend on smell even more than on sight. Even the slightest human scent on a trap will alarm the prey and cause it to avoid the area. Removing the scent from a trap is difficult but masking it is relatively easy. Use the fluid from the gall and urine bladders of previous kills. Do not use human urine. Mud, particularly from an area with plenty of rotting vegetation, is also good. Use it to coat your hands when handling the trap and to coat the trap when setting it. In nearly all parts of the world, animals know the smell of burned vegetation and smoke. It is only when a fire is actually burning that they become alarmed. Therefore, smoking the trap parts is an effective means to mask your scent. If one of the above techniques is not practical, and if time permits, allow a trap to weather for a few days and then set it. Do not handle a trap while it is weathering. When you position the trap, camouflage it as naturally as possible to prevent detection by the enemy and to avoid alarming the prey.

f. Traps or snares placed on a trail or run should use canalization. To build a channel, construct a funnel-shaped barrier extending from the sides of the trail toward the trap, with the narrowest part nearest the trap. Canalization should be inconspicuous to avoid alerting the prey. As the animal gets to the trap, it cannot turn left or right and continues into the trap. Few wild animals will back up, preferring to face the direction of travel. Canalization does not have to be an impassable barrier. You only have to make it inconvenient for the animal to go over or through the barrier. For best effect, the canalization should reduce the trail's width to just slightly wider than the targeted animal's body. Maintain this constriction at least as far back from the trap as the animal's body length, then begin the widening toward the mouth of the funnel.

(1) Use a treadle snare against small game on a trail (Figure 11-6). Dig a shallow hole in the trail. Then drive a forked stick (fork down) into the ground on each side of the hole on the same side of the trail. Select two fairly straight sticks that span the two forks. Position these two sticks so that their ends engage the forks. Place several sticks over the hole in the trail by positioning one end over the lower horizontal stick and the other on the ground on the other side of the hole. Cover the hole with enough sticks so that the prey must step on at least one of them to set off the snare. Tie one end of a piece of cordage to a twitch-up or to a weight suspended over a tree limb. Bend the twitch-up or raise the suspended weight to determine where you will tie a 5 centimeter or so long trigger. Form a noose with the other end of the cordage.

(2) Route and spread the noose over the top of the sticks over the hole. Place the trigger stick against the horizontal sticks and route the cordage behind the sticks so that the tension of the power source will hold it in place. Adjust the bottom horizontal stick so that it will barely hold against the trigger. As the animal places its foot on a stick across the hole, the bottom horizontal stick moves down, releasing the trigger and allowing the noose to catch the animal by the foot. Because of the disturbance on the trail, an animal will be wary. You must therefore use canalization.

Figure 11-6. TREADLE SNARE

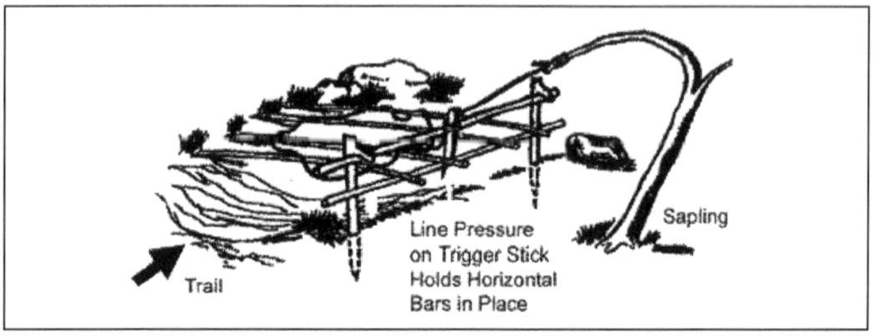

g. Trapping game can be accomplished through the use of snares, traps, or deadfalls. A snare is a noose that will slip and strangle or hold any animal caught in it. You can use inner core strands of parachute suspension lines, wire, bark of small hardwood saplings as well as hide strips from previously caught animals to make snares.

(1) The drag noose snare, Figure 11-7, is usually the most desirable in that it allows you to move away from the site, plus it is one of the easiest to make and fastest to set. It is especially suitable for catching rabbits. To make the drag noose snare, make a loop in the string using a bowline or wireman's knot. (When using wire, secure the loop by intertwining the end of the wire with the wire at the top of the loop). Pull the other end of the string (or wire) through the loop to form a noose that is large enough for the animal's head but too small for its body; tie the string (or attach the wire) to a sturdy branch. The branch should be large enough to span the trail and rest on the bush or support (two short forked sticks) you have selected. A snared animal will dislodge the drag stick, pulling it until it becomes entangled in the brush. Any attempt to escape tightens the noose, strangling or holding the animal.

Figure 11-7. DRAG NOOSE SNARE

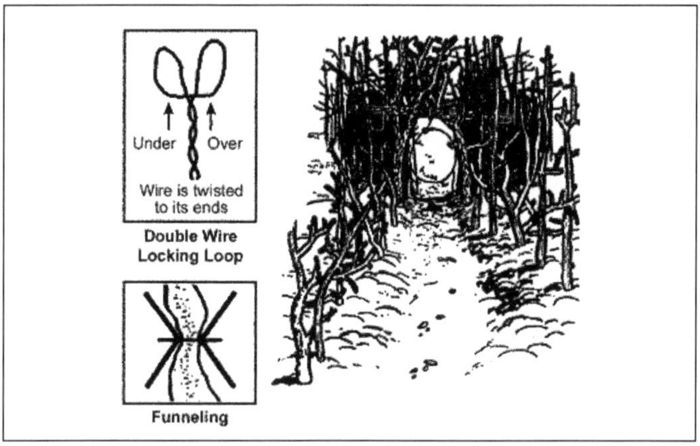

(2) Another type snare is the locking type snare loop (Figure 11-8) that will lock when pulled tight, ensuring the snared animal cannot escape. Use lightweight wire to make this snare, such as, trip wire or the wire from a vehicle or aircraft electrical system. To construct this snare, cut a piece of wire twice the length of the desired snare wire. Double the wire and attach the running ends to a securely placed object, such as the branch of a tree. Place a stick about 1/2 inch in diameter through the loop end of the wire; holding the wire taut, turn the stick in a winding motion so that the wire is twisted together. You should have four to five twists per inch. Detach the wire from the branch and then remove the loop from the stick; make a Figure Eight in the l/2-inch loop by twisting the loop over itself then fold the Figure Eight so the small loops are almost overlapping; run the loose wire ends through these loops. This forms a stiff noose that is strong. Tie the loose end to the stick (for a drag noose square) or branch you are using to complete the snare. This is an excellent snare for catching large animals.

Figure 11-8. FORMING A LOCKING-TYPE SNARE LOOP

(3) Another means of obtaining game is the use of the deadfall trap (Figure 11-9 and Figure 11-10).

Figure 11-9. TRIGGER WITH DEADFALL

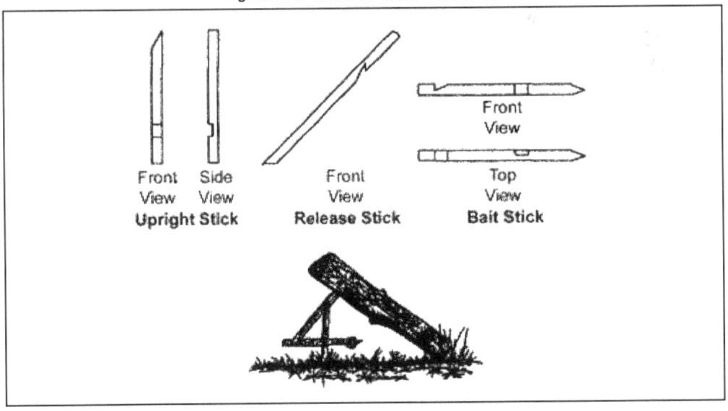

Figure 11-10. TRIP-STRING DEADFALL TRAP

h. Once you have obtained your fish or game, you must clean/butcher and cook/store it. Improper cleaning storing can result in inedible fish and game.

(1) ***Fish***. You must know how to tell if fish are free of bacterial decomposition that makes the fish dangerous to eat. Although cooking may destroy the toxin from bacterial decomposition, do not eat fish that appear spoiled.

(a) Signs of spoilage are:

- A peculiar odor.
- A suspicious color. (Gills should be red or pink. Scales should be a pronounced-not faded shade of gray).
- A dent remaining after pressing the thumb against the flesh.
- A slimy rather than moist or wet body.
- A sharp or peppery taste.

(b) Eating spoiled or poisoned fish may cause diarrhea, nausea, cramps, vomiting, itching; paralysis, or a metallic taste in the mouth. These symptoms appear suddenly 1 to 6 hours after eating. If you are near the sea, drink sea water immediately upon on set of such symptoms and force yourself to vomit.

(c) Fish spoil quickly after death, especially on a hot day, so prepare fish for eating as soon as possible after you catch them.

(d) Cut out the gills and large blood vessels that lie next to the backbone. (You can leave the head if you plan to cook the fish on a spit).

(e) Gut fish that are more than 4 inches long cut along the abdomen and scrape out the intestines.

(f) Scale or skin the fish.

(g) You can impale a whole fish on a stick and cook it over an "open fire." However, boiling the fish with the skin on is the best way to get the most food value. The fats and oil are under the skin, and by boiling the fish, you can save the juices for broth. Any of the methods used for cooking plant food can be used for cooking fish. Fish is done when the meat flakes off.

(h) To dry fish in the sun, hang them from branches or spread them on hot rocks. When the meat has dried, splash it with sea water, if available, to salt the outside. Keep seafood only if it is well dried or salted.

(2) ***Snakes***. All poisonous and nonpoisonous fresh water and land snakes are edible.
To prepare snakes for eating use the following steps (Figure 11-11):

(a) Grip the snake firmly behind the head and cut off the head with a knife.

(b) Slit the belly and remove the innards. (You can use the innards for baiting traps and snares).

(c) Skin the snake. (You can use the skin for improvising, belts, straps, or similar items).

CAUTION

Take extreme care in securing snakes--the bite of some poisonous snakes can be fatal. Even after a snake's head is cut off, its reflex action can cause it to bite, injecting poison. The best time to capture snakes is in the early morning or late evening when temperatures are low and they move slow. Kill or use a long stick to pin down its head and capture it. To pick up a snake, place the index finger on the top rear of its head with your thumb and middle finger on either side of the head behind the jaws. Keep your index finger on top of snake's head to prevent it from turning inside its skin and biting you.

Figure 11-11. CLEANING A SNAKE

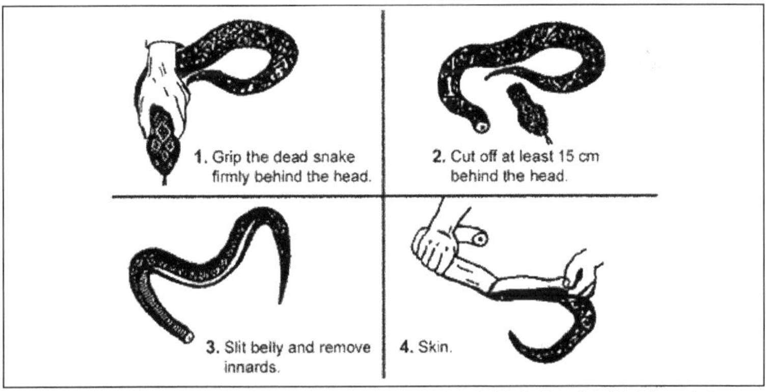

(3) ***Fowl***. Your first step after killing a fowl for eating or preserving is to pluck its feathers. If plucking is impractical, you can skin the fowl. Keep in mind, however, that a fowl cooked with the skin on retains more food value. Waterfowl are easier to pluck while dry, but other fowl are easier to pluck after scalding. After you pluck the fowl--

(a) Cut off its neck close to the body.

(b) Cut an incision in the abdominal cavity and clean out the insides. Save the neck, liver, and heart for stew. Thoroughly clean and dry the entrails to use for cordage.

(c) Wash out the abdominal cavity with fresh clean water. You can boil fowl or cook it on a spit over a fire. You should boil scavenger birds such as vultures and buzzards for at least 20 minutes to kill any parasites. Use the feathers from fowl for insulating your shoes clothing, or bedding. You can also use feathers for fish lures.

(4) **Medium-Sized Mammals**. The game you trap or snare will generally be alive when you find it and therefore dangerous. Be careful when you approach a trapped animal. Use a spear or club to kill it so you can keep a safe distance from it. After you kill an animal, immediately bleed it by cutting its throat. If you must drag the carcass any distance, do so before you cut off the hide so that the carcass is protected from dirt and debris that might contaminate it. Clean the animal near a stream if possible so that you can wash and cool the carcass and edible parts. Fleas and parasites will leave a cooled body so if the situation allows, wait until the animal cools before cleaning and dressing the carcass. To skin and dress the animal (Figure 11-12 and 11-13).

Figure 11-12. SKINNING AND BUTCHERING LARGE GAME

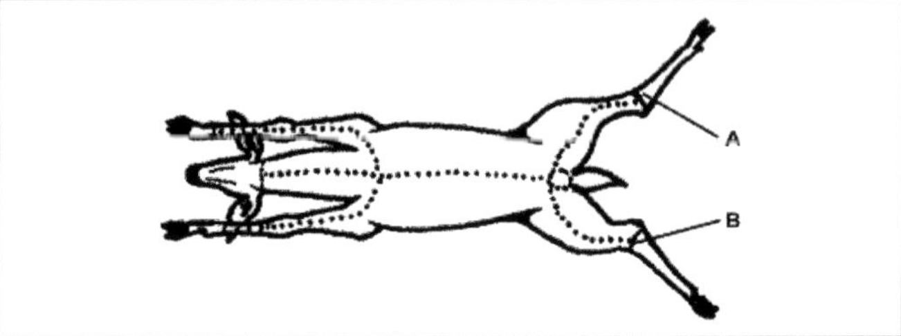

Figure 11-13. SKINNING SMALL GAME

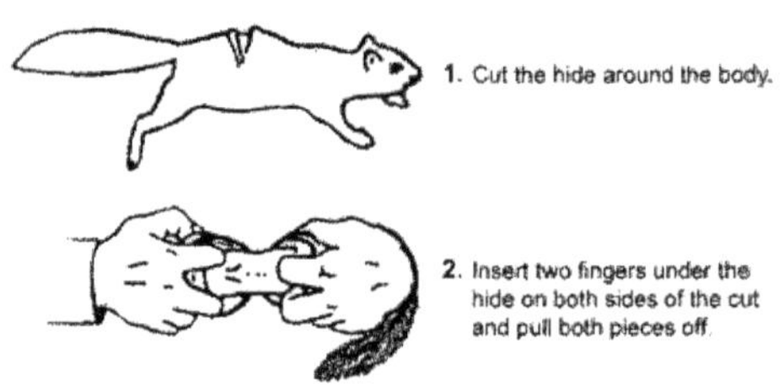

(a) Place carcass, belly up, on a slope if available. You can use rocks or brush to support it.

(b) Remove genitals or udder.

(c) Remove musk glands to avoid tainting meat.

(d) Split hide from tail to throat. Make the cut shallow so that you do not pierce the stomach.

(e) Insert your knife under the skin, taking care not to cut into the body cavity. Peel the hide back several inches on each side to keep hair out of the meat.

(f) Open the chest cavity by splitting the sternum. You can do this by cutting to one side of the sternum where the ribs join.

(g) Reach inside and cut the windpipe and gullet as close to the base of the skull as possible.

(h) With the forward end of the intestinal tract free, work your way to the rear lifting out internal organs and intestines. Cut only where necessary to free them.

(i) Carefully cut the bladder away from the carcass so that you do not puncture the bladder (urine can contaminate meat). Pinch the urethra tightly and cut it beyond the point you are pinching.

(j) Remove the bladder.

(k) From the outside of the carcass, cut a circle around the anus.

(l) Pull the anus into the body cavity and out of the carcass.

(m) Lift or roll the carcass to drain all blood.

Note: Try to save as much blood as you can as it is a vital source of food and salt. Boil the blood.

(n) Remove the hide, and make cuts along the inside of the legs to just above the hoof or paw. Peel the skin back, using your knife in a slicing motion to cut the membrane between the skin and meat. Continue this until the entire skin is removed.

(o) Most of the entrails are usable. The heart, liver, and kidneys are edible. Cut open the heart and remove the blood from its chambers. Slice the kidneys and if enough water is available, soak or rinse them. In all animals except those of the deer family, the gall bladder (a small, dark-colored, clear-textured sac) is attached to the liver.

(p) Sometimes, the sac looks like a blister on the liver. To remove the sac, hold the top portion of it and cut the liver around and behind the sac. If the gall bladder breaks and gall gets on the meat, wash it off immediately so the meat will not become tainted. Dispose of the gall.

(q) Clean blood splattered on the meat will glaze over and help preserve the meat for a short time. However, if an animal is not bled properly, the blood will settle in the lowest part of its body and will spoil in a short time. Cut out any meat that becomes contaminated.

(r) When temperatures are below 40 degrees, you can leave meat hanging for several days without danger of spoilage. If maggots get on the meat, remove the maggots and cut out the discolored meat. The remaining meat is edible. Maggots, which are the larvae of insects, are also edible.

(s) Blood, which contains salts and nutrients, is a good base for soups.

(t) Thoroughly clean the intestines and use them for storing or smoking food or lashings for general use. Make sure they are completely dry to preclude rotting.

(u) The head of most animals contains a lot of meat, which is relatively easy to get. Skin the head, saving the skin for leather. Clean the mouth thoroughly and cut out the tongue. Remove the outer skin from the tongue after cooking. Cut or scrape the meat from the head. If you prefer, you can roast the head over an open fire before cutting off the meat. Eyes are edible. Cook them but discard the retina (this is a plastic like disc). The brain is also edible; in fact, some people consider it a delicacy. The brain is also used to tan leather, the theory being that the brain of an animal is adequate to tan its hide.

(v) Use the tendons and ligaments of the body of large animals for lashings.

(w) The marrow in bones is a rich-food source. Crack the bones and scrap out the marrow, and use bones to make weapons.

(x) If the situation and time allow, you should preserve the extra meat for later use. If the air is cold enough, you can freeze the meat. In warmer climates however, you will need to use a drying or smoking process to preserve it. One night of heavy smoking will make meat edible for about 1 week. Two nights will make it remain edible for 2 to 4 weeks. To prepare meat for drying or smoking, cut it with the grain in quarter inch strips. To air dry the meat, hang it in the wind and hot sun out the reach of animals; cover it so that blow flies cannot land on it.

(y) To smoke meat, you will need an enclosed area – for instance, a teepee (Figure 11-14) or a pit. You will also need wood from deciduous trees, preferably green. Do not use conifer trees such as pines, firs, spruces, or cedars, as the smoke from these trees give the meat a disagreeable taste.

(z) When using the para-teepee or other enclosed area with a vent at the top, set the fire in the center and let it burn down to coals, then stoke it with green wood. Place the strips of meat on a grate or hang them from the top of the enclosure so that they are about 2 feet above the smoking coals. To use the pit method of smoking meat dig, a hole about 3 feet (1 meter) deep and 1 1/2 feet (1/2 meter) in diameter. Make a fire at the bottom of the hole. After it starts burning well, add chipped green wood or small branches of green wood to make it smoke. Place a wooden grate about 1 1/2 feet (1/2 meter) above the fire and lay the strips of meat on the grate. Cover the pit with poles, boughs, leaves, or other material.

11-7. **SHELTERS**. A shelter can protect you from the sun, insects, wind, rain, snow, hot or cold temperatures, and enemy observation. In some areas your need for shelter may take precedence over your need for food, possibly even your need for water. After determining your shelter site, you should keep in mind the type of shelter (protection) you need. You need to know how to make different types of shelters. Only two are described in this handbook. Additional information is available in FM 21-76.

a. **Planning Considerations**.

- How much time and effort are needed to build the shelter?
- Will the shelter adequately protect you from the elements (rain, snow, wind, sun, and so on)?
- Do you have tools to build it? If not, can you improvise tools from materials in the area?
- Do you have the type and amount of manmade materials needed to build it? If not, are there sufficient natural materials in the area?

(f) If you plan to use the lean-to for more than one night, or if you expect rain, make a center support to the lean-to. You can do this by stretching a rope between two upright poles or trees that are in line with the center of the poncho.

(g) Tie another rope to the poncho hood; pull it upward so that it lifts the center of the poncho, and tie it firmly to the rope stretched between the two uprights.

(h) Another method is to cut a stick to place upright under the center of the lean-to. This method, however, will restrict your space and movements in the shelter.

(i) To give additional protection from wind and rain, place boughs, brush, your rucksack, or other equipment at the sides of the lean-to.

(j) To reduce heat loss to the ground, place some type of insulating material, such as leaves or pine needles, inside your lean-to.

Note: When at rest, as much as 80 percent of your body heat can be lost to the ground.

(l) To increase your security from enemy observation, lower the silhouette of the lean-to by making two modifications.

- Secure the support lines to the trees knee-high rather than waist-high.
- Use two knee-high sticks in the two center grommets (sides of lean-to), and angle the poncho to the ground, securing it with sharpened sticks as above.

(2) ***Field Expedient Lean-To***. If you are in a wooded area and have sufficient natural materials, you can make an expedient lean-to (Figure 11-16) without the aid of tools or with only a knife. You need more time to make it than the shelter previously mentioned, but it will protect you from most environmental elements. You will need two trees, (or two upright poles), about 6 feet apart; one pole about 7 feet long and 1 inch in diameter. Five to eight poles about 10 feet long and 1 inch in diameter for beams, cord or vines for securing, the horizontal support to the trees and other poles, saplings, or vines to crisscross the beams. To make this lean-to:

Figure 11-16. FIELD-EXPEDIENT LEAN-TO

(a) Tie the 7-foot pole to the two trees at point about waist to chest high. This is your horizontal support. If there is a fork in the tree, you can rest the pole in it instead of tying the pole in place. If a standing tree is not available, construct a bipod using an Y-shaped sticks or two tripods.

(b) Place one end of the beams (10-foot poles) one side of the horizontal support. As with all lean-to type shelters, make sure the backside of the lean-to is placed into the wind.

(c) Crisscross sapling or vines on the beams.

(d) Cover the framework with brush, leaves, pine needles, or grass, starting at the bottom and working your way up like shingling.

(f) If you plan to use the lean-to for more than one night, or if you expect rain, make a center support to the lean-to. You can do this by stretching a rope between two upright poles or trees that are in line with the center of the poncho.

(g) Tie another rope to the poncho hood; pull it upward so that it lifts the center of the poncho, and tie it firmly to the rope stretched between the two uprights.

(h) Another method is to cut a stick to place upright under the center of the lean-to. This method, however, will restrict your space and movements in the shelter.

(i) To give additional protection from wind and rain, place boughs, brush, your rucksack, or other equipment at the sides of the lean-to.

(j) To reduce heat loss to the ground, place some type of insulating material, such as leaves or pine needles, inside your lean-to.

Note: When at rest, as much as 80 percent of your body heat can be lost to the ground.

(l) To increase your security from enemy observation, lower the silhouette of the lean-to by making two modifications.

- Secure the support lines to the trees knee-high rather than waist-high.
- Use two knee-high sticks in the two center grommets (sides of lean-to), and angle the poncho to the ground, securing it with sharpened sticks as above.

(2) ***Field Expedient Lean-To***. If you are in a wooded area and have sufficient natural materials, you can make an expedient lean-to (Figure 11-16) without the aid of tools or with only a knife. You need more time to make it than the shelter previously mentioned, but it will protect you from most environmental elements. You will need two trees, (or two upright poles), about 6 feet apart; one pole about 7 feet long and 1 inch in diameter. Five to eight poles about 10 feet long and 1 inch in diameter for beams, cord or vines for securing, the horizontal support to the trees and other poles, saplings, or vines to crisscross the beams. To make this lean-to:

Figure 11-16. FIELD-EXPEDIENT LEAN-TO

(a) Tie the 7-foot pole to the two trees at point about waist to chest high. This is your horizontal support. If there is a fork in the tree, you can rest the pole in it instead of tying the pole in place. If a standing tree is not available, construct a bipod using an Y-shaped sticks or two tripods.

(b) Place one end of the beams (10-foot poles) one side of the horizontal support. As with all lean-to type shelters, make sure the backside of the lean-to is placed into the wind.

(c) Crisscross sapling or vines on the beams.

(d) Cover the framework with brush, leaves, pine needles, or grass, starting at the bottom and working your way up like shingling.

(e) Place straw, leaves, pine needles, or grass inside the shelter for bedding.

(f) In cold weather, you can add to the comfort of your lean-to by building a fire-reflector wall (Figure 11-16). Drive four stakes about 4 feet long into the ground to support the wall. Stack green logs on top of one another between the support stales. Bind the top of the support stakes so the green logs will stay in place. Fill in the spaces between the logs with twigs or small branches. With just a little more effort, you can have a drying rack. Cut a few 3/4-inch diameter poles. The length depends on distance between the lean-to support and the top of the fire-reflector wall. Lay one end of the poles on the lean-to horizontal support and the other ends on top of the reflector wall. Place and tie into place smaller sticks across these poles. You now have a place to dry clothes, meat, or fish.

11-8. **FIRES**. A fire can full fill several needs. It can keep you warm, it can keep you dry: you can use it to cook food, to purify water, and to signal. It can also cause you problems when you are in enemy territory: it creates smoke, which can be smelled and seen from a long distance: It causes light which can be seen day or night and it leaves signs of your presence. Remember you should always weigh your need for a fire against your need to avoid enemy protection. When operating in remote areas you should always take a supply of matches in a waterproof case and keep them on your person.

a. **Selection**. When selecting a site to build a fire, you should consider the following:

(1) The area (terrain and climate) in which you are operating.
(2) The material and tools available.
(3) How much time you have.
(4) Why you need a fire.
(5) The nearness of the enemy.

b. **Preparation**. If you are in a wooded or brush-covered area, clear brush away, and scrape the surface soil from the spot you selected. The cleared circle should be at least 3 feet (1 meter) in diameter so that there is little chance of the fire spreading. To prepare the site for a fire, ensure that it is dry and that it--look for a dry spot that--

(1) Offers protection from the wind.
(2) Is suitably placed in relation to your shelter (if any).
(3) Concentrates the heat in the direction you desire.
(4) Has a supply of wood or other fire burning material.

c. **Dakota Fire Hole**. In some situations, an underground fireplace will best meet your needs. It conceals the fire to some extent and serves well for cooking food. To make an underground fireplace or Dakota fire hole (Figure 11-17):

Figure 11-17. DAKOTA FIRE HOLE

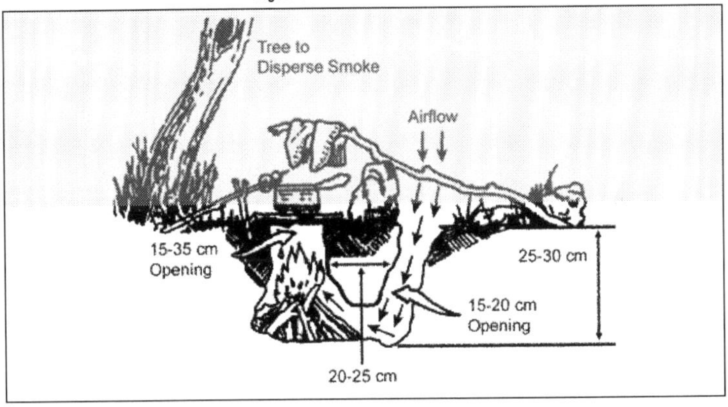

(1) Dig a hole in the ground.

(2) On the upwind side of this hole, poke one large connecting hole for ventilation.

d. **Aboveground fire**. If you are in a snow covered or wet area, you can use green logs to make a dry base for your fire (Figure 11-18). Trees with wrist-size trunks are easily broken in extreme cold. Cut or break several green logs and lay them side by side on top of the snow. Add one or two more layers, laying the top layer logs in a direction opposite those of the layer below it.

Figure 11-18. BASE FOR FIRE IN SNOW COVERED AREA

e. **Methods**. There are several methods for laying a fire for quick fire making. Three easy methods are Tepee, lean-to, and cross-ditch. Tepee (Figure 11-19). Arrange tinder and a few sticks of kindling in the shape of a cone. Fire the center. As the cone burns away, the outside logs will fall inward, feeding the heart of the fire. This type of fire burns well even with wet wood.

(1) ***Lean-To*** (Figure 11-19). Push a green stick into the ground at a 30 degree angle. Point the end of the stick in the direction of the wind. Place some tinder (at least a handful) deep inside this lean-to stick. Light the tinder. As the kindling catches fire from the tinder, add more kindling.

Figure 11-19. METHODS FOR LAYING A FIRE

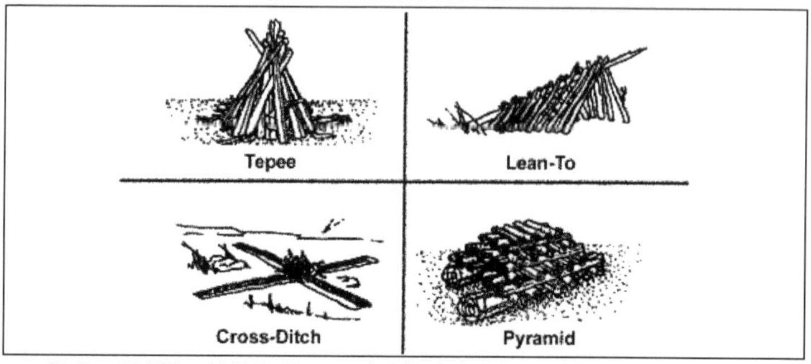

(2) ***Cross-Ditch*** (Figure 11-19). Scratch a cross about 1 foot in size in the ground. Dig the cross 3 inches deep. Put a large wad of tinder in the middle of the cross. Build a kindling pyramid above the tinder. The shallow ditch allows air to sweep under the fire to provide a draft.

NOTES

Chapter 12
FIRST AID

Patrolling, more than some other types of missions, puts Rangers in harms way. CASEVAC planning is vital. Also, because trained medical personnel might be unavailable at the initial point of injury, everyone must know how to diagnose and treat injuries, wounds, and common illnesses. The unit should also have a plan for handling KIAs.

12-1. **LIFESAVING STEPS**. Whatever the injury, (1) open the airway and restore breathing; (2) stop the bleeding and protect the wound; (3) check, treat and monitor for shock; and (4) MEDEVAC the casualty.

12-2. **CARE UNDER FIRE**. When still under fire, (1) maintain situational awareness; (2) return fire; (3) protect the casualty; (4) Identify and control severe bleeding with bandage or tourniquet; (5) and move the casualty to cover.

12-3. **PRIMARY SURVEY**. Use the ABC's to help you remember how to identify and deal with life-threatening injuries, airway complications, breathing problems, or uncontrolled hemorrhaging (bleeding).

A	Airway. Open airway by patient position or with airway adjuncts.
B	Breathing. Seal open chest wounds with occlusive dressing.
C	Circulation. Identify uncontrolled bleeding and control with pressure or tourniquet. Start IV if needed.
D	Disability. Determine Level of consciousness.
E	Exposure. Fully expose patient. (Environment dependent)

12-4. **AIRWAY MANAGEMENT**. The *airway* is most often obstructed at the base of the tongue.

a. If this happens, open the airway using the chin lift (nontrauma, Figure 12-1) or jaw thrust (trauma, Figure 12-2).

Figure 12-1. CHIN LIFT

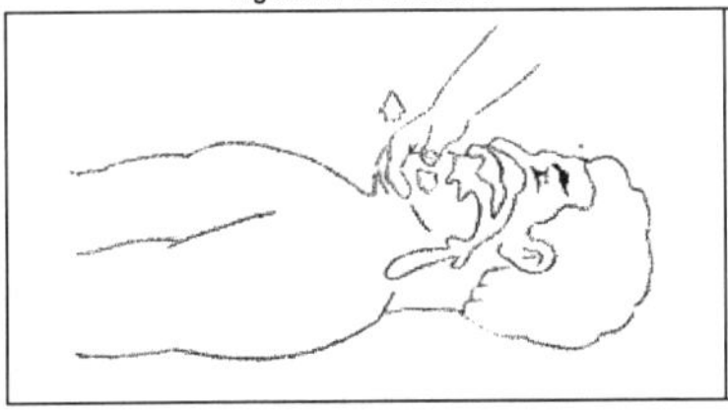

Figure 12-2. JAW THRUST

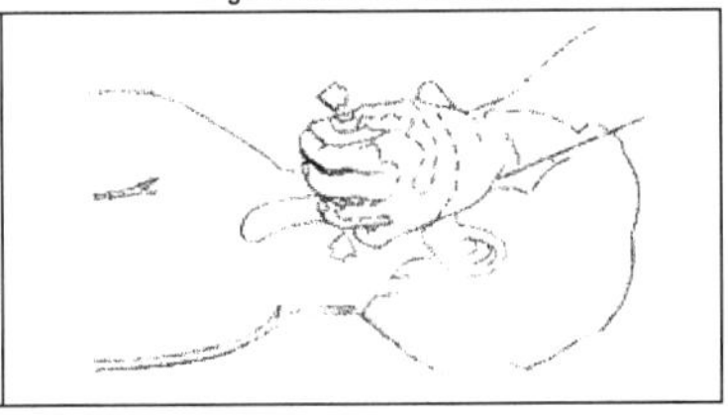

b. **Remove** debris (teeth, blood clots, bone) from the oral cavity; use suction if you have it; and place airway adjuncts to allow the victim to breathe through their nose (Figure 12-3) or mouth (Figure 12-4).

Figure 12-3. NASAL AIRWAY

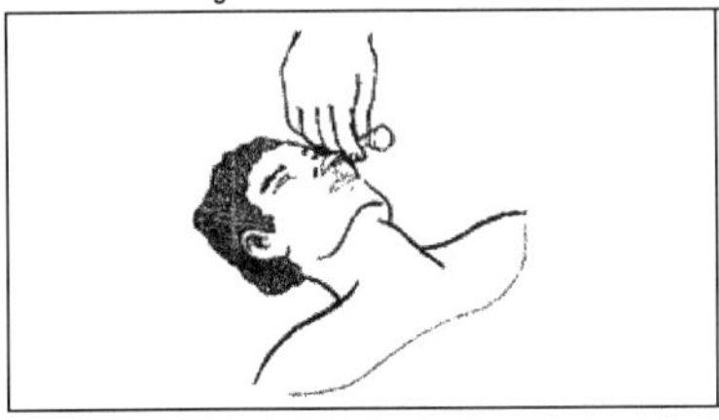

Figure 12-4. MOUTH AIRWAY

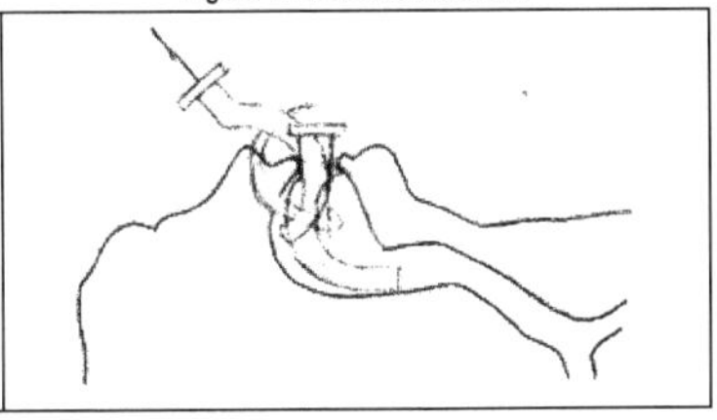

12-5. **BREATHING**. If the patient is having difficulty breathing, (1) expose the chest and identify open chest injuries; (2) place occlusive dressing (plastic and tape, or a COTS chest dressing) over open entry and exit chest wounds; and (3) place the patient on his injured side, or position him where he can breathe most comfortably.

12-6. **BLEEDING**. Quickly identify and control bleeding. (1) Apply a standard or pressure dressing. (2) If this does not control the bleeding, immediately apply a tourniquet. (3) Reassess dressings frequently to ensure bleeding has been controlled.

12-7. **SHOCK**. Shock causes an inadequate flow of oxygen to the tissues. Shock can cause bleeding (hemorrhagic) or not (nonhemorrhagic). Signs and symptoms of shock can include increased pulse, increased respiration; and decreased level of consciousness. (1) Open the victim's airway; (2) restore his breathing; (3) control the bleeding; (4) initiate an IV or a saline lock; and (5) monitor his condition.

12-8. **EXTREMITY INJURIES**. Identify and control bleeding. If you suspect a fracture, splint it as it lies. Do not reposition injured extremity.

12-9. **ABDOMINAL INJURIES**. Identify and control bleeding. Treat for shock. (1) If internal organs are exposed, cover them with dry, sterile dressing. Do not place organs back in wound area. Do not handle internal organs. (2) Place patient in position of comfort. Flex knees to relax abdomen. (3) Do not give anything by mouth to the patient.

12-10. **BURNS**. Remove patient from burn source. Then (1) remove all clothing and jewelry from the area; (2) cover with dry, sterile dressings, and ensure fingers and toes have dressings between them before covering entire area; and (3) Evacuate immediately any casualties with burns of the face, neck, hands, genitalia, or over 20 percent of his body surface.

12-11. **WEATHER (HEAT AND COLD) INJURIES**. Tables 12-1, 12-2, and 12-3 show first aid for (respectively) heat injuries, cold injuries, and environmental injuries.

Table 12-1. HEAT INJURIES

INJURY	*SIGNS/SYMPTOMS*	*FIRST AID*
Heat Cramps	Casualty experiences muscle cramps in arms, legs and/or stomach, may also have wet skin and extreme thirst.	1. Move the casualty to a shaded area and loosen clothing. 2. Allow casualty to drink 1 quart of cool water slowly per hour. 3. Monitor casualty and provide water as needed. 4. Seek medical attention if cramps persist.
Heat Exhaustion	Casualty experiences loss of appetite, headache, excessive sweating, weakness or faintness, dizziness, nausea, muscle cramps. The skin is moist, pale, and clammy.	1. Move the casualty to a cool, shaded area and loosen clothing. 2. Pour water on casualty and fan to increase cooling effect of evaporation. 3. Provide at least one quart of water to replace lost fluids. 4. Elevate legs. 5. Seek medical aid if symptoms continue.
Heatstroke (Sunstroke)	Casualty stops sweating (hot, dry skin), may experience headache, dizziness, nausea, vomiting, rapid pulse and respiration, seizures, mental confusion. Casualty may suddenly collapse and lose consciousness. THIS IS A MEDICAL EMERGENCY!	1. Move casualty to a cool, shaded area, loosen clothing, and remove outer clothing if the situation permits. 2. Immerse in cool water. If cool bath is not available, massage arms and legs with cool water. Fan casualty to increase the cooling effect of evaporation. 3. If conscious, slowly consume one quart of water. 4. SEEK MEDICAL AID AND EVACUATE AS SOON AS POSSIBLE. Perform any lifesaving measures.

Table 12-2. COLD INJURIES

INJURY	*SIGNS/SYMPTOMS*	*FIRST AID*
Chilblain	Red, swollen, hot, tender, itching skin. Continued exposure may lead to infected (ulcerated bleeding) skin lesions.	1. Area usually responds to locally applied warming (body heat). 2. Do Not rub or massage area. 3. Seek medical treatment.
Immersion (trench) foot	Affected parts are cold, numb, and painless. As parts warm, they may be hot, with burning and shooting pains. Advanced stage: skin pale with bluish cast: pulse decreases, blistering, swelling, heat hemorrhages, and gangrene may follow.	1. Gradual warming by exposure to warm air. 2. DO NOT massage or moisten skin. 3. Protect affected parts from trauma. 4. Dry feet thoroughly: avoid walking. 5. Seek medical treatment.
Frostbite	***SUPERFICIAL*** Redness, blisters in 24-36 hours and sloughing of the skin. ***DEEP*** Preceded by superficial frostbite; skin is painless, pale- yellowish, waxy, "wooden" or solid to touch, blisters form in 12-36 hours	***SUPERFICIAL*** 1. Keep casualty warm; gently warm affected parts. 2. Decrease constricting clothing, increase exercise and insulation. ***DEEP*** 1. Protect the part from additional injury. 2. Seek medical treatment as fast as possible.
Snow Blindness	Red, scratchy, or watery eyes; headache; increased pain in eyes with exposure to light.	1. Cover the eyes with a dark cloth. 2. Seek medical treatment.
Dehydration	Similar to heat exhaustion.	1. Keep warm, loosen clothes. 2. Replace lost fluids, rest, and additional medical treatment.
Hypothermia	Casualty is cold, shivers uncontrollably until shivering stops. Rectal (core) temp less 95° F can affect consciousness. Uncoordinated movements, shock, and coma may occur as body temperature drops.	***MILD HYPOTHERMIA*** 1. Warm body evenly and without delay. (Heat source must be provided.) 2. Keep dry, protect from elements. 3. Warm liquids may be given to conscious casualty only. 4. Be prepared to start CPR. 5. Seek medical treatment immediately. ***SEVERE HYPOTHERMIA*** 1. Quickly stabilize body temperature. 2. Attempt to prevent further heat loss. 3. Handle the casualty gently. 4. Evacuate to nearest medical treatment facility as soon as possible.

Table 12-3. ENVIRONMENTAL INJURIES

TYPE	*FIRST AID*
Snake bite	1. Get the casualty away from the snake. 2. Remove all rings and bracelets from the affected extremity. 3. Reassure the casualty and keep him quiet. 4. Apply constricting band(s) 1 to 2 finger widths close to the bite. You should be able to slip 1 finger between the band and skin. ***ARM OR LEG BITE*** - Place one band above and one band below the bite site. ***HAND OR FOOT BITE*** - Place one band above the wrist or ankle. 5. Immobilize the affected limb below the level of the heart. 6. Kill the snake, if possible, (without damaging its head or endangering yourself) and send it with the casualty. 7. Seek medical treatment immediately.
Brown recluse or black widow, spider bite	1. Keep the casualty calm. 2. Wash the area. 3. Apply ice or a freeze pack, if available. 4. Seek medical treatment.
Tarantula bite, scorpion sting, ant bite	1. Wash the area. 2. Apply ice or a freeze pack, if available. 3. Apply baking soda, calamine lotion, or meat tenderizer to the bite site to relieve pain and itching. 4. If site of bite(s) or sting(s) is on the face, neck (possible airway blockage), or genital area, or if reaction is severe, or if the sting is by the dangerous Southwestern scorpion, keep the casualty as quiet as possible, administer the epinephrine auto-injector (EAI) if needed, and then seek immediate medical aid.
Wasp or bee sting	1. If the stinger is present, remove by scraping with a knife or finger nail. DO NOT squeeze venom sack on stinger, more venom may be injected. 2. Wash the area. 3. Apply ice or freeze pack, if available. 4. If allergic signs or symptoms appear, be prepared to administer the EAI, and then seek medical assistance.
Human and other animal bites	1. Cleanse the wound thoroughly with soap or detergent solution. 2. Flush bite well with water. 3. Cover bite with a sterile dressing. 4. Immobilize injured extremity. 5. Transport casualty to a medical treatment facility. 6. For animal bites, without endangering yourself or damaging the animal's head, kill the animal and send its head with the casualty. 7. For human bites, try to extract some of the attacker's saliva from the wound and send that in a sealed, identified container with the casualty.

Table 12-3. ENVIRONMENTAL INJURIES (CONTINUED)

Poison Ivy, Oak, Sumac	1. Gently clean affected area. Clean areas 2-3 times daily. Wash clothing. 2. Apply topical anti-itch lotion or ointment as needed, and cover. 3. Avoid scratching or itching the area. 4. Observe for signs of infection (increasing redness, tenderness, warm to touch). 5. Seek medical attention if rash persists or signs of infection develop.

12-12. **POISONOUS PLANT IDENTIFICATION**

Poison plants include, among others, poison ivy, oak, and sumac, as well as a few more such as stinging nettles which are not discussed here.

a. **Poison Ivy**. Poison ivy is grows in two forms, a vine or shrub. The compound leaves of poison ivy consist of three pointed leaflets; the middle leaflet has a much longer stalk than the two side ones. The leaflet edges can be smooth or toothed but are rarely lobed. The leaves vary greatly in size, from 8 to 55 mm (0.31" to 2.16") in length. They are reddish when they emerge in the spring, turn green during the summer, and become various shades of yellow, orange, or red in the autumn. Small greenish flowers grow in bunches attached to the main stem close to where each leaf joins it. Later in the season, clusters of poisonous, berrylike drupes form. They are whitish, with a waxy look.

b. **Poison Oak**. Poison oak is a widespread deciduous shrub throughout mountains and valleys of North America, generally below 5,000 feet elevation. It commonly grows as a climbing vine with aerial (adventitious) roots that adhere to the trunks of oaks and sycamores. Poison oak also forms dense thickets in chaparral and coastal sage scrub. Poison oak leave typically have three leaflets (sometimes five), the terminal one on a slender rachis (also called a stalk or petiolule). Eastern poison ivy often has a longer rachis and the leaflet margins tend to be less lobed and serrated (less "oak-like"). Like many members of the sumac family (Anacardiaceae) new foliage and autumn leaves often turn brilliant shades of pink and red.

c. **Poison Sumac**. Poison Sumac is a woody perennial shrub or small tree growing from 5 - 25 feet tall that grows in peat bogs and swamps. To identify Poison Sumac, look for the fruit that grows between the leaf and the branch. Leaves are arranged in an alternate pattern on the vine. There are about 7-13 leaflets forming a feather-like appearance. The foliage has brilliant orange or scarlet coloring in the fall.

Figure 12-5. POISONOUS PLANTS

12-13. **FOOT CARE.** Use moleskin to prevent blisters prior to movement or foot march. Drain large blisters. Clean area, puncture with needle, drain blister. Place moleskin over area. Observe for signs of infection. Keep feet as clean and dry as possible. Use foot powder and change socks. Let feet air dry as mission permits.

12-14. **CASUALTY RESCUE AND TRANSPORT SYSTEM.** Remove the hoistable CRTS litter from its pack, and place the litter on the ground. Unfasten the retainer strap, step on the foot end of the litter, and unroll the litter completely to the opposite end. Then--

a. Bend the CRTS litter in half and backroll it. Repeat with the opposite end of the litter. The litter will now lay flat. Lift the sides of the litter, and fasten the four cross-straps to the buckles directly opposite the straps.

b. Place the CRTS litter next to the casualty. Insure the head end of the stretcher is adjacent to the head of casualty. Place the cross-straps under the litter. Logroll the casualty and slide the litter as far under the casualty as possible. Gently roll the casualty down on the litter. Slide the casualty to the center of the litter. Be sure to keep his spinal column as straight as possible.

c. Pull the straps out from under the CRTS litter and fasten it to the buckles. Position the foot end of the litter at the head of the casualty. Have one rescuer straddle the litter and support the casualty's head, neck, and shoulders.

d. Grasp the foot straps of the litter, and slide the litter under the casualty. Center the casualty on the litter, and feed foot straps thru the unused grommets at the foot end of the litter. Fasten the straps to the buckles.

12-15. **HYDRATION AND ACCLIMATIZATION**

a. Table 12-4 shows strategies for minimizing dehydration and increasing acclimatization and good hydration practices.

Table 12-4. HYDRATION MANAGEMENT AND ACCLIMATIZATION

STRATEGY	*SUGGESTIONS FOR IMPLEMENTATION*
Start early	1. Start at least 1 month prior to School 2. Be flexible and patient: performance benefits take longer than the physiological benefits
Mimic the training environment climate	1. In warm climates, acclimatize in the heat of day. 2. In temperate climates workout in a warm room wearing sweats.
Ensure adequate heat stress	1. Induce sweating. 2. Work up to 100 minutes of continuous physical exercise in the heat. Be patient. The first few days, you may not be able to go 100 minutes without resting. 3. Once you can comfortably exercise for 100 minutes in the heat, then continue for at least 7-14 days with added exercise intensity (loads or training runs).
Teach yourself to drink and eat	1. Your thirst mechanism will improve as you become heat acclimatized, but you will still under-drink if relying on thirst sensation. 2. Heat acclimatization will increase your water requirements. 3. Dehydration will negate most benefits of physical fitness and heat acclimatization. 4. You will sweat out more electrolytes when not acclimatized, so add salt to your food, or drink electrolyte solutions during the first week of heat acclimatization. 5. A convenient way to learn how much water your body needs to replace is to weigh yourself before and after the 100 minutes of exercise in the heat. For each pound lost, you should drink about one-half quart of fluid. 6. Do not skip meals, as this is when your body replaces most of its water and salt losses.

12-16. **WORK, REST, AND WATER CONSUMPTION.** Table 12-5 shows a work, rest, and water consumption table. The guidance applies to the average size, heat acclimated Soldier wearing ACU, not hot weather gear, except as specified otherwise. The work and rest times and fluid replacement volumes shown will help the Soldier sustain his performance and hydration for at least 4 hours of work in the specified heat category. Fluid needs can vary based on individual differences (±1 quart per hour).

a. "NL" means that there is no limit to work time per hour. "Rest" means minimal physical activity such as sitting or standing, preferably in the shade.

b. Consume no more than 1.5 quarts of fluid per hour, and no more than 12 quarts per day.

c. If you are wearing body armor in a humid climate, then add 5° F to the WBGT. if wearing MOPP 4 clothing, add 10° F to the WBGT.

d. Work categories include easy, moderate, and hard.

1) ***Easy Work***. This includes, for example, maintaining weapons; walking on hard surfaces at 2.5 mph with a load weighing no more than 30 pounds; participating in marksmanship training; and participating in drills or ceremonies.

2) ***Moderate Work***. This includes, for example, walking in loose sand at 2.5 mph (maximum) or under with no load; walking on a hard surface at 3.5 mph (maximum) with a load weighing no more than 40 pounds; performing calisthenics; patrolling; or conducting individual movement techniques such as the low or high crawl.

3) ***Hard Work***. This includes, for example, walking on a hard surface at 3.5 mph with a load weighing 40 or more pounds; walking in loose sand at 2.5 mph while carrying a load; and conducting field assaults.

Table 12-5. WORK, REST, AND WATER CONSUMPTION TABLE

HEAT CATEGORY	WBGT INDEX, F°	*EASY WORK*		*MODERATE WORK*		*HARD WORK*	
		WORK/ REST	**WATER INTAKE (QT/H)**	**WORK/ REST**	**WATER INTAKE (QT/H)**	**WORK/ REST**	**WATER INTAKE (QT/H)**
1	78° to 81.9°	NL	0.50	NL	0.75	40/20	0.75
2 (Green)	82° to 84.9°	NL	0.50	50/10	0.75	30/30	1.00
3 (Yellow)	95° to 87.9°	NL	0.75	40/20	0.75	30/30	1.00
4 (Red)	88° to 89.9°	NL	0.75	30/30	0.75	20/40	1.00
5 (Black)	90° or more	50/10 min	1.00	20/40	1.00	10/50	1.00

NOTES

Chapter 13
DEMOLITIONS

This chapter introduces Rangers to the characteristics of explosives, to initiation systems, MDI components, detonation systems, safety considerations, expedient explosives, breaching charges, and timber cutting charges (FM 5-250). The two categories of explosives are low and high (Table 13-1).

- *Low explosives* have a detonating velocity up to 1,300 feet per second, which produces a pushing or shoving effect.
- *High explosives* have a detonating velocity of 3,280 to 27,888 feet per second, which produces a shattering effect.

Table 13-1. CHARACTERISTICS OF US DEMOLITIONS EXPLOSIVES

NAME	*APPLICATIONS*	*DETONATION VELOCITY*		*RE FACTOR**	*FUME TOXICITY*	*WATER RESISTANCE*
		MIN/SEC	*FT/SEC*			
AMMONIUM NITRATE	CRATERING CHG	2,700	8.800	0.42	DANGEROUS	POOR
PETN	DET CORD BLASTING CAPS DEMOLITION CHGS	8,300	27,200	1.66	SLIGHT	EXCELLENT
RDX	BLASTING CAPS COMPOSITION EXPLOSIVE	8,350	27,400	1.60	DANGEROUS	EXCELLENT
TRINITROTOLUENE (TNT)	DEMOLITION CHG COMPOSITION EXPLOSIVE	6,900	22,600	1.00	DANGEROUS	EXCELLENT
TETRYL	BOOSTER CHG COMPOSITION EXPLOSIVE	7,100	23,300	1.25	DANGEROUS	EXCELLENT
NITROGLYCERIN	COMMERCIAL DYNAMITE	7,700	25,200	1.50	DANGEROUS	GOOD
BLACK POWDER	TIME FUSE	400	1,300	0.55	DANGEROUS	POOR
AMATOL 80/20	BURSTING CHG	4,900	16,000	1.17	DANGEROUS	POOR
COMPOSITION A3	BOOSTER CHG BURSTING CHG	8,100	26,500	---	DANGEROUS	GOOD
COMPOSITION B	BURSTING CHG	7,800	25,600	1.35	DANGEROUS	EXCELLENT
COMPOSITION C4 (M112)	CUTTING CHG BREACHING CHG	8.040	26,400	1.34	SLIGHT	EXCELLENT
COMPOSITION H6	CRATERING CHG	7,190	23,600	1.33	DANGEROUS	EXCELLENT
TETRYTOL 75/25	DEMOLITION CHG	7,000	23,000	1.20	DANGEROUS	EXCELLENT
PENTOLITE 50/50	BOOSTER CHG BURSTING CHG	7,450	24,400	---	DANGEROUS	EXCELLENT

NAME	APPLICATIONS	DETONATION VELOCITY		RE FACTOR*	FUME TOXICITY	WATER RESISTANCE
		MIN/SEC	FT/SEC			
M1 DYNAMITE	DEMOLITION CHG	6,100	20,000	0.92	DANGEROUS	FAIR
DET CORD	PRIMING DEMOLITION CHG	6,100 TO 7,300	20,000 TO 24,000	---	SLIGHT	EXCELLENT
SHEET EXPLOSIVE M118 AND M186	CUTTING CHG	7,300	24,000	1.14	DANGEROUS	EXCELLENT
BANGALORE TORPEDO M1A2	DEMOLITION CHG	7,800	25,600	1.17	DANGEROUS	EXCELLENT
SHAPED CHARGES M2A3, M2A4, AND M3A1	CUTTING CHG	7,800	25,600	1.17	DANGEROUS	EXCELLENT
* TNT = 1.00 RELATIVE EFFECTIVENESS						

13-1. **INITIATING (PRIMING) SYSTEMS**. The best way to prime demolition systems is with modernized demolition initiators (MDI). These consist of blasting caps attached to various lengths of time fuse or shock tube. MDI can be used with a fuse igniter and detonating cord to create many firing systems. In the absence of MDI, field expedient methods may be used.

a. **Shock Tube**.

(1) Thin, plastic tube of extruded polymer with a layer of special explosive material on the interior surface.

(2) Explosive material propagates a detonation wave which moves along the shock tube to a factory crimped and sealed blasting cap.

(3) Detonation is normally contained within the plastic tubing. However, burns may occur if the shock tube is held.

(4) Advantages of shock tube.

(a) The shock tube offers the instantaneous action of electric initiation without the risk of accidental initiation caused by radio transmitters, static electricity, and so on.

(b) Extremely reliable.

(c) May be extended using left over pieces from previous operations.

b. **Blasting Caps**. Five MDI blasting caps are available which replace the M6 electric and M7 Non-electric blasting cap. Three of these are high-strength caps and two are low-strength. High-strength blasting caps can be used to prime all standard military explosives (including detonating cord) or to initiate the shock tube of other MDI blasting caps.

(1) ***M11***.

- Factory crimped to 30 feet of shock tube.
- A movable "J" hook is attached for quick and easy attachment to detonating cord.
- A red flag is attached 1 meter from the blasting cap and a yellow flag 2 meters from the blasting cap.

(2) ***M14***.

- Factory crimped to 7.5 feet of time fuse.
- May be initiated using a fuse igniter or match.
- Burn-time for total length is about five minutes.
- Yellow bands indicate calibrated one-minute time intervals.

(3) ***M15***.

- Two blasting caps factory crimped to 70 feet of shock tube.
- Each blasting cap has delay elements to allow for staged detonations.
- Low-strength blasting caps. Used as a relay device to transmit a shock tube detonation impulse from an initiator to a high strength-blasting cap.

(4) ***M12***. Factory crimped to 500 feet of shock tube on a cardboard spool.

(5) ***M13***. Factory crimped to 1,000 feet of shock tube.

c. **Matches**. If fuse igniter is unavailable, light the time blasting fuse with a match. Split the fuse at the end (Figure 13-1), and place the head of an unlit match in the powder train. Light the inserted match head with a flaming match, or rub the abrasive on the match box against it. It may be necessary to use two match heads during windy conditions. Burn time will increase with altitude and colder temperatures.

d. **M81 Fuse Igniter**. This is used to ignite time blasting fuse or to initiate the shock tube of MDI blasting caps.

Figure 13-1. LIGHTING TIME FUSE WITH A MATCH

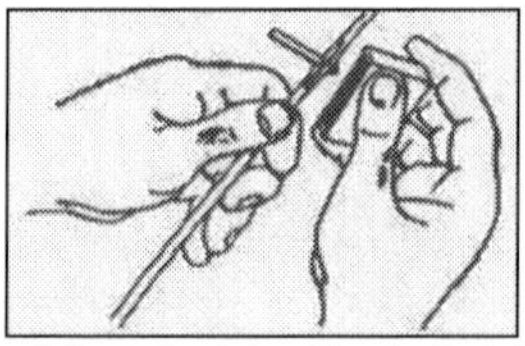

Notes: Low strength blasting caps cannot reliably set off explosives. They should only be used to set off additional shock tubes. The M60 fuse igniter may still be used to ignite time blasting fuse. However, it will not reliably initiate the shock tube.

13-2. **DETONATION (FIRING) SYSTEMS**. The two types of firing systems are MDI alone, or MDI plus detonating cord.

a. **MDI Alone**. An MDI firing system is one in which the initiation set, transmission and branch lines are constructed using MDI components and the explosive charges are primed with MDI blasting caps. Construct the charge in the following manner.

(1) Emplace and secure explosive charge (C4, TNT, cratering charge, and so on.) on target.

(2) Place a sandbag or other easily identifiable marker over the M11, M14, or M15 blasting cap to be used.

(3) Connect to an M12 or M13 transmission line if desired.

(4) Connect blasting cap with shock tube to an M14 cap with time fuse. Cut the time blasting fuse to desired delay time.

(5) Prime the explosive charge by inserting the blasting cap into the charge.

(6) Visually inspect firing system for possible misfire indicators such as cracks, bulges, or corrosion.

(7) Return to the firing point and secure a fuse igniter to the cut end of the time fuse.

(8) Remove the safety cotter pin from the igniter's body.

(9) Actuate the charge by grasping the igniter body with one hand while sharply pulling the pull ring.

b. **MDI and Detonating Cord**. Construct the charge using the above steps for MDI stand-alone system. Incorporate detonating cord branch lines into the system using the "J" hooks of the M11 shock tube. Taping the ends of the Detonation Cord reduces the effect of moisture on the system.

13-3. **SAFETY**. MDI is not recommended for below ground use, except in quarry operations with water-gel or slurry explosives. Use detonating cord when it is necessary to bury primed charges.

a. Do not handle misfires downrange until the required 30 minute waiting period for both primary and secondary initiation systems has elapsed and other safety precautions have been accomplished.

b. Never yank or pull hard on the shock tube. This may actuate the blasting cap.

c. Do not dispose of used shock tubes by burning because of potentially toxic fumes given off from the burning plastic.

d. Do not use M1 dynamite with the M15 blasting cap. The M15 delay-blasting cap should be used only with water-gel or slurry explosives.

e. Always use protective equipment when handling demolitions. Minimum protection consists of leather gloves, ballistic eye protection, and helmet.

13-4. **EXPEDIENT EXPLOSIVES--IMPROVISED SHAPED CHARGE**. An improvised shaped charge (Figure 13-2) concentrates the energy of the explosion released on a small area, making a tubular or linear fracture in the target.

a. The versatility and simplicity of these charges make them effective against targets, especially those made of concrete or those with armor plating.

(1) Bowls, funnels, cone shaped glasses, (champagne glasses with stem removed) used as cones. Champagne or cognac bottles are excellent.

(2) Charge characteristics.

(a) *Cavity Liners.* These are made of copper, tin, or zinc. If none is available, cut a cavity out of the plastic explosive.

(b) *Cavity Angle.* This will work with 30 to 60 degree angles. The cavity angle in most high-explosive antitank (HEAT) ammunition is 42 to 45 degrees.

(c) *Explosive Height (In Container).* 2 x the height of the cone measured from the base of the cone to the top of the explosive.

(d) *Standoff.* Normal standoff is one and one half the cone's diameter.

(e) *Detonation Point.* The exact top center of the charge is the detonation point. Cover the blasting cap with a small amount of C4 if any part of the blasting cap is exposed.

b. Remove the narrow neck of a bottle or the stem of a glass by wrapping it with a piece of soft, absorbent twine or by soaking the string in gasoline and lighting it. Place two bands of adhesive tape, one on each side of the twine, to hold the twine firmly in place. The bottle or stem must be turned continuously with the neck up, to heat the glass uniformly.

c. A narrow band of plastic explosive placed around the neck and burned gives the same result. After the twine or plastic has burned, submerge the neck of the bottle in water and tap it against some object to break it off. Tape the sharp edge of the bottle to prevent cutting hands while tamping the explosive in place.

d. Do not immerse the bottle in water before the plastic has been completely burned, or it could detonate.

Figure 13-2. IMPROVISED SHAPED CHARGE

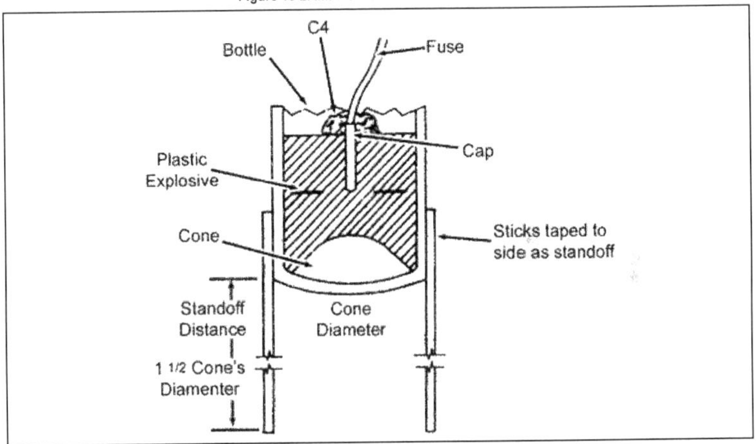

13-5. **EXPEDIENT EXPLOSIVES--PLATTER CHARGE**. This device turns a metal plate into a powerful blunt-nosed projectile (Figure 13-3). The plate should be steel, preferably round but square will work, and should weigh from 2 to 6 pounds.

a. The weight of the explosive should equal the weight of the platter.

b. Uniformly pack the explosive behind the platter. You will only need a container if the explosives fail to remain firmly against the platter. You can use tape to anchor the explosives, if needed.

c. Prime the charge at the exact, rear center of the charge. If any part of the blasting cap is exposed, cover the blasting cap with a small quantity of C4.

d. Aim charge at the direct center of the target, and ensure that the charge is on the opposite side of the platter from the target. Effective range is 35 yards for a small target. With practice, you might hit a 55-gallon drum at 25 yards 90 percent of the time. A gutted fuse igniter can serve as an expedient aiming device.

Figure 13-3. PLATTER CHARGE

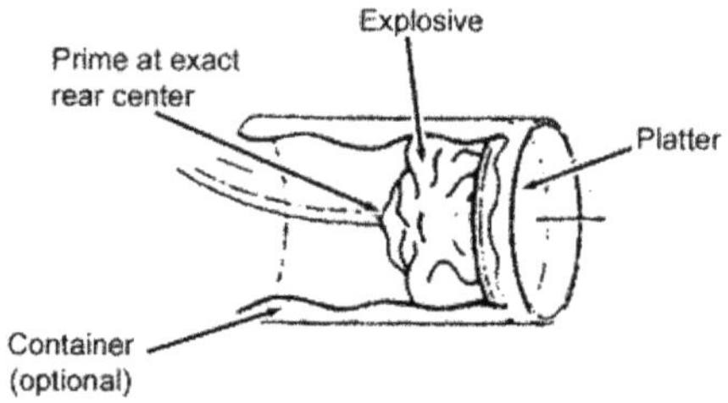

13-6. **EXPEDIENT EXPLOSIVES--GRAPESHOT CHARGE**. To use this antipersonnel fragmentation mine (Figure 13-4)--

a. **Hole**. Create a hole in the center, bottom of the container, for the blasting cap.

b. **Explosives**. Place explosives evenly on the bottom of the container. Remove all voids and air pockets by pressing the C4 into place using a non-sparking instrument.

c. **Buffer**. Place buffer material directly over top of the explosives.

d. **Projectiles**. Place projectiles over top of the buffer materials, then cover to prevent spilling from movement.

e. **Aim**. Aim at target from about 100 feet. Use a small amount of C4 on any exposed portion of the blasting cap.

Figure 13-4. GRAPESHOT CHARGE

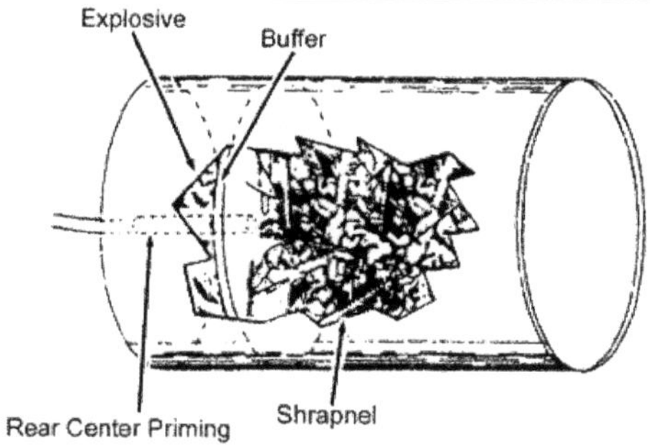

13-7. **DEMOLITION KNOTS.** Several knots are used in demolitions. Figures 13-5 and 13-6 show a few simple knots that can join demolitions to detonation cord.

Figure 13-5. VARIOUS JOINING KNOTS USED IN DEMOLITIONS

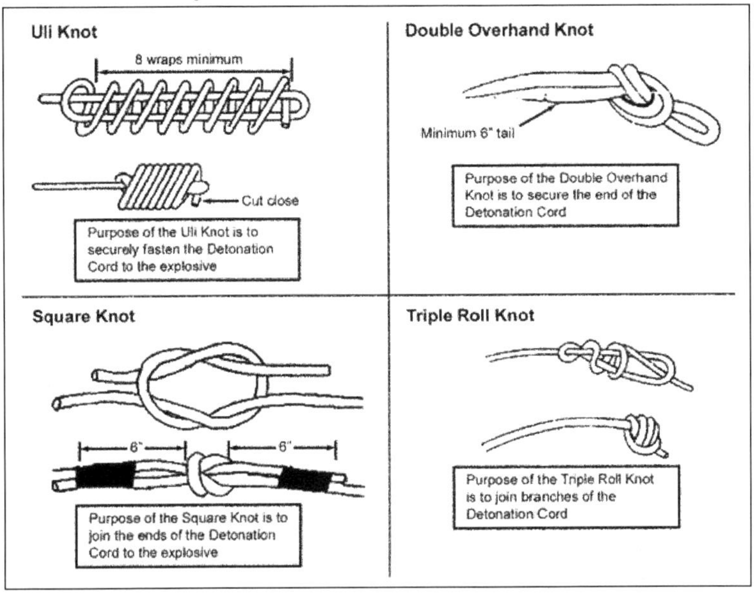

Figure 13-6. BRITISH JUNCTION

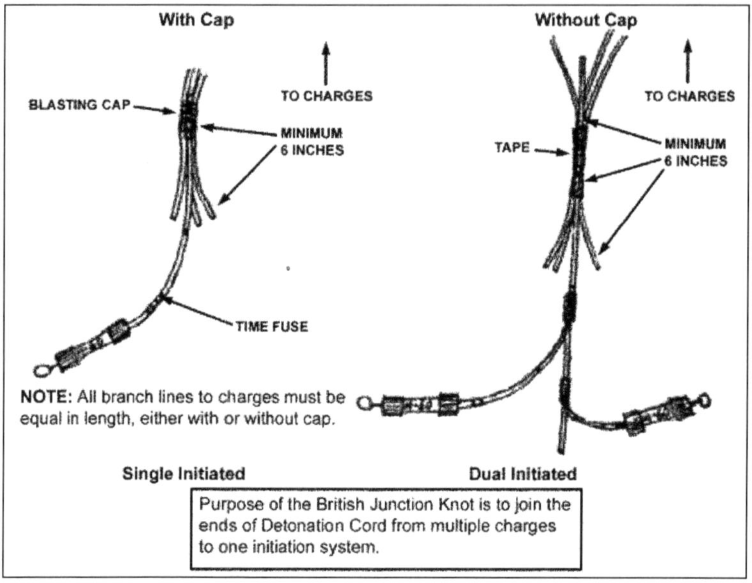

13-8. **MINIMUM SAFE DISTANCES**. Rangers must remain especially aware of their situation when using demolitions. (Table 13-2 shows minimum safe distances for employing demolitions.) For charges over 500 pounds, use the formulas shown in Figure 13-7.

Table 13-2. MINIMUM SAFE DISTANCE FOR PERSONNEL IN OPEN (BARE CHARGE)

EXPLOSIVE WEIGHT (LB)	SAFE DISTANCE		EXPLOSIVE WEIGHT (LB)	SAFE DISTANCE	
	FEET	METERS		FEET	METERS
27 OR LESS	985	300	175	1,838	560
30	1,021	311	200	1,920	585
35	1,073	327	225	1,999	609
40	1,123	342	250	2,067	630
45	1,168	356	275	2,136	651
50	1,211	369	300	2,199	670
60	1,287	392	325	2,258	688
70	1,355	413	350	2,313	705
80	1,415	431	375	2,369	722
90	1,474	449	400	2,418	737
100	1,526	465	425	2,461	750
125	1,641	500	500	2,625	800
150	1,752	534	---	---	---

Figure 13-7. MINIMUM SAFE DISTANCES FOR CHARGES OVER 500 POUNDS

$$\textit{safe distance (meters)} = 100\sqrt[3]{\textit{pounds of explosive}}$$

$$\textit{safe distance (feet)} = 300\sqrt[3]{\textit{pounds of explosive}}$$

Notes: For charges on targets, the minimum radius of danger is 1,000 meters. Minimum safe distance when in a missile-proof shelter from the point of detonation is 100 meters.

13-9. **BREACHING CHARGES**. For Table 13-3, the left column represents the thickness of reinforced concrete wall. The remaining 7 columns show the number of packages of C4 required to breach the wall using the charge placements shown in the drawings above the columns.

a. Use Table 13-3, 13-4 and 13-5 for breaching charges.

b. Use the formula in Figure 13-8 to calculate the charges.

c. Multiply number of packages of C4 from Table 13-3 by conversion factor from Table 13-4 for materials other than reinforced concrete.

Table 13-3. BREACHING CHARGES FOR REINFORCED CONCRETE

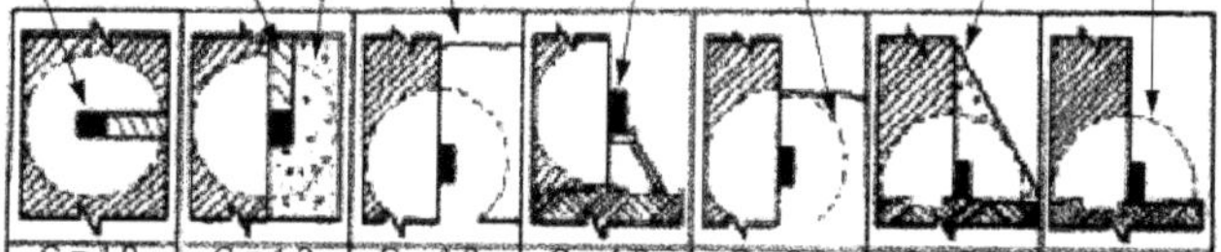

Rein-forced-Concrete Thick-ness (ft)	Placement Methods: Placed in center of mass (C = 1.0)	Tamped or stemmed (C = 1.0)	Deep water (C = 1.0)	Elevated untamped (C = 1.8)	Shallow water (C = 2.0)	Earth tamping (C = 2.0)	Ground placed, untamped (C = 3.6)
	Packages of M112 (C4)						
2.0	1	5	5	9	10	10	17
2.5	2	9	9	17	18	18	33
3.0	2	13	13	24	26	26	47
3.5	4	21	21	37	41	41	74
4.0	5	31	31	56	62	62	111
4.5	7	44	44	79	88	88	157
5.0	9	48	48	85	95	95	170
5.5	12	63	63	113	126	126	226
6.0	13	82	82	147	163	163	293
6.5	17	104	104	186	207	207	372
7.0	21	111	111	200	222	222	399
7.5	26	137	137	245	273	273	490
8.0	31	166	166	298	331	331	595

NOTE: The results of all calculations for this table have been rounded UP to the next whole package.

Table 13-4. CONVERSION FACTORS FOR MATERIALS OTHER THAN REINFORCED CONCRETE

MATERIAL	*CONVERSION FACTOR*
• EARTH	0.1
• ORDINARY MASONRY • HARDPAN • SHALE • ORDINARY CONCRETE • ROCK • GOOD TIMBER • EARTH CONSTRUCTION	0.5
• DENSE CONCRETE • FIRST-CLASS MASONRY	0.7

Table 13-5. MATERIAL FACTOR (K) FOR BREACHING CHARGES

Material	R	K
Earth	All values	0.07
Poor masonry, shale, hardpan, good timber, and earth construction	Less than 1.5 m (5 ft)	0.32
	1.5 m (5 ft) or more	0.29
Good masonry, concrete block, and rock	0.3 m (1 ft) or less	0.88
	Over 0.3 m (1 ft) to less than 0.9 m (3 ft)	0.48
	0.9 m (3 ft) to less than 1.5 m (5 ft)	0.40
	1.5 m (5 ft) to less than 2.1 m (7 ft)	0.32
	2.1 m (7 ft) or more	0.27
Dense concrete and first-class masonry	0.3 m (1 ft) or less	1.14
	Over 0.3 m (1 ft) to less than 0.9 m (3 ft)	0.62
	0.9 m (3 ft) to less than 1.5 m (5 ft)	0.52
	1.5 m (5 ft) to less than 2.1 m (7 ft)	0.41
	2.1 m (7 ft) or more	0.35
Reinforced concrete (factor does not consider cutting steel)	0.3 m (1 ft) or less	1.76
	Over 0.3 m (1 ft) to less than 0.9 m (3 ft)	0.96
	0.9 m (3 ft) to less than 1.5 m (5 ft)	0.80
	1.5 m (5 ft) to less than 2.1 m (7 ft)	0.63
	2.1 m (7 ft) or more	0.54

Figure 13-8. FORMULA FOR COMPUTING SIZES OF CHARGES TO BREACH CONCRETE, MASONRY, AND ROCK

$$P = R^3KC$$

where--

P = *TNT required, in pounds*

R = *breaching radius, in feet*

K = *material factor, which reflects the strength, hardness, and mass of the material to be demolished (Table 3-5)*

C = *tamping factor, which depends on the location and tamping of the charge (Figure 3-15)*

13-10. **TIMBER CUTTING CHARGES**. Table 13-6 shows timber-cutting charge sizes. Figures 13-9 through 13-15 show the types of charges and the formulas to use with each.

Table 13-6. TIMBER-CUTTING CHARGE SIZE

	Packages of C4 Required (1.25-lb Packages) by Timber Diameter (in)											
Charge Type	6	8	10	12	15	18	21	24	27	30	33	36
Internal	1	1	1	1	1	1	2	2	2	3	3	4
External	1	1	2	3	4	5	7	9	11	14	17	20
Abatis	---	---	---	---	---	---	---	7	9	11	14	16
NOTE: **Packages required are rounded UP to the next whole package.**												

Figure 13-9. ABATIS

Figure 13-10. FORMULA FOR FALLEN TREE OBSTACLES OR TEST SHOT

$$P = \frac{D^2}{50} = P = 0.02D^2$$

where—

P = TNT required per tree, in pounds

D = diameter or least dimension of dimensioned timber, in inches

Figure 13-11. TIMBER-CUTTING RING CHARGE

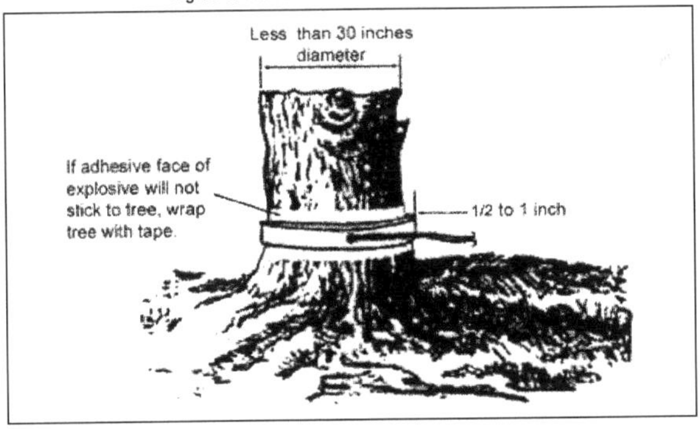

Figure 13-12. TIMBER-CUTTING CHARGE (EXTERNAL)

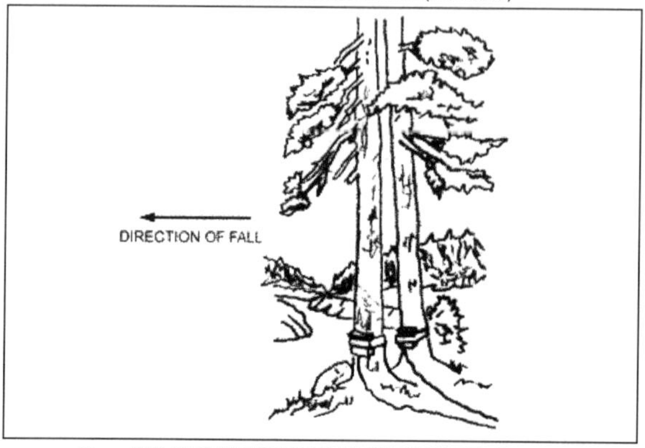

Figure 13-13. FORMULA FOR EXTERNAL TIMBER-CUTTING CHARGE

$$P = \frac{D^2}{40} \text{ or } P = 0.025D^2$$

where—

P = TNT required per target, in pounds

D = diameter or least dimension of dimensioned timber, in inches

Figure 13-14. TIMBER-CUTTING CHARGE (INTERNAL)

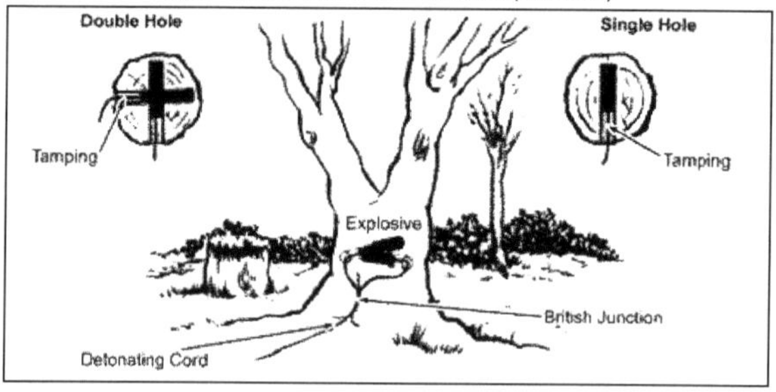

Figure 13-15. FORMULA FOR INTERNAL TIMBER-CUTTING CHARGE

$$P = \frac{D^2}{250} \text{ or } P = 0.004D^2$$

where—

P = *TNT required per tree, in pounds*

D = *diameter or least dimension of dressed timber, in inches*

NOTES

Chapter 14
RANGER URBAN OPERATIONS

Urban operations are planned and conducted in an area of operations (AO) that includes one or more urban areas. An urban area is a topographical complex where man-made construction or high population density is the dominant feature (FM 30-06.11). Urban terrain is likely to be one of the most significant future areas of operations for American forces throughout the world. Expanding urban development affects military operations as the terrain is altered. The increasing focus on stability and support operations, urban terrorism, and civil disorder emphasizes that combat in urbanized areas is unavoidable. Urban areas are the power centers, the centers of gravity, and thus the future battlefield. For further study, see FM 30-06.11, FM 90-10, FM 90-1, FM 7-8, 75th Ranger Regiment Advanced UO SOP, and Ranger Training Circular 350-1-2.

14-1. **TYPES**. Infantry units must be trained to conduct urban combat under high-intensity conditions. High-intensity urban combat requires the employment of combat power of the joint combined arms team. An Infantry unit's mission is normally to recon, isolate, penetrate, systematically clear, defend the urban area, and engage and defeat the enemy with decisive combat power. Although the changing world situation may have made urban combat under high-intensity conditions less likely for US forces, it represents the high end of the combat spectrum, and units must be trained for it. High-intensity urban operations can be casualty-intensive for both sides. With the integrated firepower of the joint, combined arms team, leaders must make every attempt to limit unnecessary destruction of critical infrastructure and casualties among noncombatants.

a. **Precision Conditions**. Infantry units train to defeat an enemy that is mixed with non-combatants in precision urban combat. Leaders plan to limit civilian casualties and collateral damage through the establishment of strict rules of engagement (ROE) and the employment of precision weapons and munitions. The ROE provides the focus for the use and restraint of combat power. The ROE may be significantly more restrictive than under high-intensity conditions.

b. **Surgical Conditions**. Operations conducted under surgical conditions include special-purpose raids, small precision strikes, or small-scale personnel seizure or recovery operations in an urban environment (for example, hostage rescue). Joint special operation forces usually conduct these operations. They may closely resemble US police operations performed by Special Weapons and Tactics (SWAT) teams. They may even involve cooperation between US forces and host nation police. Though regular units may not usually be involved in the actual surgical operation, they may support it by isolating the area, by providing security or crowd control, or providing search and rescue teams.

c. **Transitions**. Leaders must always be prepared to transition rapidly from one type of urban combat to another, and back. Real-world combat shows us that urban operations can deteriorate rapidly and without warning. A force involved in stability and support operations can suddenly find itself in a high-intensity combat situation.

14-2. **PRINCIPLES**

a. **Surprise**. Strike the enemy at a time or place or in a manner for which he is unprepared. Key to success: gives the assaulting element the advantage.

b. **Security**. Never permit the enemy to acquire unexpected advantage.

(1) Maintain during all phases of the operation.

(2) Four-dimensional battlefield (height, depth, width, subterranean).

(3) Always maintain 360 degree security (include elevated and subterranean areas).

(4) Mission is never complete as long as you remain in the urban environment. The status of actors in the urban environment does not afford the sense of security offered by "open" terrain. The key to survivability is a constant state of situational awareness.

c. **Simplicity**. Prepare clear, uncomplicated plans, and provide subordinates with concise orders to ensure thorough understanding.

(1) Always keep plans simple.

(2) Ensure everyone understands the mission and the commander's intent.

(3) Plan and prepare for the worst.

d. **Speed**. Rate of military action.

(1) Act as security.

(2) Move in a careful hurry.

(3) Smooth is fast and fast is smooth.

(4) Never move faster than you can accurately engage targets.
(5) Exercise tactical patience.

e. **Violence of Action**. Eliminate the enemy with sudden, explosive force.
(1) Combined with speed gives surprise.
(2) Prevents enemy reaction.
(3) Both physical and mental.

14-3. **METT-TC**. To effectively plan combat operations in urban environments, leaders must use Troop Leading Procedures and conduct a thorough analysis utilizing METT-TC factors. The following lists specific planning guidance that must be incorporated when planning for urban operations. For more specifics on mission planning, refer to Chapter 2, Ranger Handbook.

a. **Mission**. Know correct Task Organization to accomplish the mission (Offense, Defense, or Stability and Support Operations).

b. **Enemy**.

(1) ***Disposition***. Analyze the arrayal of enemy forces in and around your objective, known and suspected. Example: Known or suspected locations of minefields, obstacles, and strong points.

(2) ***Composition and Strength***. Analyze the enemy's task organization, troops available, suspected strength, and amount of support from local civilian populace based on intelligence estimates. Is the enemy a conventional or unconventional force?

(3) ***Morale***. Analyze the enemy's current operational status based on friendly intelligence estimates. For example, is the enemy well supplied; have they had recent success against friendly forces, taken many casualties; and what is the current weather?

(4) ***Capabilities***. Determine what the enemy can employ against your forces. Example: Enemy's weapons, artillery assets, engineer assets, air defense assets, NBC threats, thermal/NVG capabilities, close air support, armor threat, and so on.

(5) ***Probable Course(s) of Action***. Based on friendly intelligence estimates, determine how the enemy will fight within his area of operation (in and around your area of operation).

c. **Terrain**. Leaders conduct a detailed terrain analysis of each urban setting, considering the types of built-up areas and composition of existing structures. They use OCOKA when analyzing terrain, in and around the area of operation.

(1) ***Observation and Fields of Fire***. Always be prepared to conduct urban operations under limited visibility conditions. This includes the effects of reduced illumination, as well as natural and manmade obscuration. Leaders should ensure that Rangers are equipped with adequate resources, which allow them to successfully operate in the urban environment under these types of conditions.

(2) ***Cover and Concealment***. Leaders must perform a thorough analysis of peripheral as well as intra-urban areas. Leaders should identify routes to objectives, which afford assault forces with the best possible cover and concealment. Additionally, leaders should take advantage of limited visibility conditions, which would allow forces to move undetected to their final assault / breaching positions. When in the final assault position, forces should move as rapidly as tactically possible to access structures, which afford additional cover and concealment. Leaders must learn to properly employ obscurants and exercise "tactical patience" to fully take advantage of these effects. Finally, all members of the urban force must practice noise and light discipline. Soldiers must avoid unnecessary voice communications, learn the proper use of white light, and limit contact with surfaces that may alert the enemy of their presence.

(3) ***Obstacles***. There are many manmade and natural obstacles on the periphery, as well as within the urban environment. Leaders should conduct a detailed reconnaissance of routes and objectives (this must include subterranean complexes), taking into consideration route adjustments and special equipment needs.

(4) ***Key Terrain***. Analyze which buildings, intersections, bridges, LZ/PZ, airports, and elevated areas that provide a tactical advantage to you or the enemy. Additionally, the leader must identify critical infrastructure within his area of operations, which would provide the enemy with a tactical advantage on the battlefield. These may include, but are not limited to, communication centers, medical facilities, governmental facilities, and facilities that are of psychological significance.

(5) ***Avenues of Approach***. Consider roads, intersections, inland waterways, and subterranean constructions (subways, sewers, and basements). Leaders should classify areas as go, slow go, or no-go based on the navigability of the approach.

Note: Military maps may not provide enough detail for urban terrain analysis or reflect the underground sewer systems, subways, water systems, or mass transit routes.

(6) ***Troops***. Analyze your forces utilizing their disposition, composition, strength, morale, capabilities, and so on. Leaders must also consider the type and size of the objective to plan effective use of troops available.

(7) ***Time***. Operations in an urban environment have a slower pace and tempo. Leaders must consider the amount of time required to secure, clear, or seize the urban objective and stress and fatigue Rangers will encounter. Additional time must also be allowed for area analysis efforts; these may include, but are not limited to--

- Maps and urban plans Recon and analysis.
- Hydrological data analysis.
- Line-of-sight surveys.
- Long Range Surveillance and Scout reconnaissance.
- Artillery.
- Armor.
- Aviation.
- Armor.
- Engineers.

(8) ***Civilians***. Authorities such as the National Command will establish the Rules of Engagement. Commanders at all levels may provide further guidance regarding civilians occupying the area of operations (AO). Leaders must daily reiterate the ROE to subordinates, and immediately inform them of any changes to the ROE. Rangers must have the discipline to identify the enemy from noncombatants and ensure civilians understand and follow all directed commands.

Note: Civilians may not speak English, may be hiding (especially small children), or dazed from a breach. Civilians must not be given the means to resist. Rehearse how clearing/search teams will react to these variables. Never compromise the safety of your Rangers.

14-4. **CLOSE QUARTERS COMBAT.** Due to the very nature of a CQC encounter, engagements will be very close (within 10 meters) and very fast (targets exposed for only a few seconds). Most close quarter's engagements are won by who hits first and puts the enemy down. It is more important to knock a Ranger down as soon as possible than it is to kill him. In order to win a close quarters engagement, Rangers must make quick, accurate shots by mere reflex. This is accomplished by reflexive fire training. Remember, no matter how proficient you are, always fire until the enemy goes down. All reflexive fire training is conducted with the eyes open.

Note: Research has determined that, on average, only three out of ten people actually fire their weapons when confronted by an enemy during room clearing operations. Close quarters combat success for the Ranger begins with the Ranger being psychologically prepared for the close quarter's battle. The foundation for this preparedness begins with the Ranger's proficiency in basic rifle marksmanship. Survival in the urban environment does not depend on advanced skills and technologies. Rangers must be proficient in the basics.

14-5. **REHEARSALS**. Similar to the conduct of other military operations, leaders need to designate time for rehearsals. Urban operations require a variety of individual, collective, and special tasks, which are not associated with operations on less complex terrain. These task require additional rehearsal time for clearing, breaching, obstacle reduction, casualty evacuation, and support teams. Additionally, rehearsal time must be identified for rehearsals with combined arms elements. These may include, but are not limited to--

a. **Stance**. Feet are shoulder width apart, toes pointed straight to the front (direction of movement). The firing side foot is slightly staggered to the rear of the non-firing foot. Knees are slightly bent and the upper body leans slightly forward. Shoulders are not rolled or slouched. Weapon is held with the butt stock in the pocket of the shoulder. The firing side elbow is kept in against the body. The stance should be modified to ensure the Ranger maintains a comfortable boxer stance.

(1) ***Low-Carry Technique***. The butt stock of the weapon is placed in the pocket of the shoulder. The barrel is pointed down so the front sight post and day optic is just out of the field of vision. The head is always up identifying targets. This technique is safest and is recommended for use by the clearing team once inside the room.

(2) ***High-Carry Technique***. The butt stock of the weapon is held in the armpit. The barrel is pointed slightly up with the front sight post in the peripheral vision of the individual. Push out on the pistol grip and thrust the weapon forward and pull straight back into the pocket of the shoulder to assume the proper firing position. This technique is best suited for the line-up outside the door. Exercise caution with this technique always maintaining situational awareness, particularly in a multi-floored building.

Note: Muzzle awareness is critical to the successful execution of close quarter's operations. Rangers must never point their weapons or cross the bodies of their fellow Rangers at any time. Additionally Rangers should always avoid exposing the muzzle of their weapons around corners; this is referred to as "flagging."

b. **Malfunction**. If a Ranger has a malfunction with his weapon during any CQC training, he will take a knee to conduct immediate action. Once the malfunction is cleared, there is no need to immediately stand up to engage targets. Rangers can save precious seconds by continuing to engage from one knee. Whenever other members of the team see a Ranger down, they must automatically clear his sector of fire. Before rising to his feet, the Ranger warns his team members of his movement and only rises after they acknowledge him. If a malfunction occurs once committed to a doorway, the Ranger must enter the room far enough to allow those following him to enter and move away from the door. This drill must be continually practiced until it is second nature.

c. **Approach to a Building or Breach Point**. One of the trademarks of ranger operations is the use of limited visibility conditions. Whenever possible, breaching and entry operations should be executed during hours and conditions of limited visibility. Rangers should always take advantage of all available cover and concealment when approaching breach and entry points. When natural or manmade cover and concealment is not available, Rangers should employ obscurants to conceal their approach. There are times when Rangers will want to employ obscurants to enhance existing cover and concealment. Members of the breach / entry team should be numbered for identification, communication, and control purposes.

(1) The # 1 Ranger should always be the most experienced and mature member of the team, usually the team leader. The # 1 Ranger is responsible for frontal and entry and breach point security. The team leader is responsible for initiating all voice and physical commands. The team leader must exercise situational awareness at all time with respect to the task, friendly force, and enemy activity.

(2) The # 2 Ranger is directly behind the # 1 Ranger in the order of movement, and he moves through the breach point in the opposite direction from the #1 Ranger.

(3) The # 3 Ranger will simply go opposite the # 2 Ranger inside the room at least 1 meter from the door.

(4) The # 4 Ranger moves the opposite the # 3 Ranger and is responsible for rear security (and is normally the last Ranger into the room). An additional duty of the # 4 Ranger is breaching.

Note: Consider how much firepower each Ranger delivers. Where do you put the SAW gunner in the order? You must weigh the benefits of firepower against those of quick, accurate shots. If the # 4 Ranger has breaching responsibilities, it should not be the SAW gunner, because this would reduce your firepower.

d. **Actions Outside the Point of Entry**. Entry point position and individual weapon positions are important. The clearing team members should stand as close to the entry point as possible, ready to enter. Weapons are oriented in such a manner that the team provides itself with 360 degree security at all times. Team members must signal to one another that they are ready at the point of entry. This is best accomplished by sending up a "squeeze." If a tap method is used, an inadvertent bump may be misunderstood as a tap.

e. **Enter Building/Clear Room**. See Battle Drill 5, Chapter 6.

f. **Locking Down the Room**.

(1) Control the situation within the room.

(2) Use clear, concise arm and had signals. Voice commands should be kept to a minimum to reduce the amount of confusion and to prevent the enemy--who might be in the next room--from discerning what is going on. This enhances the opportunity for surprise and allows the assault force the opportunity to detect any approaching force.

(3) Physically and psychologically dominate.

(4) Establish security and report status.

(5) Cursory search of the room to include the ceiling (3 Dimensional Fight).

(6) Identify the dead using reflexive response techniques (Eye thump method).

(7) Search the room for PIR, precious cargo as per the mission and time available.

(8) Evacuate personnel.

(9) Mark room clear using chemical lights, engineer tape, chalk, paint, VS-17 panels, and so on.

14-6. **TTPS FOR MARKING BUILDINGS AND ROOMS**. Units have long identified a need to mark specific buildings and rooms during UO. Sometimes rooms need to be marked as having been cleared, or buildings need to be marked as containing friendly forces. The US Army Infantry School is currently testing a remote marking device that can be used to mark doors from as far away as across a wide street. In the past, units have tried several different field-expedient marking devices; some with more success than others. Chalk has been the most common. It is light and easily obtained but not as visible as other markings. Some of the other techniques have been to use spray paint, and paintball guns.

a. **NATO Standard Marking SOP**. The North Atlantic Treaty Organization (NATO) has developed a standard marking SOP for use during urban combat. It uses a combination of colors, shapes, and symbols. These markings can be fabricated from any material available. (Figure 14-1 shows examples.)

Figure 14-1. EXAMPLE NATO STANDARD MARKINGS

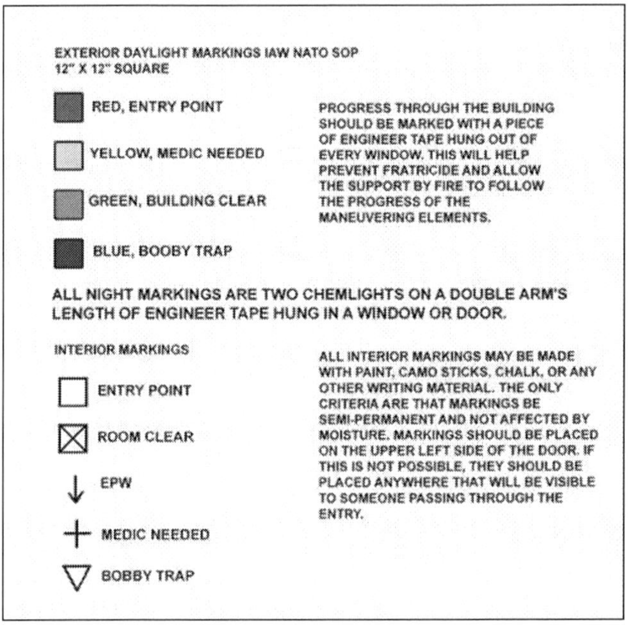

b. **Spray Paint**. Canned spray paint is easily obtained and comes in a wide assortment of colors including florescent shades that are highly visible in daylight. It cannot be removed once used. Cans of spray paint are bulky and hard to carry with other combat equipment. Paint is not visible during darkness nor does it show up well through thermal sights.

c. **Paintball Guns**. Commercial paintball guns have been purchased by some units and issued to small unit leaders. Some models can be carried in standard military holsters. They can mark a building or door from about 30 meters. The ammunition and propellant gas is not easily obtainable. The ammunition is fragile and often jams the gun if it gets wet. The available colors are not very bright, and just like spray paint, cannot be seen at night or through thermal sights.

d. **Wolf Tail**. A simple, effective, easy-to-make, lightweight device called a "wolf tail" can be fabricated to mark buildings, doorways, and windows (Figure 14-1). A unit has changed its tactical SOP to require that each Infantryman carry one of these devices in his ACU cargo pocket. Wolf tails, when used IAW a simple signaling plan understood by all members of the unit, can aid in command and control, reduce the chances of fratricide, and speed up casualty collection during urban combat.

(1) The wolf tail marking device is simple to make and versatile. It can be used together with the NATO marking scheme. Rolled up, it makes a small, easily accessible package that can be carried in the cargo pocket of the ACUs. It can be recovered easily and used again if the situation changes. All its components can be easily obtained through unit supply. It combines a variety of visual signals (colored strapping and one or more chemical lights of varying colors) with a distinctive heat signature that is easily identified through a thermal weapon sight. An infrared chemlight can be used either as a substitute for the colored chemlight(s) or in addition to them.

(2) Constructing the wolf-tail marking device requires the following material:

- A 2-foot length of nylon strap (the type used for cargo tie-downs) (engineer tape can be substituted).
- About 5 feet of 550 cord.
- A small weight such as a bolt or similar object.
- Duct tape.
- Chemical lights (colored or IR).
- Two 9-volt batteries.

(3) Assemble the items by tying or taping the cord to the small weight. Attach the other end of the cord to the nylon strapping, securing it with duct tape. Attach the 9-volt batteries in pairs to the lower end of the strapping with several wraps of duct tape, making sure that the negative terminals are opposite the positive, but not actually touching. Use more duct tape to attach the chemical lights, about 2 inches above the batteries, to the strapping.

(4) When you want to mark your position, push the batteries together firmly until the male and female plugs lock. This shorts out the battery, causing it to heat up rapidly. The hot battery is easily identified through the thermal sights of tanks or BFVs. The batteries stay visible for about 45 minutes. Activating the chemical lights provides an easily identified light source visible to the naked eye. You can use infrared chemical lights if you want them to be seen through night vision devices but not with the naked eye.

Figure 14-2. EXAMPLE OF A WOLF-TAIL MARKING DEVICE

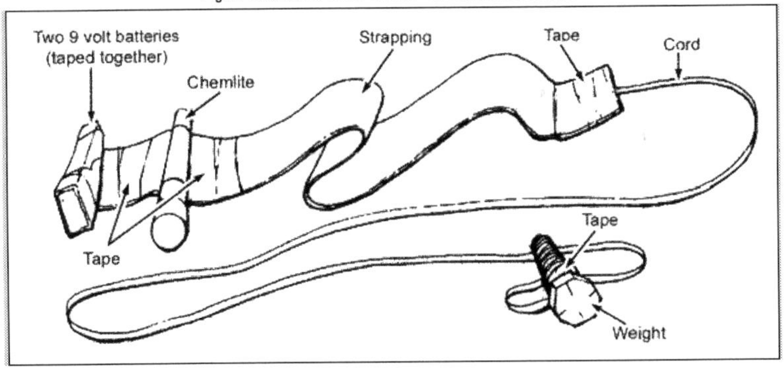

Note: One option is to place chemical lights and batteries at both ends of the wolf tail to mark the inside and outside of a building or room.

(5) Use the cord and the small weight to hold the wolf tail in position by tying or draping it out a window or hanging it on a door, wherever it is best seen by other friendly troops. Squads or platoons can vary the numbers and colors of chemical lights, or use multiple battery sets to identify precisely what unit is in which building.

(6) Medics and combat lifesavers can carry a standardized variation that can be used to clearly identify a building as containing wounded personnel needing evacuation. This could be a white strap with multiple red chemical lights, or any other easily identified combination.

14-7. **URBAN ASSAULT BREACHES**. Understanding how to employ and incorporate breaching as part of a leader's planning process is an important part of urban operations. It is imperative that elements of your patrol be skilled in the art of breaching. Whether infiltrating or exfiltrating from an objective, leaders must plan for either option. One constant disadvantage with the employment of explosives is that placement requires Rangers to expose themselves to possible enemy fires. Breach teams need to be supported by fires or obscurants and breaching operations should be performed during hours of limited visibility whenever possible. Breaching classes follow:

a. **Mechanical**. Mechanical breaching should be an important part of a leader's breaching training program because it is usually an option. Mechanical breaching is best described as gaining access by the use of tools or saws. Although most tools and saws used are recognizable and self explanatory to the individual Ranger, one must practice on various techniques to increase speed an effectiveness. This reduces fatigue and expedites the actual assault.

(1) **Tools**.

(a) Hooligan tools..........................Doors/windows of all types.
(b) Sledge hammer.......................Heavy duty doors, locks, and window frames.
(c) Picket pounder.........................Doors of all types, light walls.
(d) Bolt Cutters..............................Chain link fence, locks, and wire obstacles.
(e) Pick Ax....................................Lightweight doors and locks.
(f) Saws..Fences, light doors, locks.

b. **Ballistic**. Ballistic breaching is defined as a forced entry or exit by the use of weapons. Whether using shotguns, M16A2/M4, M249 SAW, specific considerations must be addressed:

(a) Type of round and ricochet factor.
(b) Composition of the breaching point.
(c) Composition of the floor beyond the door.
(d) Personnel behind the door (friendly/enemy).
(e) Always shoot at a 45-degree angle.

c. **Explosive**. Explosive breaching is the most viable because it is the most effective. When employing explosives during breaching operations, leaders must consider three major factors.

(1) ***Overpressure***. The amount of PSI released from the concussion of the blast.

(2) ***Missile Hazard***. Fragmentation or projectiles sent at tremendous speed from the explosion area. This occurs from either the charge or target being breached.

(3) ***Minimum Safe Distance Requirements***. Use of explosives in the urban environment must consider the presence of noncombatants and friendly forces. Additionally, the are many hazardous materials located in the urban environment, these may include chemicals as well as construction materials. There is always a risk of secondary explosions and fires, when employing explosive breaching techniques.

(4) ***Charges***. Various charges can be used for explosive breaching. Leaders must conduct extensive training on the use of the charges to get proper target feedback. Figure 14-3 shows examples of charges used for explosive breaching.

Figure 14-3. EXAMPLE CHARGES FOR EXPLOSIVE BREACHING

Charge	Target
Water Impulse................	Steel/wood doors
Flexible linear.................	Wood doors
Ranger wall breach........	Masonry/brick walls
Chain link ladder............	Chain link fence
E-Type silhouette...........	Wood doors
Brashier breacher..........	Concertina wire

Chapter 15

VEHICLE CONVOY OPERATIONS

This chapter outlines a technique for conducting vehicle convoy operations. Convoy operations present a challenge to the Ranger leader. Trucks and other combat vehicles produce a large signature on the battlefield and increase your unit's value as a target. Vehicle movement is restricted to roads and terrain that they can traverse; therefore, a sound plan must be implemented to minimize the possibility of compromise.

15-1. **PLANNING CONSIDERATIONS**. This paragraph explains how to plan a vehicle convoy IAW the eight steps of the TLP.

a. **Receive the Mission**. (PL will extract the following information from Company OPORD).

(1) Truck Support (number and type of Truck, ACL).
(2) Weather. Road conditions.
(3) Truck pickup and drop-off location / marking.
(4) Truck movement Timeline (Pick-up time, roll time, H-hour).
(5) Truck Routes (primary and alternate, Check points).
(6) Drop-off locations (primary and alternate).
(7) IED/Contact compromise and contingencies.

b. **Issue the Warning Order**. During the WARNO the PL gives a basic overview of the Truck Movement and gives specific Task to Maneuver Units.

(1) Manifest Personnel (PSG).
(2) Pick up point security and set-up (PSG).
(3) Primary and alternate routes (Security SL).

c. **Make a tentative Plan**. PL considers METT-TC.

(1) ***Mission***. Develop your Ground Tactical Plan and a then develop a truck plan that best supports it.
(2) ***Enemy***. Consider enemy patrols and activity in AO.
(3) ***Terrain***. Evaluate routes and dropoff and pickup points using OCOKA.
(4) ***Troops***.
(a) Truck ACL.
(b) Chalks and chalk leaders identified.
(c) Chalks planed with tactical integrity & self sufficiency.
(d) Key leaders and weapons are cross-loaded.
(e) Pickup point marking team identified.
(f) Pickup point security teams identified.
(5) ***Time***. PL considers the following when evaluating time available.
(a) Movement to Pickup point.
(b) Recon of Pickup point.
(c) Emplacement of security.
(d) Marking the Pickup point.
(e) Pickup point Posture time.

d. **Initiate Necessary Movement**. PL decides whether to move the platoon to the pickup location or to conduct a map reconnaissance and complete the plan.

e. **Conduct Reconnaissance**. (Is not necessary if pickup point is secured and established by Higher HQ's).

(1) Leader's Recon consists of PL, RTO, Chalk Leaders and Pickup point Security element.
(2) PL determines suitability of Pickup point.
(3) PL determines security plan (either overwatch or strong point).

f. **Complete the Plan**. Complete Truck Movement Annex on Page ??? the Ranger Handbook.

g. **Issue the OPORD**.

h. **Supervise**.

(1) Back brief from chalk leaders.
(2) Rock drill of truck movement and contingency plans.

15-2. **FIVE PHASES OF TRUCK MOVEMENT**. Each phase must support the ground tactical plan, which specifies actions in the objective area to accomplish the commander's intent for the assigned mission, be it a raid, ambush, recon or other follow-on missions.

a. **Staging Plan**.
   (1) Secure the pickup point (strongpoint or overwatch).
   (2) Mark pickup point (day/night).

b. **Loading Plan**. Each Ranger is assigned a truck.

c. **Movement Plan**.
   (1) Troops awake and alert during movement.
   (2) Platoon leader and chalk leaders tracking route progress.
   (3) Compromise and contingency plan.
   (a) React to IED.
   (b) React to ambush.
   (c) Truck breakdown.

d. **Dropoff Plan**.
   (1) Establish security of dropoff point (overwatch or strongpoint).
   (2) Dismount vehicles.
   (3) Recon and secure assembly area.
   (4) Adjust perimeter as chalks arrive.
   (5) PSG clears the trucks when last chalk departs.

e. **Ground Tactical Plan**. Execute after platoon is reconsolidated or minimum force required is assembled.

15-3. **CONVOY TECHNIQUES**. The following convoy techniques have been included for your convenience:

- Convoy warning order
- Convoy brief
- React to Ambush Near/Far
- Convoy forced to stop (Methods 1 and 2)
- Break contact
- Recovery/CASEVAC Operations

Convoy Warning Order

1. SITUATION: A brief statement of the enemy and friendly situation. Who. What, Where
2. MISSION: Who, What, When, Where, Why
3. TASK ORGANIZATION: Based on tasking from higher WARNO.
4. INITIAL TIME SCHEDULE:

When	Who	What

5. SPECIAL INSTRUCTIONS: PCC/PCI Guidance, Rehearsals, additional tasks to be accomplished.
6. SERVICE SUPPORT:

Class I: (Rations & Water)

Class III: (POL)

Class V: (Ammunition /Pyrotechnics)

Weapon System	Rounds	Type

Pyrotechnic Device	Number	Location

Class VIII: (Medical)

7. UNIFORM AND EQUIPMENT COMMON TO ALL.

CONVOY BRIEF (Modified from FM 55-30)

Movement Order No. ____

References: A. ______________________________ *(Maps, tables and relevant documents)*

B. ______________________________

TASK ORGANIZATION: (Internal organization for convoy – Manifest)

1. SITUATION:

a. Enemy Forces:

(1) Weather. General forecast.

(2) Light Data (EENT, % Illumination, MR, MS, BMNT)

(3) Discuss Enemy.

Identification of enemy (If known).
Composition / capabilities / strength / equipment
Location
(Hot Spots highlighted on map)

b. Friendly Forces:

(1) Operational support provided by higher headquarters.

Helicopter / Gunships
Quick Reaction Forces (QRFs)
MP Escorts / Rat Patrols
Fire Support elements

c. Attachments: (From outside the organization)

2. MISSION: (WHO, WHAT, WHEN, WHERE, & WHY)

3. Employment Tactics, Techniques and Procedures

a. Readiness Level
  i. Ride / Scanning (Observation)
  ii. At the Ready

b. Scanning / Sector of Fire
  i. Driver
    1. Sector of Scan is 9 – 1 clock position
    2. Observation with Mirrors
    3. Sector of Fire is 9 – 11 o'clock
  ii. TC
    4. Sector of Scan is 11 – 3 clock position
    5. Sector of Fire is 1 – 3 clock position
  iii. Other Systems are based off vehicle type and load

b. Target ID
  i. Communicate / Signal

c. Body Positioning
  i. Engage as you Train (Right or Left handed Firing)
  ii. Firing-side Shoulder Down
  iii. Maintain your Body Position
  iv. Weapon to Head

d. Rules of Engagement Concerns – Employ the appropriate systems based off the threat.

e. Point of Aim
  i. Moving Platform – Stationary Target requires aim to the rear (trail) and low dependant upon speed.

CONVOY BRIEF (Modified from FM 55-30) (Continued)

3. Employment Tactics, Techniques and Procedures

e. Point of Aim

  ii. Moving Platform – Moving Target requires aim directly on and low.

  iii. Stationary Platform – Moving Target requires aim to the front (lead) and low dependent upon speed.

f. Rate of Fire

  i. Burst Mode

  ii. Steady Suppression (ROE)

g. Magazine Awareness

  i. Serviceability / maintenance

  ii. Tracer Mix

  iii. Magazine Storage / Placement

  iv. Mounted, Reload when time is available

Dismounted, Seek cover prior to the need to Reload

REACT TO AMBUSH (NEAR)

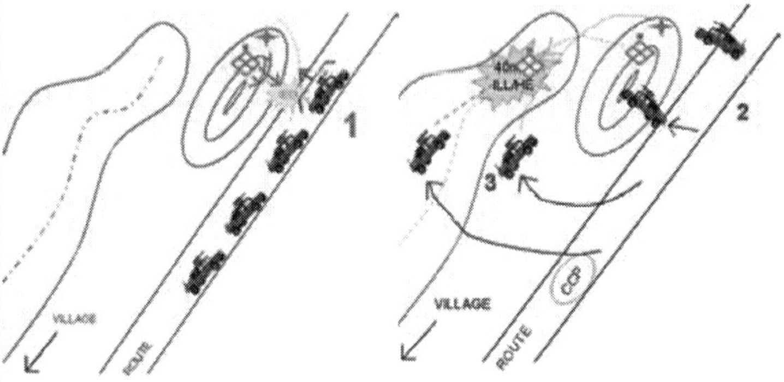

(Enemy Initiates Weapons Fire / RPG / IED)

1. Lead element of convoy reacts to contact and immediately returns fire. Lead fire team fires 40mm HE (at night also white star if NVGs drown out by vehicle headlights – leader calls it. FYI, most travel at night will be with headlights on).
2. Lead elements assualt through ambush – fire team dismounts while gun trucks use M249, MK-19, and .50 cal in close support.
3. Trail fire team and trail gun truck move to block most likely enemy escape route. Fire team uses 40mm HE/Illumination if needed. (Use of 40mm becomes less likely as units get closer to urban areas due to collateral damage risk.) Establish CCP upon action completion or enemy neutralized.

REACT TO AMBUSH (FAR)

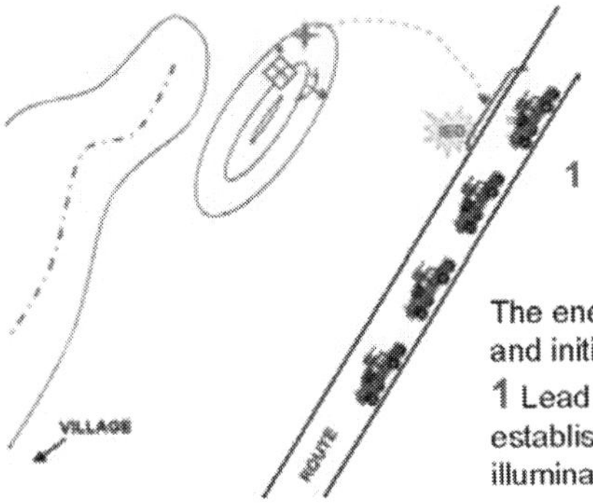

The enemy is outside 100 meters and initiates with IED/RPG

1 Lead gun truck and lead fire team establish SBF—fire team fires 40mm illumination

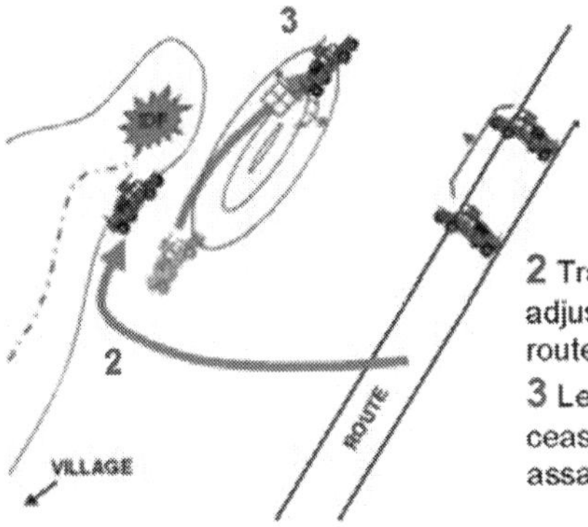

2 Trail element flanks—SL/PL adjusts IDF into enemy escape route(if in range)

3 Lead element shifts fire—arty cease loading—trail element assaults across—enemy destroyed

CONVOY FORCED TO STOP

(Weapons Fire / RPG / IED / Indirect)

1. Vehicles forced to stop. Activate vehicle turn signal to indicate direction of contact.
2. Vehicle(s) / Personnel not in direct contact, report on internal communication, identifying truck number, type of contact, clock direction and eight digit grid if available.

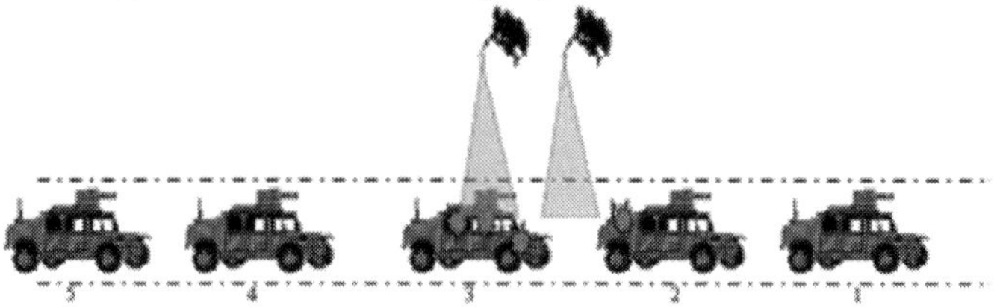

3. Personnel on vehicle(s) forced to stop dismount on the non-contact side, assume covered position(s) and provide initial base of fire.
4. The <u>entire</u> convoy halts; personnel will dismount vehicle(s) on the non-contact side and provide additional base of fire on the enemy if in range. Vehicle(s) not in contact will reposition to attack the attackers. M16's should now be on <u>Semi-Automatic</u> in order to conserve ammunition.
5. PL/CC/SC will move to better assess the situation and position the Gun Truck(s) in order to best suppress the enemy while maintaining <u>standoff</u>. Gun trucks will close with and engage the enemy.
6. Once the PL/CC/SC determines the convoy has either gained fire superiority or defeated the enemy contact, Recovery / CASEVAC operations will begin.
7. If the PL/CC/SC determines the convoy cannot gain fire superiority, leader will then conduct Break Contact procedures

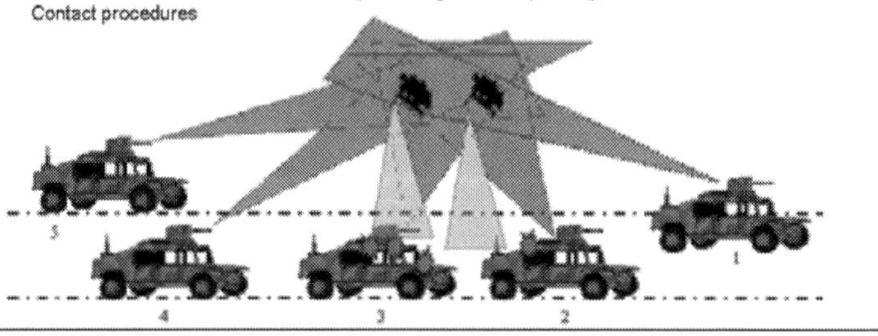

CONVOY FORCED TO STOP - METHOD 2

(Weapons Fire /RPG / IED / Indirect)

1. Vehicle(s) forced to stop. Activate vehicle turn signal to indicate direction of contact.

*** ALL PERSONNEL STAY IN VEHICLES ***

2. All vehicles immediately drive forward out of the kill zone.

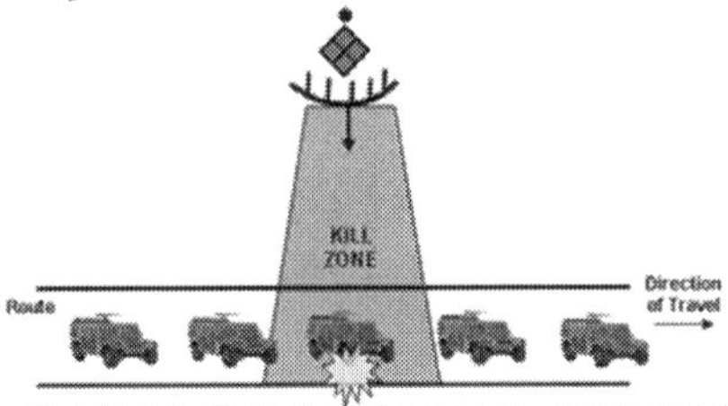

3. The vehicle(s) directly behind the disabled vehicle(s) push the disabled vehicle(s) out of the kill zone.

4. The vehicle(s) that are not disabled will establish a base of fire towards the suspected/known contact.

5. If fire superiority can be gained the PL will use the minimum force necessary to destroy the enemy.

6. If the PL determines the convoy cannot gain fire superiority, the leader will conduct break contact.

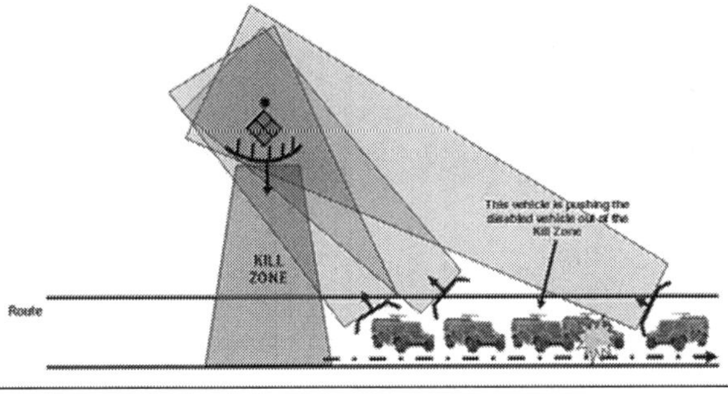

BREAK CONTACT

Always try to close with and destroy the enemy first. This way he cannot come back later on to attack you or another convoy again. However – if you must...

1. The PL/CC/SC determines the convoy cannot gain fire superiority and the decision has been made to Break Contact.
2. The PL/CC/SC designates that either Rally Point "Rear" or "Forward" will be used. If necessary, both Rally Points may be used. Communication systems and appropriate pyrotechnic signals will be used to communicate Break Contact and Rally Point.
3. Personnel will deploy obscuration measures if available. Utilizing cover and concealment, Aid & Litter team(s) will evacuate all casualty(s) under support of Gun Truck and other protective fire(s).
4. Personnel will maintain position and suppression in contact zone and assist Aid & Litter team(s) as necessary.
5. Disabled vehicle(s) will be hauled back or destroyed as directed by leaders. (Thermite or explosives)

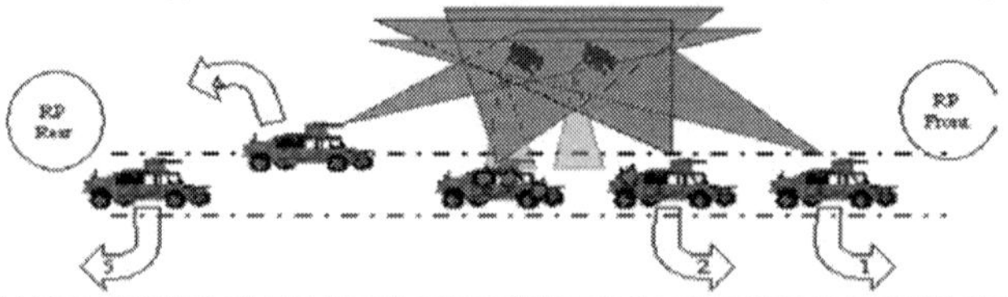

6. Vehicles will displace either backwards or forwards through the convoy lines under control of leaders. The most forward vehicle in the contact zone moves first, then the next most forward vehicle moves second. Vehicles will continue to displace. As vehicles displace, Gun Truck repositions as necessary until contact is broken.

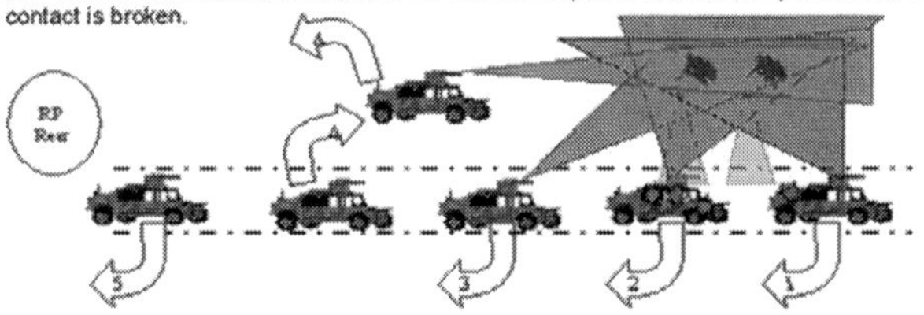

7. If Break Contact occurs with vehicles on both sides of the contact zone, displacement of vehicles will occur using an alternating displacement technique.
8. Upon occupation of the Rally Point, leaders will immediately position vehicles, setting 360 degree security, and conduct Consolidation and Reorganization.
9. If the convoy vehicles get separated when not in contact with the enemy, personnel and vehicles stay together and move to the closest Rally Point or Check Point.

RECOVERY / CASEVAC OPERATIONS

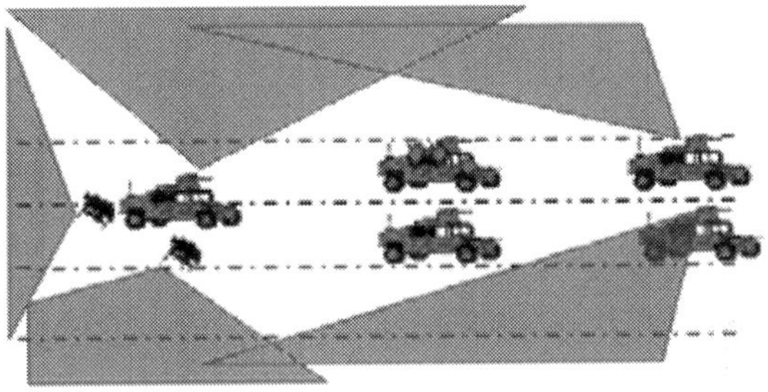

1. Once the leader assesses the enemy threat to be destroyed, neutralized, defeated and the area secured – Recovery / CASEVAC operations will begin. This keeps soldiers focused first on defeating and destroying the threat.

2. CASEVAC:

   a. Aid & Litter team will position on the safe side of the vehicle and extract casualty(s) and personnel.

   b. Treatment of casualty(s) will occur once they are safely removed from the contact area.

3. Vehicle Recovery Procedures:

   a. Recovery team will position on the safe side of the disabled vehicle.

   b. TC will dismount and assess the disabled vehicle.

   c. If determined the vehicle can be safely recovered, TC guide the recovery vehicle into position and conduct a hasty hook-up. TC will operate the disabled vehicle is necessary.

   d. Upon exiting the contact area, complete and correct hook-up procedures will occur.

   e. If assessment results in outside support necessary for recovery, leader will contact higher for guidance.

4. Once recovery operations are complete, the team will displace and conduct link-up with the convoy at the Rally Point.

5. Disabled vehicle(s) will be abandoned or destroyed as directed by leaders. (Thermite or explosives)

GLOSSARY

1SG	first sergeant
5 W's	Who, What, Where, When, and Why?
5-S Rule	Search, Silence, Segregate, Safeguard, and Speed to rear
AA	avenue(s) of approach
AAR	after-action review
AATF	air assault task force
ABCDE	method of identification and response to life-threatening conditions: Airway, Breathing, Circulation, Disability, Exposure
ACE	ammunition, casualties, and equipment
ACL	allowable combat load
ACP	aerial checkpoint
ACU	Army combat uniform
ADA	air defense artillery
AG	assistant gunner
ATC	air traffic controller; a mechanical belay device that locks down on itself when tension is applied in opposite directions
ALT	alternate
AMC	air mission commander
ammo	ammunition
ANCD	automated net-control device
AO	area of operations
AOO	actions on the objective
AR	automatic rifleman
ABF	attack by fire
ATL	Alpha team leader
ATM	Alpha Team
bangalore torpedo	A manually emplaced, 1.5-meter long explosive-filled tube used to breach wire and detonate simple, pressure-activated antipersonnel mines. Ten tubes clear a 1- by 15-meter lane.
belay	any action taken to stop a climber's fall or to control the rate a load descends
binos	binoculars
BMNT	begin morning nautical twilight
BN	battalion
body belay	belay that uses the belayer's body to apply friction by routing the rope around the his body
bowline on a coil	knot used to secure a climber to the end of a climbing rope
BP	battle position
British junction knot	knot used to join the ends of detonation cords from multiple charges to one initiation system
BTC	bridge team commander
BTL	Bravo Team leader
BTM	Bravo Team

C2	command and control
CAS	close air support
CASEVAC	casualty evacuation
CCA	close combat attack
CCIR	commander's critical information requirements
CCP	casualty collection point
CDR	commander
CDS	Camp Darby Special (map)
CLS	combat lifesaver
CO	company
COA	course of action
commo	communications
COMSEC	communications security
cordelette	short section of static rope or static cord. Also called "sling rope"
COTS	commercial off the shelf
CP	command post
CPR	cardiopulmonary resuscitation
CQC	close quarters combat
CS	combat support
CSS	*obsolete: now referred to as* sustainment
CTT	common task test
DAR	designated area of recovery
DOL	direction of landing
double figure eight knot	knot used to form a fixed loop in the end of the rope; loops are large enough to insert a carabiner
double overhand knot	knot used to secure the end of detonation cord
DP	duty position; decision point (depending on context)
DST	distance
DTG	date-time group
dynamic ropes	one of two classifications of kernmantle rope; used for climbing; *see also* static rope
DZ	drop zone
EA	engagement area
EEFI	essential elements of friendly information
EENT	end evening nautical twilight
end-of-the-rope clove hitch	intermediate anchor knot that requires constant tension
end-of-the-rope Prusik	knot used to attach a movable rope to a fixed rope; *see also* middle-of-the-rope Prusik
ENY	enemy
EPW	enemy prisoner of war
FA	field artillery
FDC	fire direction center
FFIR	friendly force information requirements

figure 8 slip knot	knot used to form an adjustable bight in the middle of a rope
FLIR	forward looking infrared
FLOT	forward line of own troops
FO	forward observer
FPF	final protective fires
FRAGO	fragmentary order
FSC	fire support coordinator
FSO	fire support officer
GOTWA	G - Where leader is GOING O - Others he is taking with him T - Time he plans to go W - WHAT to do if the leader does not return in time A - The unit's and the leader's ACTIONS on chance contact while the leader is gone
GPS	global positioning system
GRN	grenadier
HDG	heading
HE	high explosive
H-Hour	hit hour (the time the unit plans to accomplish the mission)
HI	high temperature (weather)
HPT	high-payoff target
HQ	headquarters
IAW	in accordance with
ICM	improved conventional munitions
ID	identification
IP	initial point
IR	intelligence requirements
ERRP	en route to release point
JAAT	joint air attack team
JD	Julian date
KIA	killed in action
LAW	light antiarmor weapon
LBV	load-bearing vest
LD	line of departure
LDA	linear danger area
LO	low temperature (weather)
LOGSTAT	logistical status
LP	listening post
LZ	landing zone
material factor	the strength, hardness, and mass of the material to be demolished
mb	millibar (a metric unit used to measure air pressure)

MDI	modernized demolition initiator
ME	main effort
mechanical belay	a belay that uses mechanical devices to help the belayer control the rope, as in rappelling
MEDEVAC	medical evacuation
METL	mission-essential task list
METT-TC	mission, enemy, terrain (and weather), troops (and support) available, time available, and civil considerations
middle of the rope clove hitch	knot that secures the middle of a rope to an anchor
middle-of-the-rope Prusik	knot that attaches a movable rope to a fixed rope, anywhere along the length of the fixed rope; *see also* end-of-the-rope Prusik
MG	machine gun/ner
MOPP	mission-oriented protective posture
MR	moonrise
MRE	meal, ready to eat
MS	moonset
MSD	minimum safe distance
MSL	mean sea level
munter hitch	commonly used belay that requires little equipment
NATO	North Atlantic Treaty Organization
NAV	navigation
NFA	no-fire area
NLT	no later than
NVD	night-vision device
NVG	night vision goggles
NVS	night vision system
OAKOC	observation and fields of fire, avenues of approach, key terrain, observation, and cover and concealment
OBJ	objective
OD	olive drab
OOM	order of movement
OP	observation post
OPORD	operations order
OPSKED	operational schedule
ORP	objective rally point
OT	observer-target
PB	patrol base
PCC	precombat checks
PCI	precombat inspection
PDF	principal direction of fire

PI	probability of incapacitation
PIR	priority intelligence requirements
PL	platoon leader
PLD	probable line of deployment
PLOT-CR	purpose, location, observer, trigger, communication method, resources (a format for planning fire support)
PLT	platoon
POL	petroleum, oils, and lubricants
PRI	primary
protection	a piece of equipment, natural or artificial, that is used to construct an anchor
PSG	platoon sergeant
PSI	pounds per square inch
PZ	pickup zone
R	rifleman
R&S	reconnaissance and surveillance
RACO	rear area combat operations
RAP	rocket-assisted projectile
rappel seat	a rope harness used in rappelling and climbing
RED	risk-estimate distance
REQ	required
rerouted figure 8 knot	anchor knot that also attaches a climber to a climbing rope
RFA	restrictive fire area
RFL	restrictive fire line
RFLM	rifleman
RHB	Ranger Handbook
ROE	rules of engagement
round-turn with two half hitches	a constant tension anchor knot
RP	release point
RTO	radio operator
S-2	intelligence staff officer
S-3	operations staff (and training) officer
SALUTE	Size, Activity, Location, Unit/Uniform, Time, and Equipment
SAW	squad automatic weapon
SBF	support by fire (position)
SDT	self-development test
SE	supporting effort
SEAD	suppression of enemy air defenses
SITREP	Situation Report
SITTEMP	situational template

SL	squad leader
sling rope	short section of static rope or static cord. Also called "cordelette"
SLLS	Stop, Look, Listen, Smell
SOC	succession of command
SOI	signal operating instructions
SOP	standing operating procedures
SP	start point
square knot	knot used to join two ropes of equal diameter; used to join the ends of the detonation cord to the explosive
SR	sunrise
SS	sunset
STANO	surveillance, target acquisition, and night observation
static ropes	one of two classifications of kernmantle rope; used where rope stretch is undesired, and when the rope is subjected to heavy static weight. *See also* dynamic rope
SURVIVAL	S Size up the situation, your surroundings, your physical condition, and your equipment. U Undue haste makes waste; don't be too eager to move. Plan your moves. R Remember where you are in relation to important friendly and locations and critical resources V Vanquish fear and panic. I Improvise. You can improve your situation. Use what you have. Use your Imagination. V Value your life. Remember your goal: to get out alive. Remain stubborn. Refuse to give in to problems and obstacles that face you. This will give you the mental and physical strength to endure. A Act like the natives; watch their daily routines. When, where, and how do they get food? Where they get water? L Live by your wits. Learn basic skills.
suspension traverse	used to move personnel and equipment over rivers, ravines, chasms, and up or down a vertical obstacle
TAC	tactical air controller
tamping factor	depends on the location and tamping of the charge
technical climbing	using safe and proper equipment and techniques to climb on a rock formation in parties of two or more
tensionless anchor	used to anchor rope on high-load installations such as bridging
TL	team leader
TLP	troop-leading procedures
TL	team leader
TOC	tactical operations center
triple roll knot	knot used to join branches of detonation cord
TTP	tactics, techniques, and procedures
uli knot	knot used to securely fasten detonation cord to explosive
VIXL	video image crosslink
WARNO	warning order
WBGT	wet bulb globe temperature

WFFs	warfighting functions (fire support, movement and maneuver, protection, command and control, and sustainment)
WIA	wounded in action
XO	executive officer
WSL	weapons squad leader

NOTES

INDEX

Appendix A

RESOURCES

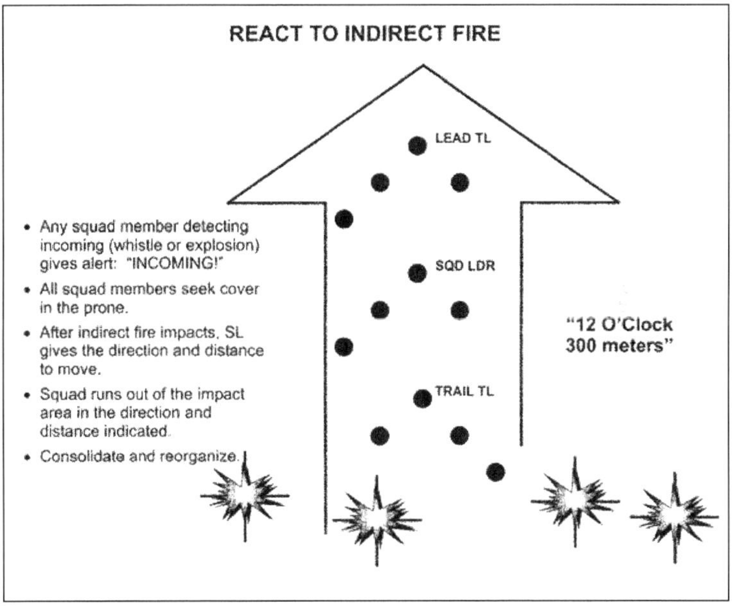

REACT TO CONTACT

- Seek nearest cover.
- Return fire (known or suspected enemy position).
- TM LDRs control fire by using fire commands.
- Report enemy situation (number/position).
- Maintain contact (visual/oral) with team members.
- SQD LDR moves to team in contact and makes an assessment of the situation.
- Factors of his assessment:
    - Can squad move out to engagement area.
    - Can squad gain and maintain suppressive fire.
    - Location of enemy.
    - Size of enemy.
    - Vulnerable flanks.
    - Covered/concealed flanking routes.
- SQD LDR determines COA (break contact, squad attack, etc.)
- Report situation to PL

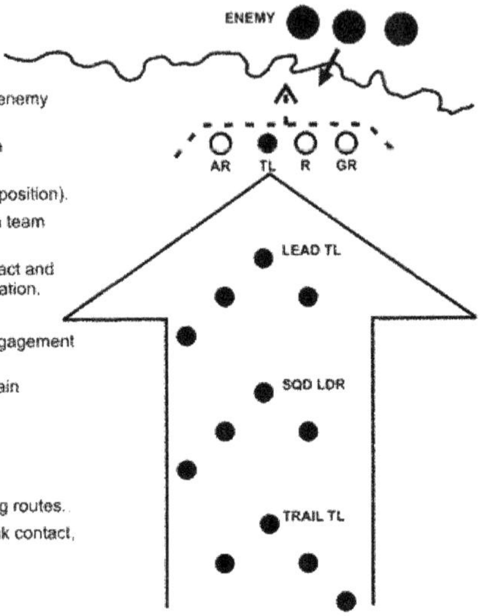

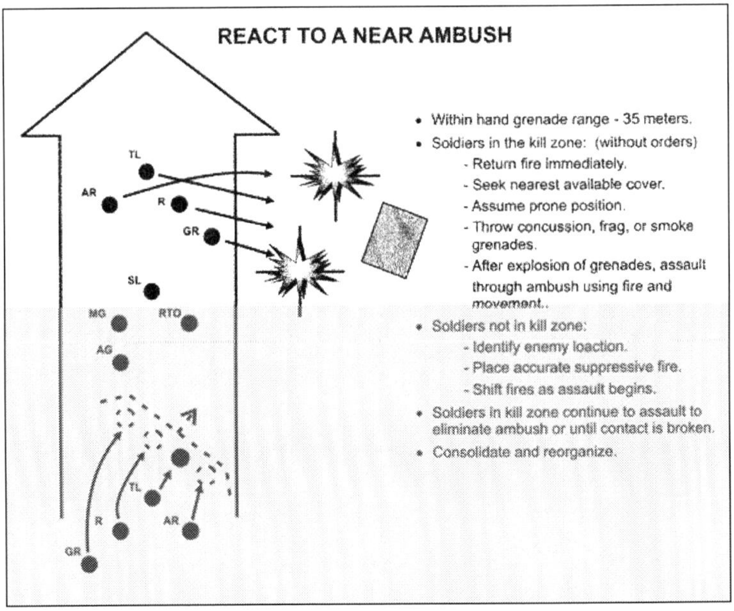
REACT TO A NEAR AMBUSH
TL
AR
R
GR
SL
MG
RTO
AG
TL
R
AR
GR
• Within hand grenade range - 35 meters.
• Soldiers in the kill zone: (without orders)
- Return fire immediately.
- Seek nearest available cover.
- Assume prone position.
- Throw concussion, frag, or smoke grenades.
- After explosion of grenades, assault through ambush using fire and movement..
• Soldiers not in kill zone:
- Identify enemy loaction.
- Place accurate suppressive fire.
- Shift fires as assault begins.
• Soldiers in kill zone continue to assault to eliminate ambush or until contact is broken.
• Consolidate and reorganize.

REACT TO A FAR AMBUSH

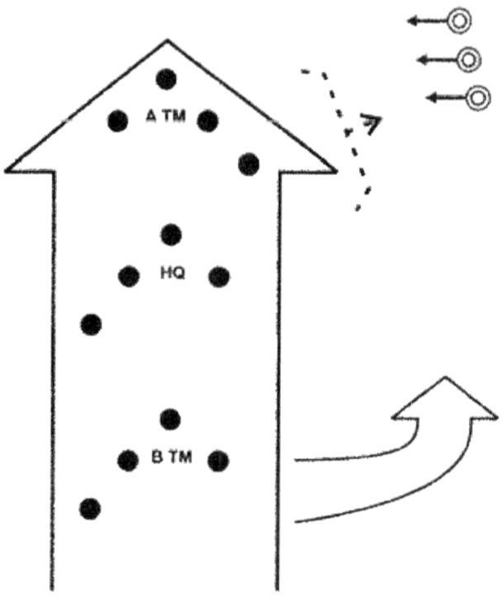

- More than 35 meters.
- TM in the kill zone: (without orders)
  - Return fire immediately.
  - Seek cover and concealment.
  - Suppress enemy (overwatch).
- SL assess situation.
  - Determines COA (flank).
- TM not in contact:
  - Moves along covered and concealed route.
  - Assaults enemy on weak flank.
  - Suppress enemy (overwatch).
- Overwatch TM continues to suppress shifts/cease fire as bounding team enters sector.
- Bounding team continues to assault through enemy.
- SL may request indirect fire.
- Consolidate and reorganize.

BREAK CONTACT

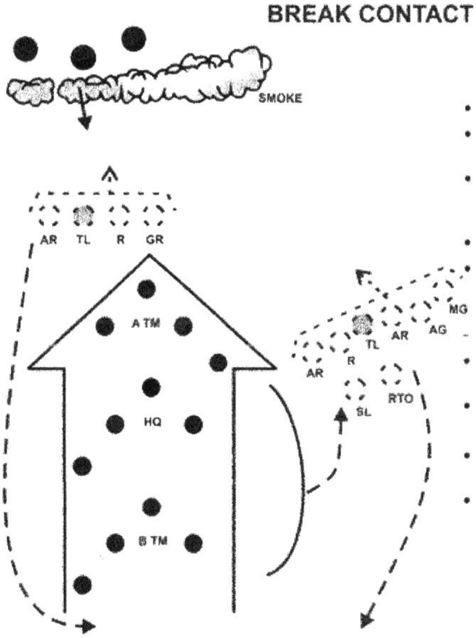

- Squad Leader orders: "Break Contact".
- Squad Leader designates SPT element and maneuver element.
- SL issues distance and direction or a terrain feature for the maneuver element.
- SBF suppresses enemy position.
- Maneuver uses smoke to mask movement.
    - Takes up overwatch position.
    - Begins to suppress enemy
- Squad Leader directs SBF to break contact.
- SBF uses smoke to screen movement.
    - Takes up overwatch position.
- Squad continues to bound away until contact is broken.
- Consolidate / reorganize.

FORMATIONS AND ORDERS OF MOVEMENT

I. Movement Formation: Fire Team Wedge; MG Team attached.

II. Three Movement Techniques utilized:

   A. Traveling technique used behind FFL when contact is not likely.

   B. Traveling Overwatch forward of the FFL when enemy contact is possible.

   C. Bounding Overwatch used forward of the FFL when enemy contact is expected.

III. Distances are based on but not dictated by visibility, terrain, and vegetation.

IV. Actions at Night: Modified Wedge

V. Actions at the Halt: Short and Long Halt (GV/LV)

VI. Leader Location: Fixed/Unfixed

FIRE TEAM WEDGE

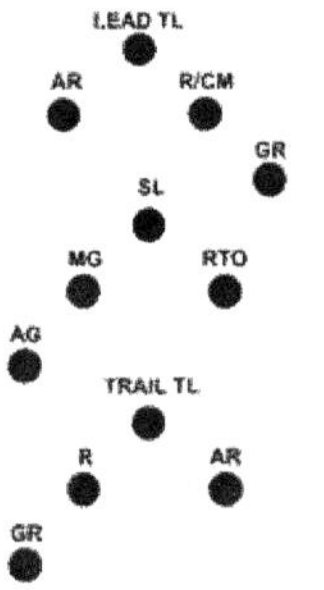

MODIFIED WEDGE

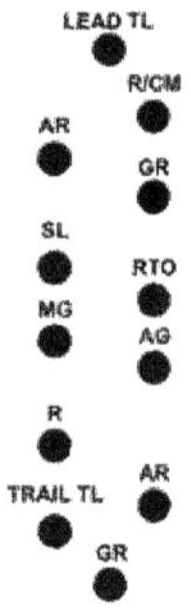

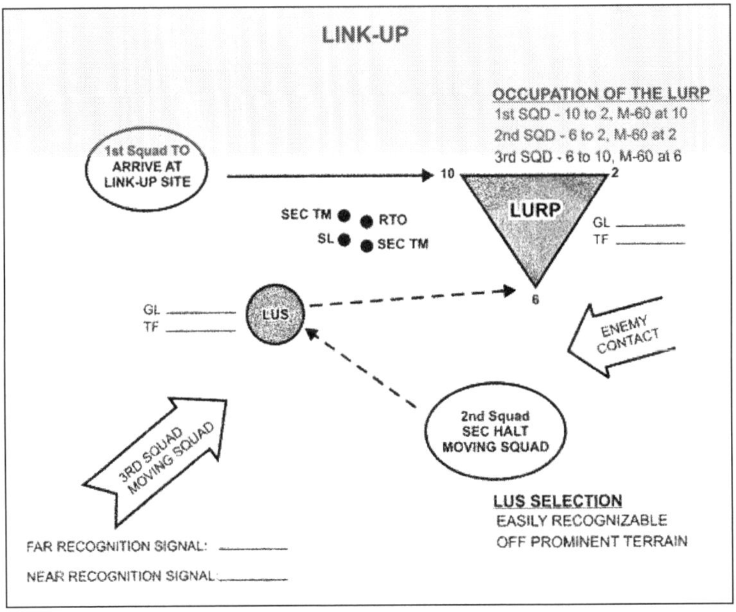
LINK-UP
OCCUPATION OF THE LURP
1st SQD - 10 to 2, M-60 at 10
2nd SQD - 6 to 2, M-60 at 2
3rd SQD - 6 to 10, M-60 at 6
1st Squad TO ARRIVE AT LINK-UP SITE
10
2
LURP
GL
TF
SEC TM
RTO
SL
SEC TM
GL
TF
LUS
6
ENEMY CONTACT
3RD SQUAD MOVING SQUAD
2nd Squad SEC HALT MOVING SQUAD
LUS SELECTION
EASILY RECOGNIZABLE
OFF PROMINENT TERRAIN
FAR RECOGNITION SIGNAL:
NEAR RECOGNITION SIGNAL:

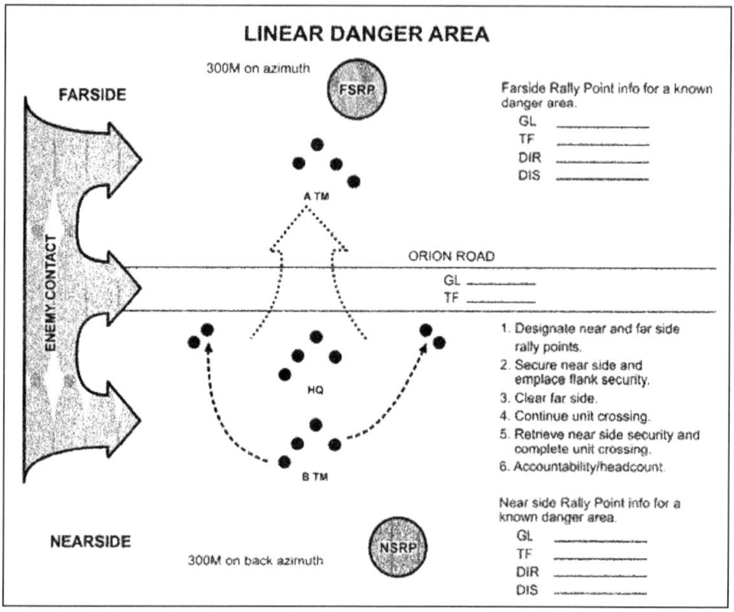
LINEAR DANGER AREA
300M on azimuth
FSRP
FARSIDE
Farside Rally Point info for a known danger area.
GL
TF
DIR
DIS
A TM
ENEMY CONTACT
ORION ROAD
GL
TF
1. Designate near and far side rally points.
2. Secure near side and emplace flank security.
3. Clear far side.
4. Continue unit crossing.
5. Retrieve near side security and complete unit crossing.
6. Accountability/headcount.
HQ
B TM
Near side Rally Point info for a known danger area.
GL
TF
DIR
DIS
NEARSIDE
300M on back azimuth
NSRP

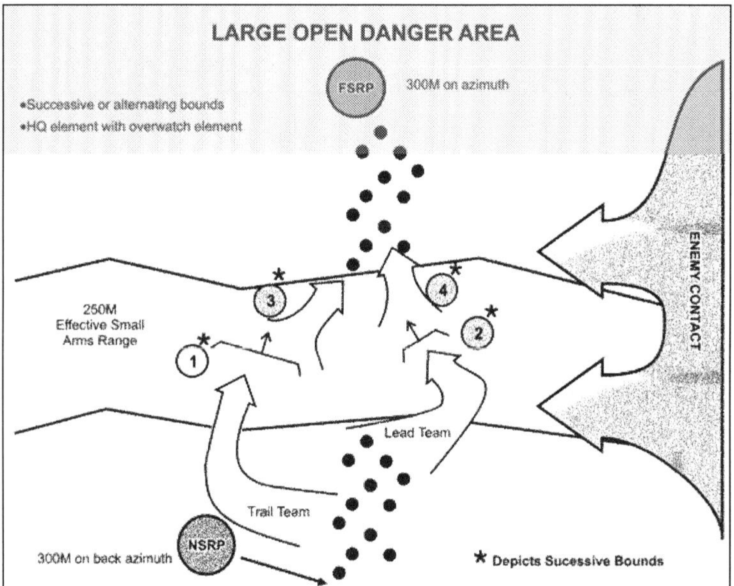

LARGE OPEN DANGER AREA
FSRP
300M on azimuth
•Successive or alternating bounds
•HQ element with overwatch element
ENEMY CONTACT
250M
Effective Small
Arms Range
1
2
3
4
Lead Team
Trail Team
NSRP
300M on back azimuth
* Depicts Sucessive Bounds

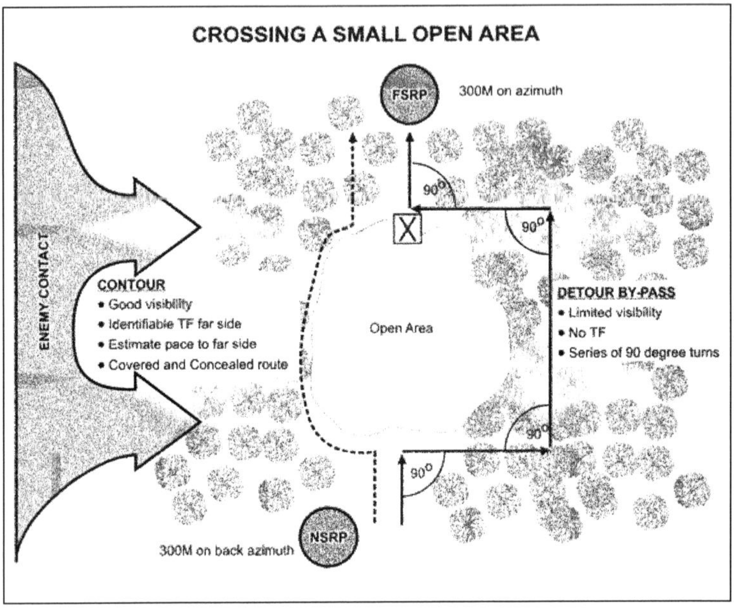
CROSSING A SMALL OPEN AREA
FSRP
300M on azimuth
90°
90°
ENEMY CONTACT
CONTOUR
• Good visibility
• Identifiable TF far side
• Estimate pace to far side
• Covered and Concealed route
Open Area
DETOUR BY-PASS
• Limited visibility
• No TF
• Series of 90 degree turns
90°
90°
300M on back azimuth
NSRP

SQUAD ATTACK

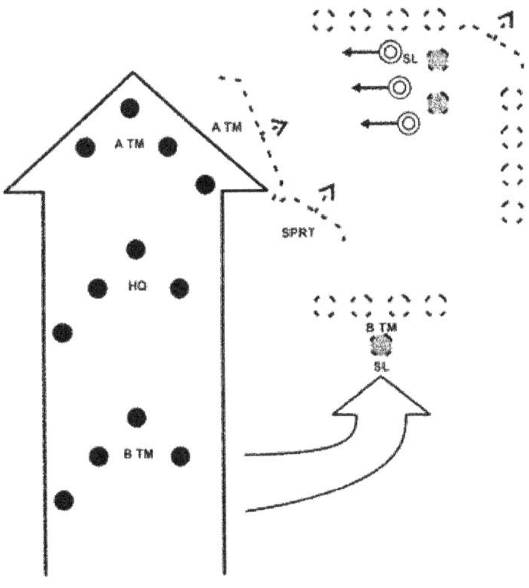

- React to contact.
- TM in the kill zone: (without orders)
  - Return fire.
  - Seek cover and concealment.
  - Suppress enemy calls out 3 D's becomes overwatch.
- SL assess situation.
  - Determines COA (flank/attack).
- TM not in contact: (with SL)
  - Moves along covered and concealed route.
  - Assaults enemy on weak flank.
- Overwatch TM continues to suppress shifts/cease fire as bounding team enters sector.
- Bounding team continues to assault through enemy.
- SL may request indirect fire.
- Consolidate and reorganize.

RAID BOARDS (LEFT)
GENERAL INFO

Raid References:
- RHB Chapter 5
- FM 3-21.8

Purposes:
1. Destroy
2. Liberate
3. Collect Intel

Planning Considerations:
- Mission
- Enemy
- Troops
- Terrain / OAKOC
- Time
- Civilians

STUDENT INPUT WRITTEN HERE

Definition:
A surprise attack on a fixed position or installation ending in a planned withdrawal.

Characteristics:
1. Surprise - Gain
2. Coordinated fires - Maintain
3. Violence of action - Retain
4. Planned withdrawal

INITIATIVE

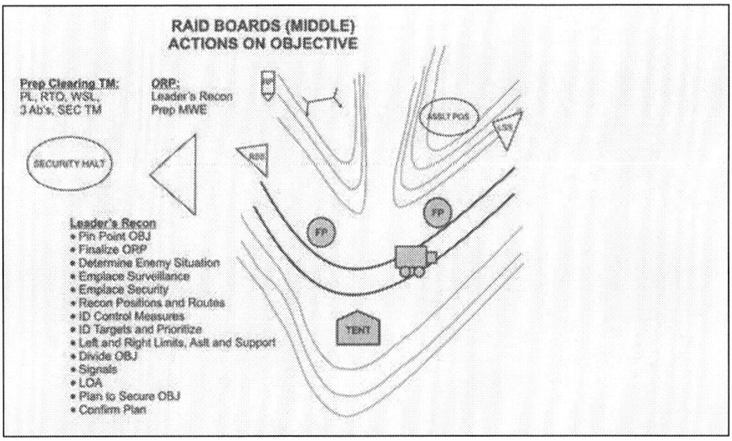
RAID BOARDS (MIDDLE)
ACTIONS ON OBJECTIVE
Prep Clearing TM:
PL, RTO, WSL,
3 Ab's, SEC TM
ORP:
Leader's Recon
Prep MWE
ASSLT POS
SECURITY HALT
FP
FP
TENT
Leader's Recon
• Pin Point OBJ
• Finalize ORP
• Determine Enemy Situation
• Emplace Surveillance
• Emplace Security
• Recon Positions and Routes
• ID Control Measures
• ID Targets and Prioritize
• Left and Right Limits, Aslt and Support
• Divide OBJ
• Signals
• LOA
• Plan to Secure OBJ
• Confirm Plan

RAID BOARDS (MIDDLE)
TASK ORGANIZATION

ELEMENT	WHO	WHY	WHAT
HQ	PL, APL RTO, FO, MEDIC	C2/FACILITATE SIGNALS, PEQ-2, STANO	1x119, ICOM, SKEDCO, PLUGGER
ASSLT 1	SQUAD (+)(-)	DESTROY	ICOM, STANO, M18A1, AT4, SPECIAL TEAM KITS
ASSLT 2	SQUAD (+)(-)	DESTROY	ICOM, STANO, M18A1, AT4, SPECIAL TEAM KITS
SUPPORT	WPN S SQD (+)(-)	SUPPRESS	3xM240B COMPLETE, ICOM, PEQ-2, PVS-4, AT, STANO, SIGNALS, MAP
SECURITY	SQUAD (+)(-)	DELAY	ICOMs, M18A1, AT, STANO, MAP SIGNALS

RAID BOARDS (RIGHT)
SOP'S

CONTINGENCIES

1. COMPROMISE
   a. ORP
   b. LEADERS RECON
   c. OCCUPATION

2. MASS CAL

3. COUNTERATTACK

SOP's

1. EPW SEARCH

2. AID LITTER

3. MED EVAC

4. CCP

5. WITHDRAW PLAN

AMBUSH SOPS (LEFT)

References:
- RHB Chapter 5
- FM 3-21.8

Ambush:
A surprise attack from a covered and concealed position on a moving or temporarily halted target.

Types:
- Point
- Area

Fundamentals:
1. Surprise - Gain
2. Fire superiority - Maintain
3. Violence of action - Retain

Purposes:
1. Disrupt/Destroy
2. Collect Intelligence
3. Block or Deny Access
4. Canalize

Categories:
Deliberate: has specific target at a predetermined time and location.

Hasty: Platoon makes visual contact with the enemy and has time to establish an ambush without being detected.

Planning Considerations:
- Mission - Task/Purpose
- Enemy - MPCOA, MDCOA
- Terrain - Map Recon/Ldrs Recon
- Time - 1/3 2/3 Rule
- Troops - Task Org
- Civilians -

AMBUSH BOARDS (MIDDLE)

HQ	PL, PL RTO, PSG, MEDIC, FO	COMMAND AND CONTROL	PRC-119s, ICOM, NVGs, BINO's, SKEDCO, PEQ-2, SIGNALS, MAP
SECURITY	MINIMUM OF 3-2 MAN TEAMS	EARLY WARNING, SEALS OFF OBJ, LFS, RFS, ORP SECURITY	COMMO, NVGs, CLAYMORES, AT WPNS, BINO's, SIGNALS, MAP
SUPPORT	WSL, 3 GUN TEAMS COMPLETE +/-	SUPPRESSIVE FIRE ON KILL ZONE, SEC ON OBJ DURING RECON/REORG	MG's, COMMO, NVGs, AT WPNS, BINO's, SKEDCO, PEQ-2, SIGNALS., MAP
ASSAULT	2 SQUADS +/-	MAIN EFFORT, BLOCK, DESTROY, CANALIZE, CAPTURE PIR	NVGs, BINO's, CLAYMORE, AT, FLEX CUFFS, POLELESS LITTER, MARKINGS, SIGNALS

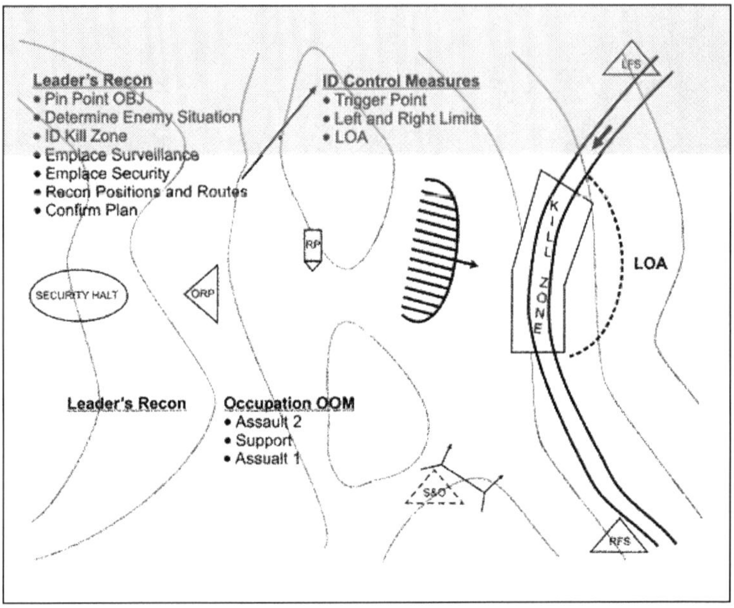

Leader's Recon
• Pin Point OBJ
• Determine Enemy Situation
• ID Kill Zone
• Emplace Surveillance
• Emplace Security
• Recon Positions and Routes
• Confirm Plan
ID Control Measures
• Trigger Point
• Left and Right Limits
• LOA
LFS
RP
KILL ZONE
LOA
SECURITY HALT
ORP
Leader's Recon
Occupation OOM
• Assault 2
• Support
• Assualt 1
S&O
RFS

AMBUSH BOARDS (RIGHT)
SOP'S

CONTINGENCIES

1. COMPROMISE
   a. ORP
   b. LEADERS RECON
   c. OCCUPATION
2. MASS CAL
3. COUNTERATTACK

SOP's

1. EPW SEARCH
2. AID LITTER
3. MASS CAL
4. MED EVAC
5. CCP

NOTES

Appendix B

QUICK REFERENCE CARDS

CASUALTY FEEDER CARD

NAME ____________________ ROS # ___________

COMPANY ________ ALLEGERIES______________

LOCATION/EVENT ______________________________

TIME	______________	TIME	______________
TEMP	______________	TEMP	______________
PULSE	______________	PULSE	______________
RESP	______________	RESP	______________
B/P	______________	B/P	______________

MENTAL STATUS ______________________________

A & O x 1 2 3

NPA OPA ET Tube

SAO^2 ______________ GLUC __________________

IV _____ AC FOREARM HAND OTHER

1000-1000-500-500-500- NS RL D_5W

NOTES: ______________________________________

____

TROOP-LEADING PROCEDURES

1. Receive mission.
2. Issue warning order.
3. Make a tentative plan.
4. Start movement.
5. Reconnoiter.
6. Complete plan.
7. Issue plan.
8. Supervise.

FIRE REQUEST (FM 6-30)

1. Identification.
2. Warning order.
3. Target location.
4. Target description.
5. Method of engagement.
6. Method of control.

ESTIMATE OF SITUATION (FM 7-10)

1. Conduct a detailed mission analysis.
2. Analyze the situation and develop courses of action.
3. Analyze courses of action (wargame).
4. Compare courses of action.
5. Make a decision.

GTA 07-01-038
INFANTRY LEADERS' REFERENCE CARD
January 1995

HEADQUARTERS, DEPARTMENT OF THE ARMY

References. FM 3-3, FM 6-30, FM 6-121, FM 7-8, FM 7-10, FM 7-90

(Supersedes GTA 7-1-31)

OPERATION ORDER (FM 7-10)

Task Organization
1. Situation
   a. Enemy
   b. Friendly
   c. Attachments/detachments
2. Mission
   - Who, what, where, when, and why
   - Task and purpose
3. Execution.
   a. Concept of operation.
      (1) Maneuver.
      (2) Fire support.
      (3) Engineer.
   b. Tasks to maneuver units.
   c. Tasks to combat support units.
   d. Coordinating instructions.
4. Service support.
   a. General.
   b. Materiel and services.
   c. Casualty evacuation.
   d. Miscellaneous.
5. Command and signal.
   a. Command.
   b. Signal.

SPOT REPORT

1. Size.
2. Activity.
3. Location.
4. Unit/uniform.
5. Time.
6. Equipment.

SHELL REPORT (FM 6-121)

1. Observer identification.
2. Location (coded).
3. Azimuth to flash or sound.
4. Time (from and to).
5. Area shelled.
6. Nature of fire.
7. Type rounds received (artillery, mortars, etc.).
8. Damage (coded).

WEAPONS (FM 7-8)

TYPE	MAX EFF RANGE (m)
M16A2	580 (pt) 800 (area) 200 (mov)
M203	150(pt) 350 (area)
M249	600 (pt) 800 (area)
M136 (AT4)	300
M47 (Dragon)	1,000 (sta) 100 (mov)
MK19	1500 (pt) 2,212 (area)
M3 RAAWS	700 (sta) 60 (mov)
M60 MG	1,100 (600 grazing)
.50 Caliber MG	1,800 (1,000 grazing)
TOW	3,000 (plng purposes)
TOW 2	3,750
105-mm	11,500
105-mm Tank	*2 to 2.5 km
120-mm Tank	*2 to 2.5 km
25-mm BFV	2,200
155-mm M109A3	18,100
M198	24,000
8-in Howitzer	22,900

WEAPONS (MORTAR) HE ONLY (FM 7-90)

	MIN	MAX
60-mm	70m	3,500m
81-mm (M252)	80m	5,800m
81-mm (M29A1)	73m	4,790m
4.2-inch	770m	6,840m
120-mm	200m	7,200m

FPFs (FM 7-90)

GUNS	MORT	WIDTH		DEPTH
2	60-mm	60m	x	30m
4	81 (M252)	150m	x	50m
4	81 (M29A1)	140m	x	40m
6	4.2-in	240m	x	40m
6	120-mm	360m	x	60m
Plt	155-mm	200m	x	50m
Btry	155-mm	400m	x	50m

*Optimum engagement ranges

MEDEVAC REQUEST

1. Requesting unit identification.
2. Location.
3. Number of patients by type (litter or ambulatory).
4. Type of injuries.
5. Special equipment needed.
6. Tactical situation.

AIRCRAFT REQUEST

1. Identification.
2. Precedence/priority.
3. Target description.
4. Target location.
5. Target time/date.
6. Desired ordnance/results.
7. Final control.

DELIBERATE ATTACK CONSIDERATIONS

1. Reconnoiter, pinpoint objective/enemy positions/obstacles.
2. Determine weak points; designate supporting positions.
3. Assign platoon/squad objectives—identify the decisive point.
4. Determine main attack, supporting attack, reserve.
5. Assign breach, support, assault missions.
6. Designate fire control measures.
7. Coordinate indirect/direct fires and CAS to time of attack.
8. Control measures during attack.
9. Secure (ground and air).
10. Consolidate and reorganize.

NBC 1 OBSERVER'S INITIAL OR FOLLOW-UP REPORT

Instructions

1. Line items D and H are mandatory for NBC 1 reports.
2. Line items A, E, G, I, K, L, M, S, Y, and ZA are optional for NBC 1 reports.
3. Line Items B, C, F, PAR, and PBR are reported if data is available.

Section I. Chemical or Biological Only

A. Strike serial number, if known (assigned by NBCE).
B. Position of observer.
C. Azimuth of attack from observer (state degrees or mils).
D. Date and time attack started (Zulu, local, or letter zone).
E. Time attack ended, if known.
F. Location of attack (UTM or place) (state actual or estimated).
G. Means of delivery, if known.
H. Type of agent and height of burst, if known.
I. Type and number of munitions or aircraft (state which).
K. Description of terrain (bare, scrubby vegetation, wooded, urban, or unknown).
S. Date and time contamination detected (Zulu, local, or letter zone).
Y. Representative downwind direction—4 digits (state degrees or mils), wind speed—3 digits (data kmph or knots).
ZA. Temperature (centigrade)—2 digits, cloud cover—1 digit, significant weather phenomena—1 digit, air stability—1 digit.
ZB. Remarks.

Section II. Nuclear Only

A. Strike serial number, if known (assigned by NBCE).
B. Position of observer.
C. Azimuth of attack from observer (state degrees or mils and grid or magnetic).
D. Date and time attack started (Zulu, local, or letter zone).
F. Location of attack (UTM or place) (state actual or estimated).
G. Means of delivery, if known.
H. Type of burst (state air, surface, or unknown).
J. Flash-to-bang time (seconds).
K. Crater diameter (meters), if known.
L. Cloud width at H +5 minutes (degrees or mils).
M. Cloud angle(top or bottom) or cloud height(top or bottom) at H +10 minutes (state degrees, mils, meters, or feet).
PAR. Location of radioactive cloud outline (UTM).
PAB. Downwind direction of radioactive cloud (state degrees or mils).
ZB. Remarks.

DEFENSE PLANNING CONSIDERATIONS

1. Establish security (OP/patrols, PWs, M8).
2. Position key weapons:
   a. Coordinate with units on left and right.
   b. Establish FPF or PDF for machine gun.
   c. Ensure mutual support between machine guns.
   d. Cover armor approaches with antiarmor systems.
   e. Establish fire control measures.
3. Prepare positions:
   a. Check sectors of fire.
   b. Check overhead cover and view positions from enemy vantage.
   c. Position in depth and achieve mutual support between positions.
   d. Select/prepare alternate and supplementary positions.
4. Integrate indirect fires, CAS, and obstacles with direct and indirect fire.
5. Check communications and establish emergency signals.
6. Designate ammunition, supply, PW, and casualty points.

IED / UXO

Procedures when IEDs are found

Security - Maintain 360 degrees. Scan close in and far out, up high and down low.

Always - Scan you immediate surroundings for more IEDs.

Move - Move away. Vary safe distances, but plan for 300m minimum safe distance and adapt to your METT-TC.

Attempt - To confirm suspected IEDs using optics while staying back as far as possible.

Cordon - Off the area. Direct people out of danger area. Do not allow anyone to enter except for EOD. Question, search, and detain suspects as defined by your Existing ROE.

Report - Your situation using the 9 line IED / UXO spot report format.

This could be your hand if you try to dispose of UXOs or IEDs. The enemy has developed anti-handling to catch you when you try defusing. Leave it to experts!

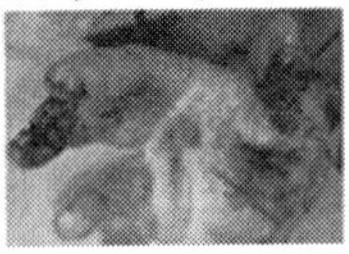

Call EOD - Don't be a Hero!

IED / UXO Report

LINE 1. DATE-TIME-GROUP: When the item was discovered.

LINE 2. REPORT ACTIVITY AND LOCATION: Unit and grid location of the IED/UXO.

LINE 3. CONTACT METHOD: Radio frequency, call sign, POC, and telephone number.

LINE 4. TYPE OF ORDNANCE: Dropped, projected, placed, or thrown. Give the number of items, if more than one.

LINE 5. NBC CONTAMINATIONS: Be as specific as possible.

LINE 6. RESOURCES THREATENED: Equipment, facilities, or other assets that are threatened.

LINE 7. IMPACT ON MISSION: Short description of current tactical situation and how the IED/UXO affects the status of the mission.

LINE 8. PROTECTIVE MEASURES: Any measures taken to protect personnel and equipment.

LINE 9. RECOMMENDED PRIORITY: Immediate, Indirect, Minor, No Threat.

Priority

Immediate: Stops unit's maneuver and mission capability or threatens critical assets vital to the mission.

Indirect: Stops the unit's maneuver and mission capability or threatens critical assets important to the mission.

Minor: Reduces the unit's maneuver and mission capability or threatens non-critical assets of value.

No Threat: Has little or no effect on the unit's capabilities or assets.

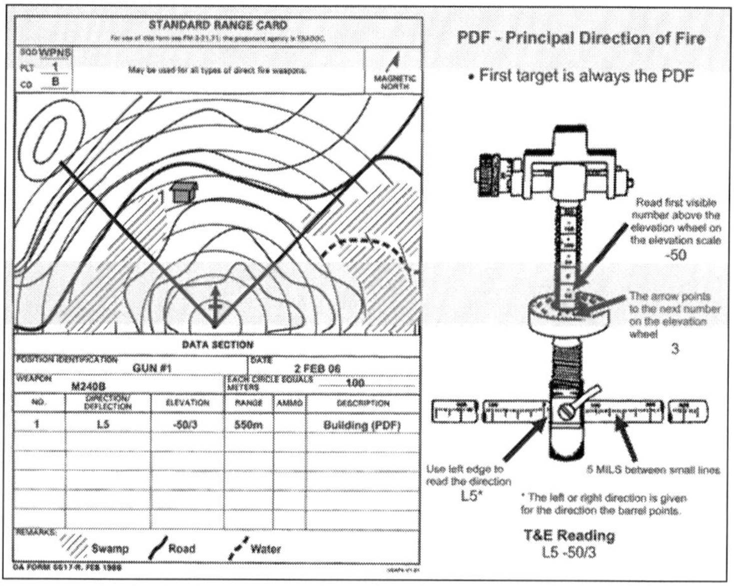
STANDARD RANGE CARD
SQD WPNS
PLT 1
CO B
May be used for all types of direct fire weapons.
MAGNETIC NORTH
1
DATA SECTION
POSITION IDENTIFICATION GUN #1
DATE 2 FEB 06
WEAPON M240B
EACH CIRCLE EQUALS 100 METERS
NO. | DIRECTION/DEFLECTION | ELEVATION | RANGE | AMMO | DESCRIPTION
1 | L5 | -50/3 | 550m | | Building (PDF)
REMARKS:
Swamp
Road
Water
DA FORM 5517-R, FEB 1986
PDF - Principal Direction of Fire
• First target is always the PDF
Read first visible number above the elevation wheel on the elevation scale
-50
The arrow points to the next number on the elevation wheel
3
Use left edge to read the direction
L5*
5 MILS between small lines
* The left or right direction is given for the direction the barrel points.
T&E Reading
L5 -50/3

STANDARD RANGE CARD

For use of this form see FM 3-21.71; the proponent agency is TRADOC.

SQD WPNS
PLT 1
CO B

May be used for all types of direct fire weapons.

MAGNETIC NORTH

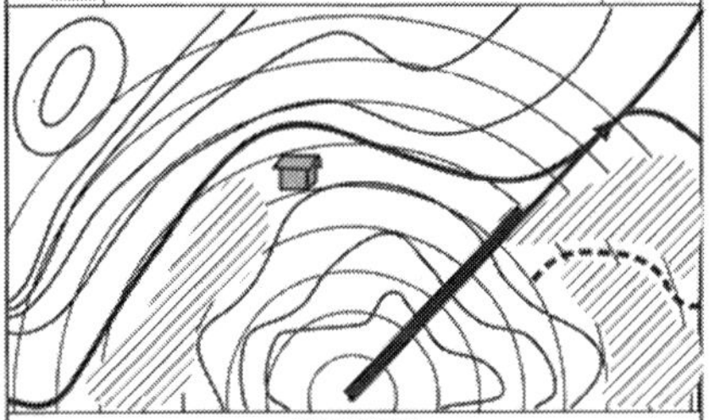

DATA SECTION

POSITION IDENTIFICATION: GUN #1
DATE: 2 FEB 06
WEAPON: M240B
EACH CIRCLE EQUALS 100 METERS

NO.	DIRECTION/ DEFLECTION	ELEVATION	RANGE	AMMO	DESCRIPTION
1	R450	-50/10	600m		FPL

REMARKS: Swamp Road Water

DA FORM 5517-R, FEB 1986

FPL - Final Protective Line

- FPL is always target #1
- FPL will always be metal-to-metal
- Represented by thick line
  - Break in thick line for dead space out to 600 meters
  - The gap is equal to the width of the dead space

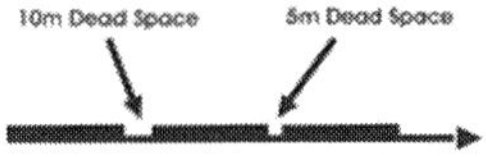

FPL - Final Protective Line

STANDARD RANGE CARD

For use of this form see FM 3-21.71; the proponent agency is TRADOC.

SQD ______
PLT ______
CO ______

May be used for all types of direct fire weapons.

MAGNETIC NORTH

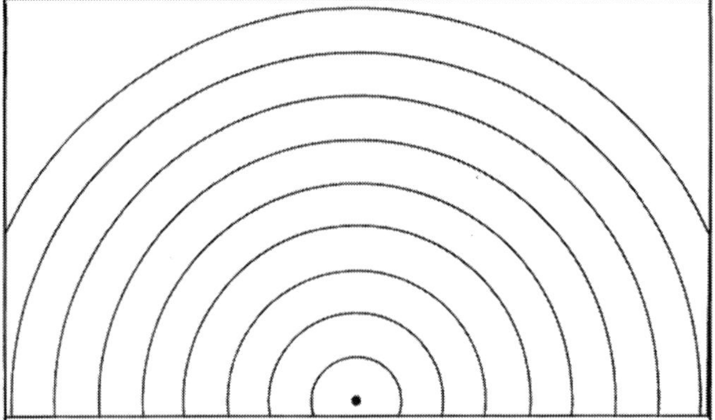

DATA SECTION

POSITION IDENTIFICATION | DATE

WEAPON | EACH CIRCLE EQUALS ______ METERS

NO.	DIRECTION/ DEFLECTION	ELEVATION	RANGE	AMMO	DESCRIPTION

REMARKS:

DA FORM 5517-R, FEB 1986

USAPA V1.01

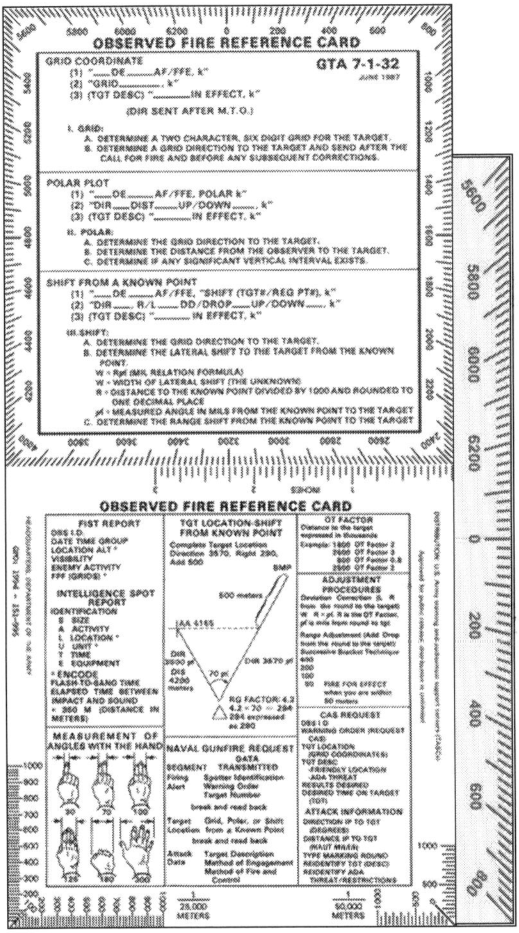
OBSERVED FIRE REFERENCE CARD
GTA 7-1-32
JUNE 1987
GRID COORDINATE
(1) "___OE___AF/FFE, k"
(2) "GRID___, k"
(3) (TGT DESC) "___IN EFFECT, k"
(DIR SENT AFTER M.T.O.)
I. GRID:
A. DETERMINE A TWO CHARACTER, SIX DIGIT GRID FOR THE TARGET.
B. DETERMINE A GRID DIRECTION TO THE TARGET AND SEND AFTER THE CALL FOR FIRE AND BEFORE ANY SUBSEQUENT CORRECTIONS.
POLAR PLOT
(1) "___OE___AF/FFE, POLAR k"
(2) "DIR___DIST___UP/DOWN___, k"
(3) (TGT DESC) "___IN EFFECT, k"
II. POLAR:
A. DETERMINE THE GRID DIRECTION TO THE TARGET.
B. DETERMINE THE DISTANCE FROM THE OBSERVER TO THE TARGET.
C. DETERMINE IF ANY SIGNIFICANT VERTICAL INTERVAL EXISTS.
SHIFT FROM A KNOWN POINT
(1) "___OE___AF/FFE, "SHIFT (TGT#/REG PT#), k"
(2) "DIR___, R/L___DD/DROP___UP/DOWN___, k"
(3) (TGT DESC) "___IN EFFECT, k"
III. SHIFT:
A. DETERMINE THE GRID DIRECTION TO THE TARGET.
B. DETERMINE THE LATERAL SHIFT TO THE TARGET FROM THE KNOWN POINT.
W = Rø (MIL RELATION FORMULA)
W = WIDTH OF LATERAL SHIFT (THE UNKNOWN)
R = DISTANCE TO THE KNOWN POINT DIVIDED BY 1000 AND ROUNDED TO ONE DECIMAL PLACE
ø = MEASURED ANGLE IN MILS FROM THE KNOWN POINT TO THE TARGET
C. DETERMINE THE RANGE SHIFT FROM THE KNOWN POINT TO THE TARGET
INCHES
OBSERVED FIRE REFERENCE CARD
FIST REPORT
OBS I.D.
DATE TIME GROUP
LOCATION ALT *
VISIBILITY
ENEMY ACTIVITY
FPF (GRIDS) *
INTELLIGENCE SPOT REPORT
IDENTIFICATION
S SIZE
A ACTIVITY
L LOCATION *
U UNIT *
T TIME
E EQUIPMENT
* ENCODE
FLASH-TO-BANG TIME
ELAPSED TIME BETWEEN IMPACT AND SOUND × 350 M (DISTANCE IN METERS)
TGT LOCATION-SHIFT FROM KNOWN POINT
Complete Target Location
Direction 3570, Right 290, Add 500
BMP
500 meters
AA 4165
DIR 3500 ø
DIS 4200 meters
DIR 3570 ø
70 ø
RG FACTOR: 4.2
4.2 × 70 = 294
294 expressed as 290
OT FACTOR
ADJUSTMENT PROCEDURES
FIRE FOR EFFECT when you are within 50 meters
CAS REQUEST
OBS I.D.
WARNING ORDER (REQUEST CAS)
TGT LOCATION (GRID COORDINATES)
TGT DESC
-FRIENDLY LOCATION
-ADA THREAT
RESULTS DESIRED
DESIRED TIME ON TARGET (TOT)
ATTACK INFORMATION
DIRECTION IP TO TGT (DEGREES)
DISTANCE IP TO TGT (NAUT MILES)
TYPE MARKING ROUND
REIDENTIFY TGT (DESC)
REIDENTIFY ADA THREAT/RESTRICTIONS
MEASUREMENT OF ANGLES WITH THE HAND
30
70
100
120
180
300
NAVAL GUNFIRE REQUEST
SEGMENT DATA TRANSMITTED
Firing Alert Spotter Identification
Warning Order
Target Number
break and read back
Target Location Grid, Polar, or Shift from a Known Point
break and read back
Attack Data Target Description
Method of Engagement
Method of Fire and Control
1/25,000 METERS
1/50,000 METERS

9 Line MEDEVAC

LINE ITEM	EXPLANATION
1. Location of Pickup Site.	Encrypt grid coordinates. When using *DRYAD Numeral Cipher, the same SET line* will be used to encrypt grid zone letters and coordinates. To preclude misunderstanding, a statement is made that grid zone letters are included in the message (unless unit SOP specifies its use at all times).
2. Radio Frequency, Call Sign, Suffix.	Encrypt the frequency of the radio at the pickup site, *not* a relay frequency. The call sign (and suffix if used) of person to be contacted at the pickup site may be transmitted in the clear.
3. No. of Patients by Precedence.	Report only applicable info & encrypt brevity codes. A = Urgent, B = Urgent-Surg, C = Priority, D = Routine, E = Convenience. (If 2 or more categories reported in same request, insert the word "*break*" between each category).
4. Spec Equipment.	Encrypt applicable brevity codes. A = None, B = Hoist, C = Extraction equipment, D = Ventilator.
5. No. of Patients by Type.	Report only applicable information and encrypt brevity code. If requesting MEDEVAC for both types, insert the word "*break*" between the litter entry and ambulatory entry: L + # of Pnt - Litter; A + # of Pnt - Ambul (sitting).
6. Security Pickup Site (Wartime).	N = No enemy troops in area, P = Possibly enemy troops in area (approach with caution), E = Enemy troops in area (approach with caution), X = Enemy troops in area (armed escort required).
6. Number and type of Wound, Injury, Illness (Peacetime).	Specific information regarding patient wounds by type (gunshot or shrapnel). Report serious bleeding, along with patient blood type, if known.
7. Method of Marking Pickup Site.	Encrypt the brevity codes. *A* = Panels, *B* = Pyrotechnic signal, *C* = Smoke Signal, *D* = None, *E* = Other.
8. Patient Nationality and Status.	Number of patients in each category need not be transmitted. Encrypt only applicable brevity codes. A = US military, B = US civilian, C = Non-US mil, D = Non-US civilian, E = EPW.
9. NBC Contamination (Wartime).	Include this line only when applicable. Encrypt the applicable brevity codes. N = nuclear, B = biological, C = chemical.
9. Terrain Description (Peacetime).	Include details of terrain features in and around proposed landing site. If possible, describe the relationship of site to a prominent terrain feature (lake, mountain, tower).

Reference: FM 8-10-6, *Medical Evacuation in a Theater of Operations, pages 7-7 through 7-9.*

*FM 4-02.2

Field Manual
No. 4-02.2

Headquarters
Department of the Army
Washington, DC, 8 May 2007

MEDICAL EVACUATION

Contents

Page

Distribution Restriction: Approved for public release; distribution is unlimited.

***This publication supersedes Chapters 1–7 and Appendixes A, B, D–F, K, L, and N of FM 8-10-6 dated 14 April 2000, and FM 8-10-26 dated 16 February 1999.**

Page

Figures

Tables

Preface

This field manual (FM) provides doctrine, as well as techniques and procedures for conducting medical evacuation and medical regulating operations. Medical evacuation encompasses both the evacuation of Soldiers from the point of injury (POI) or wounding to a medical treatment facility (MTF) staffed and equipped to provide essential care in theater and further evacuation from the theater to provide definitive, rehabilitative, and convalescent care in the continental United States (CONUS) and the movement of patients between MTFs or to staging facilities. Medical evacuation entails the provision of en route medical care; supports the joint health service support (JHSS) system; and links the continuum of care. In addition, it discusses the difference between medical evacuation and casualty evacuation (CASEVAC), as well as coordination requirements for and the use of nonmedical transportation assets to accomplish the CASEVAC mission. This publication is intended for use by medical commanders and their staffs, command surgeons, and nonmedical commanders involved in medical evacuation operations.

Users of this publication are encouraged to submit comments and recommendations to improve this publication. Comments should include the page, paragraph, and line(s) of the text where the change is recommended. The proponent for this publication is the United States (US) Army Medical Department (AMEDD) Center and School (USAMEDDC&S). Comments and recommendations should be forwarded, in letter format, directly to **Commander, USAMEDDC&S, ATTN: MCCS-FCD-L, 1400 East Grayson Street, Fort Sam Houston, Texas 78234-5052** or by using the e-mail address: medicaldoctrine@amedd.army.mil.

This publication applies to the Active Army, the Army National Guard (ARNG)/Army National Guard of the United States (ARNGUS), and the U.S. Army Reserve (USAR), unless otherwise stated.

Unless this publication states otherwise, masculine nouns and pronouns do not refer exclusively to men.

Use of trade or brand names in this publication is for illustrative purposes only and does not imply endorsement by the Department of Defense (DOD).

The staffing and organization structure presented in this publication reflects those established in base tables of organization and equipment (TOE) and are current as of the publication print date. However, such staffing is subject to change to comply with manpower requirements criteria outlined in Army Regulation (AR) 71-32. Those requirements criteria are also subject to change if the modification table of organization and equipment (MTOE) is significantly altered.

This publication implements the following North Atlantic Treaty Organization (NATO) International Standardization Agreements (STANAGs) and American, British, Canadian, and Australian (ABCA) Quadripartite Standardization Agreements (QSTAGs):

TITLE	STANAG	QSTAG
Stretchers, Bearing Brackets, and Attachment Supports	2040	
Medical Employment of Air Transport in the Forward Area	2087	
Medical and Dental Supply Procedures	2128	
Minimum Labeling Requirements for Medical Materiel		436
Documentation Relative to Medical Evacuation, Treatment, and Cause of Death of Patients	2132	470
Road Movements and Movement Control	2454	

TITLE	STANAG	QSTAG
Orders for the Camouflage of the Red Cross and Red Crescent on Land in Tactical Operations	2931	
Aeromedical Evacuation	3204	

When amendment, revision, or cancellation of this publication is proposed which will affect or violate the international agreements concerned, the preparing agency will take appropriate reconciliatory action through international standardization channels. These agreements are available on request from the Standardization Documents Order Desk, 700 Robins Avenue, Building 4, Section D, Philadelphia, Pennsylvania 19111-5094.

The AMEDD is in a transitional phase with terminology. This publication uses the most current terminology; however, other FM 4-02-series and FM 8-series may use the older terminology. Changes in terminology are a result of adopting the terminology currently used in the joint and/or NATO and ABCA Armies publication arenas. Also, the following terms are synonymous and the current terms are listed first:

- Medical logistics (MEDLOG); health service logistics (HSL); and combat health logistics (CHL).
- Roles of care, echelons of care, and level of care.
- Combat and operational stress control (COSC) and combat stress control (CSC).
- Behavioral health (BH) and mental health (MH).
- Chemical, biological, radiological and nuclear (CBRN) and nuclear, biological, and chemical (NBC).
- Stability operations; stability, security, transition, and reconstruction (SSTR) operations; and stability operations and support operations.

Introduction

The Army Health System (AHS) is a complex system of interrelated and interdependent systems which provides a continuum of medical treatment from the POI or wounding through successive roles of health care to definitive, rehabilitative, and convalescent care in the CONUS, as required. *Medical evacuation* is the system which provides the vital linkage between the roles of care necessary to sustain the patient during transport. This is accomplished by providing en route medical care and emergency medical intervention, if required, and to enhance the individual's prognosis and to reduce long-term disability.

Medical evacuation occurs at the tactical, operational, and strategic levels and requires the synchronization and integration of service component medical evacuation resources and procedures with the DOD worldwide evacuation system operated by the United States Transportation Command (USTRANSCOM).

Army medical evacuation is a multifaceted mission accomplished by a combination of dedicated ground and air evacuation platforms synchronized to provide direct support (DS), general support (GS), and area support within the joint operations area (JOA). At the tactical level, organic or DS medical evacuation resources locate, acquire, treat, and evacuate Soldiers from the POI or wounding to an appropriate MTF where they are stabilized, prioritized, and, if required, prepared for further evacuation to an MTF capable of providing required essential care within the JOA.

Although the most recognized mission of Army medical evacuation assets is the evacuation and provision of en route medical care to battlefield wounded, the essential and vital functions of medical evacuation resources encompass many additional missions and tasks that support the JHSS system. Medical evacuation resources are used to transfer patients between MTFs within the JOA and from MTFs to United States Air Force (USAF) mobile aeromedical staging facilities (MASFs) or aeromedical staging facilities (ASFs); emergency movement of Class VIII, blood and blood products, medical personnel and equipment; and serve as messengers in medical channels.

Medical regulating provides the interface with the DOD worldwide medical evacuation system by determining the patient's destination (the MTF best suited to provide the required care) and scheduling the means to transport the patient with the required en route medical care. Formal medical regulating begins at Role 3, however technological advances in information management (IM)/information technology (IT) are permitting this capability to be used at Role 1 and Role 2 MTFs in some situations.

Chapter 1

Overview of Army Health System Operations and Medical Evacuation

The AHS is the Army component of the military health system (MHS). Its capabilities are focused on delivering health care across the continuum of military operations—from the POI or wounding, through the JOA, to the CONUS support base. The two missions of the AHS are to provide health service support (HSS) (casualty care, medical evacuation, and medical logistics) and force health protection (FHP) (casualty prevention). The AHS is focused on promoting wellness, preventing casualties due to disease and nonbattle injuries (DNBIs), and providing timely and effective casualty care and management. Medical evacuation is the key factor in ensuring the continuity of care provided to our Soldiers by providing en route medical care during evacuation, facilitating the transfer of patients between MTFs to receive the appropriate specialty care, and ensuring that scarce medical resources (personnel, equipment, and supplies [to include blood]) can be rapidly transported to areas of critical need on the battlefield.

SECTION I — ARMY HEALTH SYSTEM

1-1. The provision of AHS is governed by well established and time-tested principles and rules which ensure the care provided to our Soldiers is timely and effective. For an in-depth discussion of these principles, rules, and roles of medical care, refer to FM 4-02.

1-2. The AHS is comprised of 10 medical functions. They are—

- Medical command, control, communications, computers, and intelligence.
- Medical treatment.
- Medical evacuation.
- Hospitalization.
- Dental services.
- Preventive medicine (PVNTMED) services.
- Combat and operational stress control (COSC).
- Veterinary services.
- Medical logistics.
- Medical laboratory support.

1-3. Army health system resources are arrayed across the battlefield in successive levels of support. These successive levels have increased medical care capabilities at each higher level. Medical evacuation and the provision of en route medical care ensures an uninterrupted continuum of care is maintained while Soldiers are moved through the roles of medical care to the MTF best suited to treat the patient's specific injuries.

PRINCIPLES OF ARMY HEALTH SYSTEM

1-4. The principles of AHS provide a framework in which medical planners can ensure that comprehensive plans are developed to support the tactical commander's operation plan (OPLAN). The principles of AHS are conformity, continuity, control, proximity, flexibility, and mobility. For this publication, the discussion of these principles has been focused toward the medical evacuation mission. For a general discussion of how these principles relate to the overall AHS mission, refer to FM 4-02.

CONFORMITY

1-5. Participating in the development of the OPLAN or the operation order (OPORD) ensures that the medical planner conforms to strategic, operational, and tactical plans. This is the most fundamental element for effectively providing AHS and ensures medical influence over the execution of medical evacuation operations. Only by participating in the orders process and developing a medical evacuation plan, will the medical planner ensure that medical evacuation support is arrayed on the battlefield in the right place at the right time and synchronized across operational commands to maximize responsiveness and effectiveness.

CONTINUITY

1-6. En route medical care provided during medical evacuation must be effective and continuous to prevent interruptions in the continuum of care. An interruption in medical treatment may result in an increase in morbidity, mortality, and disability. No patient is evacuated any farther than his physical condition or the military situation requires.

1-7. Medical evacuation resources provide the linkage between the roles of care within the JOA. They also provide interface with other deployed elements of the MHS operated by other services to enhance and facilitate the continuum of care from the POI to the CONUS support base.

CONTROL

1-8. Medical planners must ensure medical control is exercised over the execution of ground medical evacuation operations and that medical influence (technical and operational supervision) is exercised over the execution of aeromedical evacuation (AE) operations. Furthermore, medical planners must ensure the medical evacuation system is responsive to changing requirements and tailored to effectively support the forces within an assigned area of operations (AO). Since medical evacuation resources are limited, it is essential that medical control and influence be retained at the highest level consistent with the tactical situation.

1-9. A thorough and comprehensive medical evacuation plan is essential to establishing and maintaining control of medical evacuation operations characterized by decentralized execution of the plan. The medical evacuation plan complies with the combatant command guidance and intent and maximizes the use of scarce medical evacuation resources. When directed by the combatant command, Army air and ground ambulances may support operations conducted by other services, allied and coalition partners, and the host nation (HN).

PROXIMITY

1-10. The location of medical evacuation assets in support of combat operations is dictated by orders and the tactical situation (mission, enemy, terrain and weather, troops and support available, time available, and civil considerations [METT-TC]). Accurately determining time and distance factors and the availability of evacuation resources are critical to determining the disposition of evacuation assets. The speed with which medical evacuation is initiated is extremely important in reducing morbidity, mortality and disability. Medical evacuation time must be minimized by the effective disposition of resources, ensuring close proximity of both supported elements and MTF. Medical evacuation assets cannot be located so far forward that they interfere with the conduct of combat operations. Conversely, they must not be located so far to the rear that medical treatment is delayed due to lengthy evacuation routes.

1-11. Medical evacuation resources, both ground and air, are arrayed on the battlefield to best support both the tactical commander and the AHS. Depending upon the situation, evacuation resources may be placed in a DS role to support maneuvering forces or GS, during stability operations which are centrally located to accomplish an area support mission.

FLEXIBILITY

1-12. Changes in tactical plans or operations may require redistribution or reallocation of medical evacuation resources. Therefore, the medical evacuation plan must be designed to ensure flexibility and agility as well as enhance the ability to rapidly task-organize and relocate medical evacuation assets to meet changing battlefield requirements. Medical planners must also ensure medical control and influence is exercised through the orders process and facilitates the synchronization of air and ground evacuation assets to rapidly clear the battlefield.

MOBILITY

1-13. Medical evacuation assets must have the same mobility and survivability capability (such as armor protection) as the forces supported. This mobility and survivability ensures that medical evacuation resources can rapidly respond and that evacuation routes do not become too lengthy. Medical evacuation assets also enhance the mobility of forward deployed MTFs by rapidly evacuating their patients to the next role of medical care.

BATTLEFIELD RULES

1-14. The AMEDD has developed a set of battlefield rules to aid in establishing priorities and to resolve competing priorities within AHS activities. These rules are intended to guide the medical planner to resolve system conflicts encountered in designing and coordinating AHS operations. Although medical personnel seek always to provide the full scope of AHS services and support in the best possible manner, during every combat operation there are inherent possibilities of conflicting support requirements. The planner or operator applies these rules to ensure that the conflicts are resolved appropriately. These battlefield rules are depicted in Table 1-1.

Table 1-1. Army Medical Department Battlefield Rules

AMEDD BATTLEFIELD RULES
BE THERE (MAINTAIN A MEDICAL PRESENCE WITH THE SOLDIER)
MAINTAIN THE HEALTH OF THE COMMAND
SAVE LIVES
CLEAR THE BATTLEFIELD OF CASUALTIES
PROVIDE STATE-OF-THE-ART MEDICAL CARE
ENSURE EARLY RETURN TO DUTY OF THE SOLDIER

BE THERE

1-15. Ensure that medical evacuation assets are in close proximity to supported elements to enhance response time, increase Soldier confidence and be a combat multiplier. This is accomplished by complementing organic medical evacuation assets with medical evacuation assets placed in DS, GS, and area support roles.

Maintain the Health of the Command

1-16. Ensure that medical evacuation assets are in close proximity to the MTF. The closer evacuation assets are to an MTF, the more rapidly care may be rendered.

Save Lives

1-17. Ensure medical evacuation assets are the primary means of evacuating patients. Medical evacuation assets provide en route medical care that provides a continuum of care that is instrumental in preserving life and reducing long-term disability. The use of CASEVAC should only be used in extreme emergencies or when the medical evacuation system is overwhelmed. Refer to paragraphs 1-27 through 1-33 for a discussion on the differences between medical evacuation and CASEVAC operations.

Provide State-of-the Art Medical Care

1-18. Ensure sufficient medical evacuation assets are available to facilitate the movement of patients between MTFs (intratheater evacuation) and the emergency movement of Class VIII, blood and blood products, and medical personnel and equipment. This capability permits the medical planner to maximize the use of scarce medical resources and exercise economy of force without sacrificing state-of-the-art medical care.

Early Return To Duty of The Soldier

1-19. Ensure seamless integration with the intertheater medical evacuation system. This allows rapid evacuation and continuous medical treatment throughout the continuum of care resulting rapid return to duty (RTD) rates.

SECTION II — MEDICAL EVACUATION

PURPOSE

1-20. An efficient and effective medical evacuation system—

- Minimizes mortality by rapidly and efficiently moving the sick, injured, and wounded to an MTF.
- Serves as a force multiplier as it clears the battlefield enabling the tactical commander to continue his mission with all available combat assets.
- Builds the morale of Soldiers by demonstrating that care is quickly available if they are wounded.
- Provides en route medical care that is essential in improving the prognosis and reducing disability of the wounded, injured, or ill Soldiers.
- Provides medical economy of force.
- Provides connectivity of the AHS as appropriate to the MHS.

Tactical	Operational	Strategic
Corps Division BCT Emergency Class VIII Resupply Emergency Movement of Medical Personnel/Equipment Stability Operations	Coalition Joint Class VIII Resupply Joint Blood Program DOD Civilians/Contractors Medical Personnel/Equipment Military Working Dogs Homeland Security Operations Shore-to-Ship Stability Operations	Host Nation Nongovernment Organizations Private Volunteer Organizations DOD support to Stability Operations International Disaster Relief

Figure 1-1. Army medical evacuation as it supports full spectrum operations

1-21. The AHS is established in roles of increasing capability from the POI to definitive care. On the battlefield, casualties are evacuated rearward from one role of care to the next, the sequencing of this movement is dependent upon METT-TC factors. In a contiguous battlefield, well established lines of communication (LOC), large numbers of casualties, and wide array of MTFs can result in very deliberate evacuation from one sequential role to the next higher. However in many situations, such as noncontiguous battlefield, the array of medical resources across the battlefield, the availability of medical evacuation resources, and the number of patients being evacuated may facilitate procedures that permit bypassing roles of care in order to ensure the timely treatment and care of casualties. This evacuation plan will be established by the appropriate level of command in coordination with the command surgeon to ensure the best treatment is provided to all casualties.

ATTRIBUTES

DEDICATED RESOURCES

1-22. The Army medical evacuation system is comprised of dedicated air and ground evacuation platforms. These platforms have been designed, staffed, and equipped to provide en route medical care to patients being evacuated and are used exclusively to support the medical mission. The focus of the medical evacuation mission coupled with the dedicated platforms permit a rapid response to calls for support. The dedicated nature of this mission dictates that Army medical evacuation unit's posture themselves in a "ready alert" status, ready to rapidly respond to evacuation missions and not diverted to perform any other task. Medical evacuation resources are protected under the provisions of the Geneva Conventions from intentional attack by the enemy. (For a discussion of the Geneva Conventions in relation to air and ground evacuation operations, refer to Appendix A.)

EN ROUTE MEDICAL CARE

1-23. En route care is provided on all Army medical evacuation platforms. This care is essential for minimizing mortality, enhancing survival rates, and reducing disability of wounded, injured, or ill Soldiers. Refer to paragraphs 1-27 through 1-33 for a discussion on the differences between medical evacuation and CASEVAC operations.

CONNECTIVITY

1-24. The Army provides connectivity to the MHS. The synchronized employment of medical evacuation resources provides and maintains the seamless continuum of care from the POI through successive roles of essential care within the theater. In addition to evacuating patients and providing en route medical care, medical evacuation resources provide for the emergency movement of scarce medical resources such as critical Class VIII, blood, medical personnel, and medical equipment. Further, medical evacuation resources are used to transfer patients from one MTF to another within the theater, to facilitate specialty care as well as transferring patients from an MTF to a MASF to facilitate intertheater evacuation. Figure 1-1 depicts the connectivity provided by medical evacuation platforms between the roles of care within theater. In addition to providing connectivity within the AHS, Army medical evacuation resources provide medical evacuation support and interface with MTFs of the other services deployed in the theater. In joint and multinational operations, the geographic combatant commander (GCC) may direct Army assets to provide this support and connectivity to joint, allied, and coalition forces within the JOA.

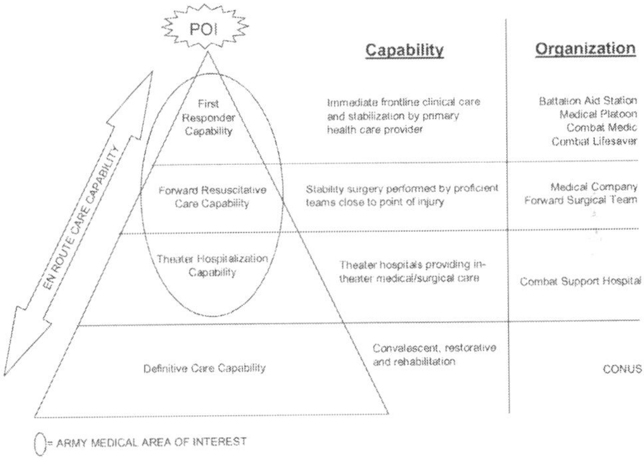

Figure 1-2. Army air and ground evacuation platforms provide connectivity to assure a seamless continuum of medical care

Medical Economy of Force

1-25. The flexibility and versatility which medical evacuation resources afford the AHS to respond to urgent requirements throughout the battlefield is essential in ensuring the provision of seamless medical care. Medical economy of force is achieved for low density, high demand medical specialties (such as a neurosurgeon), medical supplies, and medical equipment (such as computer tomography scans or magnetic resonance imaging equipment) by having the capability to move the patient to the required care over long distances and by permitting the emergency of cross-leveling on medical supplies reducing the need for large Class VIII stockpiles. By providing medical economy of force, the deployed medical footprint is reduced without negatively impacting on the care provided to the Soldier.

Force Multiplier

1-26. A highly effective, efficient, and responsive AHS increases Soldiers' confidence that if wounded in battle, they will receive timely and appropriate medical care to enhance their prognosis, speed their recovery, and reduce long-term disabilities. Medical evacuation resources locate and acquire wounded Soldiers and facilitate their entry into the AHS. These resources clear the battlefield of casualties which enables the tactical commander the ability to capitalize on and exploit battlefield opportunities with all available assets in the execution of his warfighting mission.

SECTION III — MEDICAL EVACUATION VERSUS CASUALTY EVACUATION

MEDICAL EVACUATION

1-27. Medical evacuation is performed by dedicated, standardized medical evacuation platforms, with medical professionals who provide the timely, efficient movement and en route care of the wounded, injured, or ill persons from the battlefield and/or other locations to MTFs. Medical evacuation is an AMEDD function that supports and is an integral part of the AHS. The provision of en route care on medically equipped vehicles or aircraft greatly enhances the patient's potential for recovery and may reduce long-term disability by maintaining the patient's medical condition in a more stable manner.

1-28. Medical evacuation ground/air ambulance platforms are defined as: Platforms especially for the medical evacuation mission with allocated medical equipment specifically designed for the purpose of enroute care and by trained medical personnel.

1-29. The gaining MTF in coordination with the losing MTF, is responsible for arranging for the evacuation of patients from the lower role of care. For example, Role 2 medical units are responsible for evacuating patients from Role 1 MTFs.

1-30. Medical evacuation begins when a request is initiated and continues until the patient is released from the AHS by medical authorities.

CASUALTY EVACUATION

1-31. Casualty evacuation is a term used to refer to the movement of casualties aboard nonmedical vehicles or aircraft.

WARNING

Casualties transported in this manner may not receive proper en route medical care or be transported to the appropriate MTF to address the patient's medical condition. If the casualty's medical condition deteriorates during transport, or the casualty is not transported to the appropriate MTF, an adverse impact on his prognosis and long-term disability or death may result.

1-32. If dedicated medical evacuation platforms (ground and air) are available, casualties should be evacuated on these conveyances to ensure they receive proper en route medical care.

1-33. Since CASEVAC operations can reduce combat power and degrade the efficiency of the AHS, units should only use CASEVAC to move Soldiers with less severe injuries when medical evacuation assets are overwhelmed. Medical planners should ensure CASEVAC operations are addressed in the OPLAN/OPORD as a separate operation, as these operations require preplanning, coordination, synchronization, and rehearsals. The CASEVAC plan should ensure casualties with severe or life-threatening injuries are prioritized for evacuation and are evacuated on dedicated medical evacuation platforms.

1-34. When possible, nonmedical vehicles/aircraft transporting casualties should be augmented with a trauma specialist or combat lifesaver (CLS). (On nonmedical aircraft, sufficient space may not be available to permit a caregiver to accompany the casualties.) The type of en route monitoring and medical care/first aid provided is limited by the following factors:

- Skill level of the individual providing care. (The combat medic is military occupational specialty [MOS]-qualified [MOS 68W] to provide emergency medical treatment [EMT]; the CLS is trained to provide enhanced first aid.) The combat medic can provide emergency medical intervention, whereas the CLS can only monitor the casualty and ensure that the basic lifesaving first-aid tasks are accomplished.
- Medical equipment available.
- Number of casualties being transported.
- Accessibility of casualties—If nonstandard evacuation vehicle is loaded with the maximum number of casualties, the combat medic or CLS may not be able to attend to the casualties while the vehicle is moving. If the condition of a casualty deteriorates and emergency measures are required, the vehicle will have to be stopped to permit care to be given.

SECTION IV — THEATER EVACUATION POLICY

ESTABLISHING THE THEATER EVACUATION POLICY

1-35. The theater evacuation policy is established by the Secretary of Defense, with the advice of the Joint Chiefs of Staff and upon the recommendation of the theater commander. The policy establishes, in number of days, the maximum period of noneffectiveness (hospitalization and convalescence) that patients may be held within the theater for treatment. This policy does not mean that a patient is held in the theater for the entire period of noneffectiveness. A patient who is not expected to be ready for RTD within the number of days established in the theater evacuation policy is evacuated to the CONUS or other safe haven. This is done providing that the treating physician determines that such evacuation will not aggravate the patient's disabilities or medical condition. For example, a theater evacuation policy of seven days does not mean that a patient is held in the theater for six days and then evacuated. Instead, it means that a patient is evacuated as soon as possible after the determination is made that the Soldier cannot be returned to duty within seven days following admission to a Role 3 hospital.

TEMPORARY REDUCTIONS

1-36. To the degree that unplanned for increases in patients occur (due perhaps to an epidemic or heavy combat casualties), a temporary reduction in the policy may be necessary. This reduction is used to adjust the volume of patients being held in the theater hospital system. A reduction in the evacuation policy increases the number of patients requiring evacuation out-of-theater and increases the requirement for evacuation assets. This action is necessary to relieve the congestion caused by the patient increases. A decrease in the theater evacuation policy increases the evacuation asset requirements.

ESSENTIAL CARE IN THEATER AND LENGTH OF POLICY

1-37. Due to the reduction of the medical footprint within the theater, health care delivery to deployed forces is now provided under the concept of essential care in theater. This concept provides for essential and stabilizing care being rendered within the theater, with the patient being evacuated to CONUS or other safe haven for definitive, rehabilitative, and convalescent care. In current and ongoing operations, if the patient cannot be treated and returned to duty within seven days of admission to a combat support hospital (CSH), the patient is evacuated from the theater for continued care.

1-38. The time period established by the theater evacuation policy starts on the date the patient is admitted to the first hospital. The total time a patient is hospitalized in the theater (including transit time between MTFs) for a single, uninterrupted episode of illness or injury should not exceed the number of days stated in the theater evacuation policy. Though guided by the evacuation policy, the actual selection of a patient for evacuation is based on clinical judgment as to the patient's ability to tolerate and survive the movement to the next role of care.

EXCEPTION TO POLICY

1-39. An exception to the theater evacuation policy may be required with respect to SOF personnel. This exception may be required to retain low density MOS skills within the theater. Retaining these personnel within the theater for an extended period of time is possible if the medical resources are available within the theater to treat their injuries and provide for convalescence and rehabilitation. If retention within the theater would result in a deterioration of their medical condition or would adversely impact on their prognosis for full recovery, they are evacuated from the theater for definitive care.

EVACUATION POLICY CONSIDERATIONS

1-40. Paragraphs 1-41 through 1-45 discuss the various considerations of the theater evacuation policy.

Physicians and Dentists

1-41. To physicians and dentists engaged in direct patient treatment and decisions relating to patient disposition, it means that there is a maximum period which clinical staffs may complete the necessary treatment needed to return the patient to full duty within the theater. If the theater policy is seven days and full RTD can be predicted within that time, the patient is retained in the theater. If the patient cannot be returned to full duty within seven days, the patient is evacuated out-of-theater as early as clinically prudent.

Medical Planners and Staff

1-42. The medical staff computes mix, number, and distribution of hospital beds required in the theater.

1-43. The medical operator has a management tool which, when properly adjusted and used, provides the balance between patient care and tactical support requirements. The medical staff is able to tailor a medical package specifically designed to handle the patient workloads, with maximum benefit to the patients and with maximum economy of available resources.

Logisticians

1-44. The nonmedical logistician is able to estimate his total obligation to support this system.

United States Air Force Planner

1-45. The USAF planner can accurately plan the USAF AE requirements for both intra- and intertheater patient movements.

FACTORS DETERMINING THE THEATER EVACUATION POLICY

1-46. To fully understand how the theater evacuation policy affects AHS operations, the medical operator should be aware of the factors that influence the establishment of this policy.

NATURE OF THE OPERATION

1-47. A major factor in determining the theater evacuation policy is the nature of the tactical operations. For example: Will the operations be of short duration and with a low potential for conflict? Will there be operations of long duration with significant combat operations? Will chemical, biological, radiological, nuclear, and high yield explosives (CBRNE) be employed? Will only conventional weapons be used? Is a static combat situation expected? Is there a significant threat of terrorist activities? Are the majority of patients anticipated to be DNBI patients or those with combat-related trauma?

NUMBER AND TYPE OF PATIENTS

1-48. Another factor affecting the policy is the number and types of patients anticipated and the rate of patient RTD. Admission rates vary widely in different geographical areas of the world and in different types of military operations.

EVACUATION MEANS

1-49. The means (quantity and type of transportation) available for evacuation of patients from the theater to CONUS is an essential factor impacting on the evacuation policy.

AVAILABILITY OF REPLACEMENTS

1-50. The capability of CONUS to furnish replacements to the theater is another consideration. For each patient who is evacuated from the theater to CONUS, a fully trained and equipped replacement must be provided. During a small-scale conflict overseas, the CONUS replacement capability is much greater than when compared to a large-scale conflict such as World War II.

AVAILABILITY OF IN-THEATER RESOURCES

1-51. Limitations of all AHS resources (such as insufficient number and types of medical units in the division/corps to support the brigade combat teams [BCTs] and an insufficient amount of medical and nonmedical logistics) will have an impact on the theater evacuation policy. The more limitations (or shortages), the shorter the theater evacuation policy.

IMPACT OF THE EVACUATION POLICY ON ARMY HEALTH SYSTEM REQUIREMENTS

SHORTER EVACUATION POLICY

1-52. A shorter theater evacuation policy—

- Results in fewer hospital beds required in the theater and a greater number of beds required elsewhere.
- Creates a greater demand for intertheater USAF evacuation resources. (A shortened intratheater evacuation policy would likewise increase the number of airframes required in the theater.)
- Increases the requirements for replacements to meet the rapid personnel turnover which could be expected, especially in combat units. (The impact this would have on both intra- and intertheater transportation and other requirements must also be considered.)

LONGER EVACUATION POLICY

1-53. A longer theater evacuation policy—

- Results in a greater accumulation of patients and a demand for a larger AHS infrastructure in the theater. It decreases bed requirements elsewhere.
- Increases the requirements for medical logistics (medical supplies, equipment, and equipment maintenance) and nonmedical logistics support.
- Increases the requirements for hospitals, engineer support, and all aspects of base development for deployed AHS force. (It demands the establishment of a larger number of hospitals within the theater and may require medical specialty augmentation.)
- Provides for a greater proportion of patients to RTD within the theater and, thus, reduces the loss of experienced manpower.
- A longer intratheater evacuation policy may decrease the demand on the intratheater evacuation assets and system.

1-54. The concept of essential care in theater does not support longer evacuation policies as the deployed hospitals are not designed to provide definitive, rehabilitative, and convalescent care/services. If the theater evacuation policy is extended in theaters operating under the essential care in theater concept, augmentation of medical specialty resources will be required.

PATIENT STABILIZATION

1-55. The evacuation policy has no impact on the patient stabilization period for movement. This period is known as the *evacuation delay*. It is the period of time planned for between the time of patient reporting and the time of medical evacuation of the patient to the next role of care. Evacuation delays normally range from 24 to 72 hours and are designated by the Army service component command (ASCC) surgeon.

Chapter 2

Medical Evacuation Resources

This chapter discusses the mission, functions, and capabilities of medical evacuation units and elements, as specified in the unit's TOE. It also discusses the command and control (C2) headquarters to which they are assigned. The discussion of each organization also includes the basis of allocation (BOA), limitations, and dependencies.

SECTION I — MANEUVER BATTALION MEDICAL PLATOON AMBULANCES

MEDICAL PLATOON AMBULANCE SQUADS

2-1. The ground ambulance team is the basic medical evacuation element used within the BCTs and at the corps and theater AOs. These ground ambulance teams provide medical evacuation from the POI to supporting MTFs while ensuring the continuity of care en route. Ambulance squads consisting of two ambulance teams organic to the medical platoons of movement and maneuver and fires units and to the evacuation platoon of the brigade support medical companies (BSMCs).

2-2. The primary mission of the maneuver medical platoon ambulance squads is to provide ground ambulance evacuation support from supported infantry/armor companies or from POI back to a casualty collection point (CCP) or to the Role 1 MTF/battalion aid station (BAS). Maneuver medical platoon ambulance teams are assigned to heavy BCT (HBCT), infantry BCT, Stryker BCT (SBCT), reconnaissance squadrons, and reconnaissance and target acquisition (RSTA) squadrons. They also provide area support to other elements (which do not have organic medical evacuation resources) operating in their AO. Ground ambulance support provided is consistent with evacuation precedence, tactical standing operating procedures (TSOPs), and other operational considerations.

2-3. For definitive information on maneuver battalion medical platoon operations, see FM 4-02.4, FM 4-02.6, and FM 4-02.21.

MANEUVER UNIT MEDICAL PLATOON

ORGANIZATION

2-4. The medical platoon consists of a platoon headquarters section, a treatment squad, combat medic section, and an ambulance squad. The number of ambulance squads in an ambulance platoon varies and is based on the type of parent organization. The infantry, airborne, and air assault battalions' ambulance platoons have two ambulance squads equipped with high mobility multipurpose wheeled vehicle (HMMWV) ambulances. The mechanized infantry and armor combat maneuver battalions' ambulance platoons have four ambulance squads equipped with M-113 track ambulances. The SBCT infantry battalion's and RSTA squadron's ambulance platoons have two ambulance squads equipped with Stryker wheeled armored ambulances (referred to as the Stryker medical evacuation vehicle [MEV]).

Medical Platoon Headquarters

2-5. The medical platoon leader is a physician and also serves as the battalion surgeon. He is assisted by the medical operations officer (field medical assistant) in the operational, administrative, and logistical support aspects of the platoon.

Ambulance Section

2-6. The ambulance section of the medical platoon is organized into ambulance squads and is supervised by the platoon sergeant. Each squad contains a noncommissioned officer (NCO) squad leader, three combat medics, and two ambulances.

Ambulance Squad Leader

2-7. The ambulance squad leader ensures that ambulance team members perform operator maintenance on assigned vehicles. An ambulance team consists of one vehicle and two medical personnel (an aide/evacuation NCO and a combat medic).

Ambulance Team

2-8. Ambulance team members are also responsible for performing operator maintenance on assigned ambulances. Specific duties of the ambulance team are to—

- Administer EMT as required.
- Initiate the Department of Defense (DD) Form 1380, US Field Medical Card (FMC), as required.
- Prepare patients for ground and air medical evacuation.
- Provide medical evacuation of wounded or injured Soldiers from the POI to supporting MTF or ambulance exchange point (AXP).
- Provide patient care en route.
- Provide emergency movement of medical personnel and emergency delivery of blood, medical supplies, and medical equipment.
- Assist in the care of COSR casualties.
- Maintain operational readiness of the ambulance.
- Resupply company and platoon trauma specialists, when required.
- Operate the vehicle and maintain contact with supported and supporting elements.
- Find and collect the wounded.
- Serve as messengers within medical channels.

PRIMARY FUNCTIONS OF AMBULANCE TEAMS

2-9. The primary function of the ambulance team is to collect and treat the sick, injured, and wounded Soldiers on the battlefield and to provide medical evacuation support from the POI, CCP, or AXP to the supporting MTF.

SECTION II — AMBULANCE/EVACUATION PLATOON—FORWARD SUPPORT, BRIGADE SUPPORT, OR AREA SUPPORT MEDICAL COMPANY

MEDICAL COMPANIES

2-10. Medical companies provide Role 2 MTF and medical evacuation. Medical companies assigned to or in DS of a BCT are referred to as either a forward support medical company (FSMC) or a BSMC. Area support medical companies (ASMCs) are assigned to multifunctional medical battalions and operate Role 2 MTF and provide medical evacuation support for division, and theater units. In addition, all medical companies provide Role 1 area medical support to units without organic medical personnel and may augment/reinforce maneuver medical platoons/sections with treatment and medical evacuation support.

FORWARD SUPPORT MEDICAL COMPANY

2-11. The FSMC, an Army of Excellence (AOE) designed unit is organized to provide triage and management of mass casualties, advanced trauma management (ATM), initial resuscitation and stabilization, care for patients with DNBI, and battle wounded and injured Soldiers. The FSMC also provides intervention for combat and operational stress disorders to include COSR and preparation of patients for further medical evacuation. The FSMC is organized into a company headquarters, a treatment platoon, an ambulance platoon, a PVNTMED section, and a BH section.

2-12. The ambulance platoon of FSMC-Light and Airborne units employ a platoon headquarters with four-wheeled ambulance squads (eight ambulances); the FSMC-Air Assault employs a platoon headquarters with three-wheeled ambulance squads (six ambulances); the FSMC-Heavy employs a platoon headquarters with three-tracked ambulance squads (six ambulances) and two-wheeled ambulance squads (four ambulances); and the FSMC-Airborne employs a platoon headquarters with four-wheeled ambulance squads (eight ambulances). A complete discussion on the organization, mission, and functions of the FSMC is provided in FM 4-02.6.

BRIGADE SUPPORT MEDICAL COMPANY

2-13. The new BSMC is designed to support the modular BCT. The BOA for the BSMC is one per maneuver brigade supported. The mission of the BSMC is to provide HSS to all BCT units operating within the brigade AO. The BSMC locates and establishes its company headquarters and also operates a Role 2 MTF and may operate Role 1 MTFs on an area support basis for units that do not have organic medical assets. The company provides C2 for its organic and attached/operational control (OPCON) medical augmentation elements. The BSMC may be augmented with a forward surgical team (FST) providing the company surgical capability, see FM 4-02.25. The BSMC is organized into a company headquarters, a PVNTMED section, a BH section, a treatment platoon, and an evacuation platoon.

2-14. The evacuation platoon performs ground medical evacuation and en route patient care for the supported units. The evacuation platoon consists of a platoon headquarters, a GS evacuation section, and a DS evacuation section. The platoon employs five HMMWV evacuation squads (or ten evacuation teams). A complete discussion on the organization, mission, and functions of the BSMC is provided in FM 4-02.6 and Field Manual Interim (FMI) 4-90.1.

AREA SUPPORT MEDICAL COMPANY

2-15. The ASMC performs functions similar to those of the BCT medical companies. The ASMCs are employed primarily in support of division, corps, and echelons above corps (EAC) units. They are deployed to a geographical area to provide area HSS, or may be deployed to provide HSS for designated units. The ASMC also establishes its Role 2 MTF in a secure location centrally located for supported units. Medical treatment squads/teams of the ASMC may be deployed to establish Role 1 MTF and provide HSS support to concentrations of nondivisional units that do not have organic medical capabilities. The ASMC is organized into a company headquarters, a treatment and ambulance platoon, and a BH section.

2-16. The ambulance platoon performs ground medical evacuation and en route patient care for supported units. The ambulance platoon consists of a platoon headquarters, four ambulance squads (or eight ambulance teams), one HMMWV control vehicle, and eight HMMWV ambulances. A complete discussion on the organization, mission, and functions of the ASMC is provided in FM 4-02.24.

ORGANIZATIONS

2-17. The BSMC/FSMC normally establishes its Role 2 MTF in the BSA, while the ASMC normally establishes its Role 2 MTF in echelons above brigade (EAB) and echelons above division sustainment areas and at theater level. The medical company provides DS/GS ground ambulance evacuation for supported units. It coordinates/ requests air ambulance support for air evacuation of patients from supported Role 1 MTFs/BASs, POI, CCP, and AXPs. The organizational structure, number and type of ambulances assigned to an ambulance/evacuation platoon may differ depending on their location on the battlefield and the types of units supported. For example, a BSMC in support of a HBCT has a mix of armored and wheeled evacuation vehicles while a BSMC in support of the SBCT has only wheeled evacuation vehicles.

2-18. The company headquarters section is organized into a command element; a support element; a unit supply element; a medical supply and medical maintenance element; and an operations and communications element.

Treatment Platoon

2-19. The treatment platoon is composed of a platoon headquarters, treatment squads, and an area support section, which consists of an area support squad, an area support treatment squad, and an area patient holding squad.

2-20. The platoon headquarters is the C2 element of the platoon. The platoon headquarters:

- Determines and directs the disposition of patients and submits requests through the company command post (CP) for their evacuation of patients to supporting hospitals.
- Directs, coordinates, and supervises platoon operations.
- Directs the activities of the Role 2 MTF, monitors Class VIII supplies, blood usage and inventory levels, and keeps the commander informed.
- Manages platoon operations, operations security (OPSEC), communications, medical administration, organizational training, supply transportation, patient accountability, statistical reporting functions, and blood situation reporting.

2-21. The treatment squad element can contain up to four treatment squads, depending on the type company assigned. The treatment squads have the capabilities to:

- Provide emergency and routine sick call treatment services to Soldiers assigned to supported units.
- Operate as separate treatment teams (Teams A and B) for limited periods of time.
- Operate up to four treatment stations.
- Reinforce or reconstitute similar treatment squads and can operate for up to 48 hours while separated from their parent unit.

2-22. For medical companies deployed in the BSA, these squads can further be used to—

- Provide augmentation to maneuver battalion medical platoons.
- Facilitate the movement of the treatment platoon MTF from one location to another or echelons forward to establish an MTF at a new location. A treatment squad can be echeloned forward to establish an MTF at a new location. The echeloning of elements allows the old treatment site to remain operational until the new site is established.
- Augment an AXP, CCP, or positioned at designated points in the ambulance shuttle system, when required.

2-23. The area support section of the treatment platoon is composed of an area treatment squad, an area support squad, and a patient-holding squad. These squads form the medical company Role 2 MTF. The area support treatment squad provides trauma care and routine sick call services to personnel assigned to units located in the supported area of the medical company. The area support squad provides operational dental care that includes emergency and essential dental services, limited laboratory and radiological services, and blood support commensurate with Role 2 MTF. The patient-holding squad provides up to 40 patient beds (40 patient beds for heavy divisions and 20 patient beds for light divisions) for patients requiring minimal treatment. Patients held in the patient-holding area are those patients who are expected to be RTD within 72 hours from the time they are held for treatment.

2-24. The area treatment squad is identical in personnel and equipment as the treatment squads of the treatment section. It is the base medical treatment element of a Role 2 MTF. It provides sick call services and initial resuscitative treatment for supported units. For communications, the squad employs frequency modulated (FM) radios and is deployed in the medical radio and wires communications nets.

Ambulance/Evacuation Platoon

2-25. The ambulance/evacuation platoon provides C2 for ambulance platoon operations. The ambulance platoon headquarters element maintains communications to direct ground ambulance evacuation of patients. It provides ground ambulance evacuation support for supported maneuver battalions and for supported units operating in the sustainment areas. The ambulance headquarters element performs route reconnaissance and develops and issues graphic overlays to all its ambulance teams. It also coordinates and establishes AXPs for both air and ground ambulances, as required.

2-26. Ambulance squads provide ground ambulance evacuation of patients from supported BASs/unit aid stations back to the Role 2 MTF that is located in the brigade support area (BSA). An ambulance squad consists of two ambulance teams (two ambulances, wheel or tracked vehicles). Ambulance squad personnel—

- Perform EMT, evacuate patients, and provide for their continued care en route.
- Operate and maintain assigned communication and navigational equipment.
- Perform preventive maintenance checks and services (PMCS) on ambulances and associated equipment.
- Maintain supply levels for the ambulance medical equipment sets (MES).
- Ensure that appropriate property exchange of medical items (such as litters and blankets) is made at sending and receiving MTF.
- Prepare patients for ground and air medical evacuation.
- Provide medical evacuation of wounded or injured Soldiers from the POI to supporting MTF, CCP or AXP.
- Initiate the DD Form 1380, FMC, as required.
- Provide emergency movement of medical personnel and emergency delivery of blood, medical supplies, and medical equipment.
- Assist in the care of COSR casualties.
- Maintain operational readiness of the ambulance.
- Resupply company and platoon trauma specialists, when required.
- Operate the vehicle and maintain contact with supported and supporting elements.
- Find and collect the wounded.
- Serve as messengers within medical channels.

Preventive Medicine Section

2-27. Preventive medicine sections are organic to the divisions and the BSMC of the BCT. The PVNTMED section is also found in the multifunctional medical battalion (MMB) headquarters. The PVNTMED section assigned to medical companies has a primary responsibility for supervising the unit's PVNTMED program as described in AR 40-5. The section—

- Ensures measures are implemented to protect personnel against food, water, and arthropodborne diseases, as well as environmental injuries.
- Provides advice and consultation in the area of health threat assessment, FHP, environmental sanitation, epidemiology, sanitary engineering, and pest management.
- Assists the higher headquarters in determining requirements for medical intelligence assessments, particularly with respect to CBRN and disease prevalence.
- Coordinates with supporting veterinary teams for conducting and implementing food safety and quality assurance surveillance and assisting in foodborne and zoonotic disease surveillance and control.

Behavioral Health Section

2-28. Combat and operational stress is controlled through vigorous prevention, consultation, and restoration programs by the BSMC BH section. These programs are designed to maximize the RTD rate of

COSR Soldiers by providing rest/restoration within or near their unit areas. The BSMC BH section has a primary responsibility for assisting commanders in controlling combat stress and serves as a consultant to the commander, staff, and others involved with providing prevention and intervention services to unit Soldiers and their families. For additional information on the operations and functions of BSMCs, and ASMCs, refer to FMs 4-02.6 and 4-02.24.

SECTION III — MEDICAL COMPANY (GROUND AMBULANCE)

OPERATIONAL INFORMATION

2-29. The AHS is a continuum of increasing roles of care extending from the POI through the CONUS base. All sick, injured, and wounded patients must be evacuated from the battlefield in the shortest possible time to the MTFs that can provide the required treatment. Ground ambulances serve as one of the primary means of evacuating patients from the battlefield.

MISSION

2-30. The mission of the medical company, ground ambulance, TOE 08453A00, is to provide ground evacuation within the theater.

ASSIGNMENT

2-31. The medical company, ground ambulance is normally assigned or attached to MMB (TOE 08485G00) or a medical brigade (MEDBDE) for C2.

EMPLOYMENT

2-32. The medical company, ground ambulance, provides DS to BCTs and is employed in the division, corps, and EAC to provide area support. It is tactically located where it can best control its assets and execute its patient evacuation mission.

BASIS OF ALLOCATION

2-33. The BOA within the combat zone is one per division supported and one per two divisions supported within the theater.

CAPABILITIES

2-34. At Level I, the unit is capable of providing:

- A single-lift capability for evacuation of 96 litter patients or 192 ambulatory patients.
- Medical evacuation from BCT/division medical companies and ASMCs to supporting hospitals.
- Reinforcement of BCT/division medical company evacuation assets.
- Reinforcement of covering force and unassigned area operations.
- Movement of patients between hospitals and aeromedical staging squadrons/MASFs, railheads, or seaports in the division, corps, and EAC.
- Area evacuation support beyond the capabilities of the ASMC.
- Emergency movement of medical personnel and supplies.
- Preparation of patients for ground and air medical evacuation.
- Medical evacuation of wounded or injured Soldiers from the POI to supporting MTF, CCP, or AXP.
- Field feeding and vehicle refueling support when collocated with the MMB.

DEPENDENCY

2-35. This unit is dependent upon the appropriate elements of the division, corps, EAC, or ASCC for—

- Religious, financial management, legal, personnel, mortuary affairs, and administrative services.
- Viable communications systems for C2 and adequate road networks.
- Laundry, shower, and clothing repair.
- Generator, communication equipment, and communication security (COMSEC) equipment maintenance.
- Health service support to include hospitalization.

ORGANIZATION AND FUNCTIONS

2-36. This company is organized into a company headquarters section, two ambulance platoon headquarters, and two evacuation sections (Figure 2-1).

2-37. The company headquarters provides C2, communications, administration, field feeding, and logistical support (to include maintenance) for the subordinate ambulance platoons.

2-38. The ambulance platoon headquarters provides C2 for the subordinate ambulance squads.

2-39. The evacuation section operates ambulances and provides en route medical care for patients. Each evacuation section consists of 12 ambulances, each with a two-man crew. The members of the squad maintain the level of expendable Class VIII supplies in the ambulance MES by reconstituting supplies from medical companies or hospitals when they pick up or drop off patients. They are also responsible for performing operator maintenance on assigned vehicles.

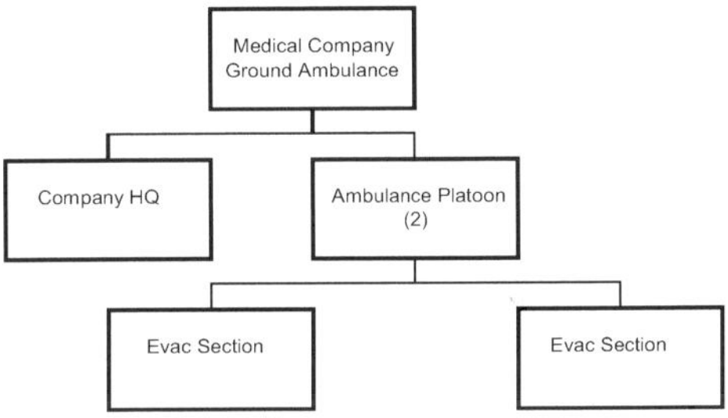

Figure 2-1. Medical company, ground ambulance

SECTION IV — MULTIFUNCTIONAL MEDICAL BATTALION

MODULAR FORCE

2-40. A modular force is a key aspect in the transformation of the Army. The modular force model is based on brigade-sized elements that are more responsive than division-sized elements and can perform joint and expeditionary-type missions. Using modular units, Army planners can tailor force structure, reduce strategic lift requirements, and create flexible forces with specialized capabilities based on everchanging mission requirements. To facilitate the synchronization and provision of AHS services and support, the AMEDD C2 organizations have been redesigned into a modular design. The MMB will replace the functional battalions (area support, evacuation, and medical logistics).

2-41. The new modular design of medical units gives the medical command (MEDCOM) the ability to—

- Assist in the employment of the optimal mix of medical capabilities within an expeditionary to campaign framework.
- Ensure seamless, state-of-the-art care, regardless of delivery location.
- Provide tested and proven systems to the battlefield and ensure the provision of the right care at the right place and time.
- Promote scalability through easily tailored, capabilities-based packages that result in improved tactical mobility, reduced footprint, and increased modularity for flexible task organization.
- Provide and enable the joint force commander (JFC) to choose among augmentation packages, thus enabling rapid synchronization of desired medical capabilities.

MULTIFUNCTIONAL MEDICAL BATTALION

MISSION

2-42. The mission of the MMB, TOE 08486G000, is to provide a scalable, flexible and modular C2, administrative assistance, logistical support, and technical supervision capability for assigned and attached medical organizations (companies, detachments, and teams) task-organized for support of deployed forces.

ASSIGNMENT

2-43. This unit is assigned to a MEDCOM or a MEDBDE.

EMPLOYMENT

2-44. The MMB is normally employed in a MEDCOM or MEDBDE AO.

BASIS OF ALLOCATION

2-45. The BOA is one per three to six subordinate company-size units plus the blood detachment.

CAPABILITIES

2-46. At Level I, this unit provides—

- Command and control, staff planning, supervision of operations, medical and general logistics support as required, and administration of the assigned and attached units conducting AHS operations in its supported AO.
- Task organization of medical assets to meet the projected patient workload.
- Advice to senior commanders in the AO on the medical aspects of their operations.
- Coordination of medical regulating and patient movement within the AO.
- Monitoring, planning, and coordinating ground and air evacuation within the battalion AO. Coordinating air evacuation support requirements and synchronizating the air evacuation plan into the overall medical evacuation plan with the supporting aviation unit.
- Consultation and technical advice on PVNTMED (medical entomology; medical, and occupational and environmental health [OEH] surveillance; and sanitary engineering), pharmacy procedures, COSC and BH, medical records administration, veterinary services, nursing practices and procedures, dental services, medical laboratory procedures, and automated medical information systems to supported units.
- Guidance for facility site selection and area preparation, as required.
- Monitoring and supervising medical logistics operations, to include Class VIII supply/resupply, medical equipment maintenance and repair support, optical fabrication and repair support, and blood management.
- Planning and coordination for Role 1 and Role 2 AHS operations, to include staff advice to units without organic medical assets within the AO.

- Unit level maintenance for wheeled vehicles and power generation equipment and wheeled vehicle recovery operations support to assigned or attached units.
- Organizational communications equipment maintenance support for the battalion.
- Food service support for staff and other assigned/attached medical units.
- Consolidated property book maintenance.
- Religious support for MMB staff, assigned/attached medical units, and casualties in MMB subordinate unit MTFs.

DEPENDENCY

2-47. This unit is dependent on the—

- Corps for legal, administrative, financial management, human resources, and transportation services, area damage control, CBRN decontamination assistance, mortuary affairs, and laundry and bath services.
- Quartermaster supply company or equivalent for Class I rations.
- Engineer company or equivalent for site selection, waste disposal, and minor construction.
- Movement control battalion or equivalent for supplemental transportation requirements.

ORGANIZATION AND FUNCTIONS

2-48. Figure 2-2 depicts the organization of the MMB. Specific information concerning the early entry and campaign elements is provided below.

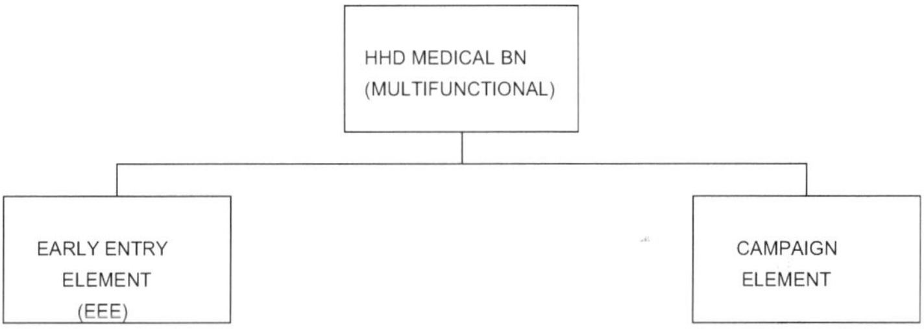

Figure 2-2. Multifunctional medical battalion

EARLY ENTRY ELEMENT, MULTIFUNCTIONAL MEDICAL BATTALION

2-49. Deployed units will begin operations immediately upon arrival in the theater avoiding the delays experienced in past conflicts. There is a need to deploy an EEE medical force to assess, plan, and coordinate AHS operations in the AO and provide medical surveillance, PVNTMED, and area medical support early on. One of the keys to providing early medical care is the development of modular medical elements. These modular AHS elements will perform specific medical functions and will allow medical planners to bring necessary medical units into theater as required by the battlefield mission.

MISSION

2-50. The mission of the EEE of the MMB, TOE 08486GA00, is to provide a scalable, flexible and modular C2, administrative assistance, logistical support, and technical supervision capability for assigned and attached medical organizations (companies, detachments, and teams) task-organized for support of deployed forces.

ASSIGNMENT

2-51. This EEE is assigned to an MMB.

EMPLOYMENT

2-52. Normally employed in a MEDCOM/MEDBDE AO. It can also be deployed to provide C2 to expeditionary forces in early entry operations and facilitates the reception, staging, onward movement, and integration (RSOI) of medical forces.

BASIS OF ALLOCATION

2-53. The BOA is one per MMB.

CAPABILITIES

2-54. At Level I, this unit provides—

- Command and control, staff planning, supervision of operations, medical and general logistics support as required, and administration of the assigned and attached units conducting AHS operations in its supported AO.
- Task organization of medical assets.
- Advice to senior commanders in the AO on the medical aspects of their operations.
- Coordination of medical regulating and patient movement within the AO.
- Monitoring, planning, and coordinating ground and air evacuation within the battalion AO. Coordinating air evacuation support requirements with the supporting aviation unit, and synchronizing the air evacuation plan into the overall medical evacuation plan.
- Consultation and technical advice on PVNTMED (medical entomology, medical and OEH surveillance, and sanitary engineering), COSC and BH, medical records administration, veterinary services, nursing practices and procedures, dental services, and automated medical information systems to supported units.
- Guidance for facility site selection and area preparation, as required.
- Monitoring and supervising medical logistics operations, to include Class VIII supply/resupply.
- Consolidated property book maintenance.
- Unit-level maintenance for wheeled vehicles and power generation equipment and wheeled vehicle recovery operations support to assigned or attached units.
- Organizational communications equipment maintenance support for the battalion.
- Food service support for staff and assigned/attached medical units.

DEPENDENCY

2-55. This unit is dependent on the—

- Corps for legal, administrative, financial management, human resources, religious support, transportation services, area damage control, CBRN decontamination assistance, mortuary affairs, and laundry and bath services.
- Quartermaster supply company or equivalent for Class I rations.
- Engineer company or equivalent for site selection, waste disposal, and minor construction.
- Movement control battalion or equivalent for supplemental transportation requirements.

ORGANIZATION AND FUNCTIONS

2-56. The battalion command section provides C2 of assigned and attached medical companies, detachments, and teams. (See Figure 2-3.)

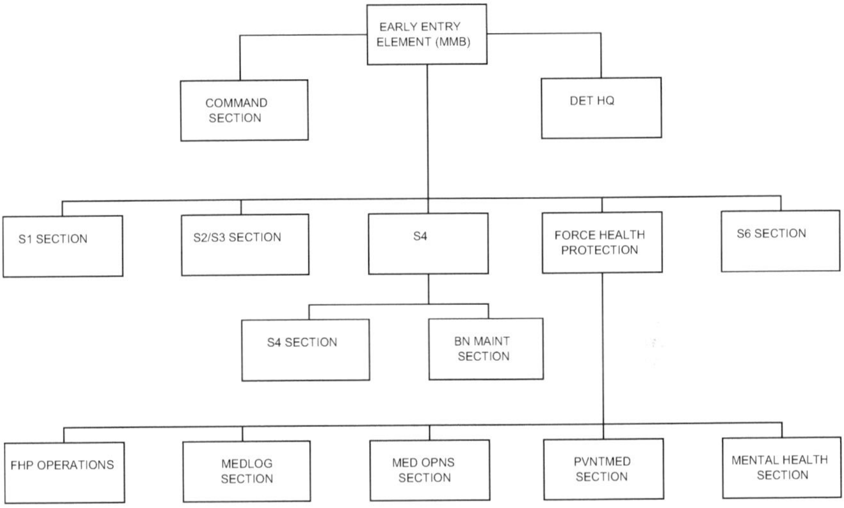

Figure 2-3. Early entry element, multifunctional medical battalion

2-57. The Adjutant (US Army) (S1) operates according to METT-TC and is the principle coordinating staff officer responsible for the delivery of human resources support. The S1 is responsible for the execution of all of the human resources (HR) core competencies to include: personnel readiness management (PRM); personnel accountability and strength reporting (PASR); personnel information management (PIM); reception, replacement, RTD, rest and recuperation (R&R), redeployment (R5) operations; casualty operations; essential personnel services (EPS); and postal operations. Morale, welfare, and recreation (MWR) operations, HR planning; and staff operations. The S1 also coordinates with elements of supporting agencies for finance, legal, religious, and administrative services.

2-58. The Intelligence Officer (US Army)/Operations and Training Officer (US Army) (S2/S3) section is responsible for security, plans and operations, deployment, relocation, and redeployment of the battalion and its assigned and attached units. The battalion's primary net control station is in this paragraph.

2-59. The Supply Officer (US Army) (S4) section is responsible for coordination, control, and management of logistics for assigned and attached units. The battalion maintenance section is under the staff supervision of the battalion S4. This section provides unit level maintenance for wheeled vehicles assigned to the HHD and assigned or attached units without unit level maintenance capability.

2-60. The FHP operations section is responsible for the planning, coordination, and execution of the AHS mission within the battalion's AO. This section supervises the operations of the medical logistics, operations, PVNTMED, and BH sections as well as coordinates medical regulation and patient movement within the AO. The medical operations section is responsible for—

- Management of area medical support (Roles 1 and 2).
- Planning, coordination, and execution of the Class VIII mission within the battalion's AO. This includes blood and medical maintenance management.

- Situational awareness for locations of ground and air ambulance assets.
- Planning and coordination for evacuation of patients using ground and air ambulance assets.
- Planning and coordination between air evacuation elements and the maneuver brigade S3. Information provided to the maneuver brigade S3 should include, but not be limited to the location of—
  - Medical treatment facilities, CCPs, and AXPs.
  - Forward air refueling equipment.
  - Location of supported units and liaison requirements. Air evacuation corridors, ground evacuation routes, and recommendations on usage.
- The PVNTMED section is responsible for the planning, coordination, and execution of the PVNTMED mission within the battalion's AO. This includes management of PVNTMED and veterinary assets.
- The BH section is responsible for the planning, coordination, and execution of the COSC mission within the battalion's AO. This section also collects and records social and psychological data.

2-61. The S6 section is responsible for all aspects of information management, automation, and communications-electronic (CE) support to assigned and attached units. The battalion's alternate net control station is in this paragraph.

2-62. The detachment headquarters provides for billeting, field feeding, discipline, security, training, and administration for personnel assigned to the EEE.

2-63. The battalion maintenance section is under the staff supervision of the battalion S4. This section provides unit level maintenance for wheeled vehicles assigned to the HHD and assigned or attached units without unit level maintenance capability.

CAMPAIGN SUPPORT ELEMENT, MULTIFUNCTIONAL MEDICAL BATTALION

Mission

2-64. The mission of the campaign support element, MMB, TOE 08486GB00, is to complete the staffing of the headquarters of the MMB to enhance the delivery of campaign quality health care to deployed forces. To provide a scalable, flexible, and modular C2, administrative assistance, logistical support, and technical supervision capability for assigned and attached medical organizations (companies, detachments, and teams) task-organized for support of deployed forces.

Assignment

2-65. This unit is assigned to an MMB.

Employment

2-66. Normally employed with the MEDBDE or a MEDCOM. If the EEE is deployed first, the campaign element falls in on this deployed module to provide a more robust C2 element and additional operational and planning capabilities, increased medical logistical staff, and a more diverse and robust clinical staff.

Basis of Allocation

2-67. The BOA is one per MMB.

CAPABILITIES

2-68. At Role 1, this unit provides—

- Augmentation of C2 in personnel, logistics, HSS, FHP, medical operations, PVNTMED, BH, automation, maintenance, religious support, and food service to the EEE.
- Unit maintenance on all organic equipment except COMSEC equipment.

DEPENDENCY

2-69. This unit is dependent on the HHD, MMB for legal, administrative, financial management, human resources, food service, transportation services, area damage control, CBRN decontamination assistance, mortuary affairs, and laundry and bath services.

ORGANIZATION AND FUNCTIONS

2-70. The command section provides C2 for assigned and attached medical companies, detachments, and teams. (See Figure 2-4.)

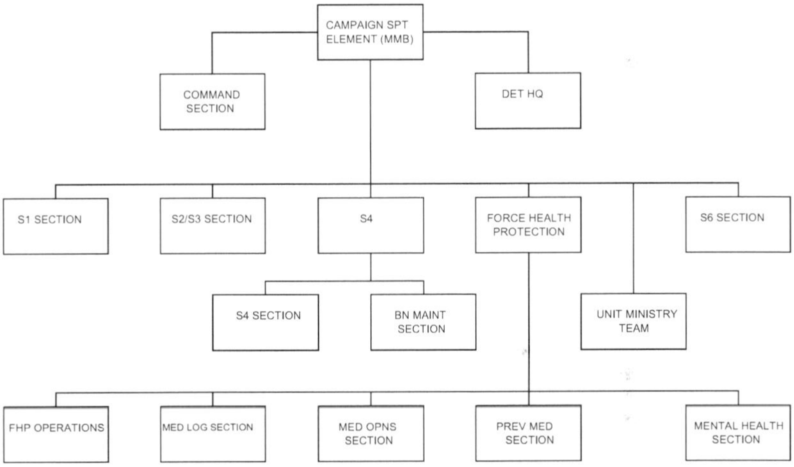

Figure 2-4. Campaign support element, multifunctional medical battalion

2-71. The S1 operates according to METT-TC and is the principle coordinating staff officer responsible for the delivery of HR support. The S1 is responsible for the execution of all of the HR core competencies to include: PRM, PASR, PIM, R5 operations, and casualty operations, EPS, postal operations, MWR operations, and HR planning and staff operations. The S1 also coordinates with elements of supporting agencies for finance, legal, religious, and administrative services.

2-72. The S2/S3 section coordinates with the EEE S2/S3 section of the MMB in the plans and operations support to subordinate organizational elements.

2-73. The S4 section is responsible for coordination, control, and management of logistics for assigned and attached units.

2-74. The FHP operations section plans and coordinates the HSS/FHP mission within the battalion's AO when collocated with the EEE. This section also provides advice on the operation of medical logistics, operations, PVNTMED, BH, medical regulating, and patient movement to supported organizations.

2-75. The medical logistics section plans and organizes Class VIII supply support (including blood and medical equipment management) in coordination with the medical logistics section of the EEE.

2-76. The medical operations section is responsible for—

- Management of area medical support (Roles 1 and 2).
- Situational awareness for locations of ground and air ambulance assets.
- Planning and coordination for evacuation of patients using ground and air ambulance assets.
- Planning and coordination between air evacuation elements and the maneuver brigade S3. Information provided to the maneuver brigade S3 should include, but not be limited to the location of—
  - Medical treatment facilities, CCPs, and AXPs.
  - Forward air refueling equipment.
  - Supported units and liaison requirements.
  - Evacuation corridors and evacuation routes and recommendations on usage.

2-77. The PVNTMED section plans and coordinates PVNTMED/veterinary support to subordinate organizations.

2-78. The BH section plans and coordinates BH support to subordinate organizations. This section also collects and records social and psychological data when collocated with the EEE.

2-79. The S6 section provides advice and support to subordinate organizations for CE support and information management.

2-80. The unit ministry team provides and coordinates religious support for all MMB staff, assigned/attached medical units and casualties in MMB subordinate unit MTFs.

2-81. The detachment headquarters provides coordination with the EEE in support of the field feeding and administration mission of the unit.

2-82. The battalion maintenance section provides unit level maintenance management support to assigned and attached units in coordination with the battalion S4 and the battalion maintenance section of the EEE.

SECTION V — GENERAL SUPPORT AVIATION BATTALION, COMBAT AVIATION BRIGADE

GENERAL SUPPORT AVIATION BATTALION

MISSION

2-83. The mission of the general support aviation battalion (GSAB) (Figure 2-5) is to provide the combat aviation brigade with aerial sustainment and maneuver support.

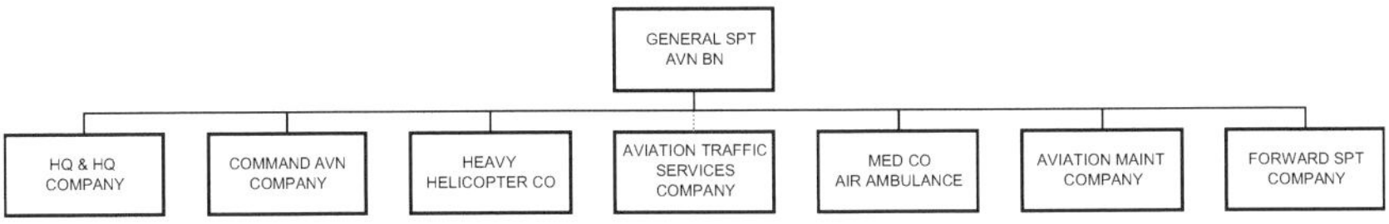

Figure 2-5. General support aviation battalion

ASSIGNMENT

2-84. Organic to the division combat aviation brigade, theater assault aviation brigade, theater general support aviation brigade, and the Army National Guard (ARNG) aviation expeditionary brigade (TOE 01300G400).

BASIS OF ALLOCATION

2-85. The BOA is one per division combat aviation brigade, one per ARNG aviation expeditionary brigade, one per theater assault aviation brigade, and three per theater GSAB (TOE 01300G400).

CAPABILITIES

2-86. This organization provides the following when all of its subordinate units are organized at Level I as shown in their respective TOEs:

- Battle command on the move.
  - Employment of Army airspace command and control (A2C2) system platforms in support of the division command group, maneuver, and aviation commanders' C2 requirements.
  - Radio relay and transport of liaison officers as required.
- Air assault.
  - Airlift for air assault operations and air movement of tactical forces (other than assault), field artillery or other fire support assets, tactical air defense (AD) systems, engineer equipment and personnel, and other resources.
  - Casualty evacuation when medical aircraft are not readily available.
  - Evacuation of downed aircraft and personnel.
- Sustainment.
  - Division sustainment with heavy-lift helicopter company.
  - General aviation support to the aviation brigade.
  - Support aircraft to the aviation support battalion.
  - Personnel recovery (PR).
  - Downed aircraft recovery team operations.
  - Casualty evacuation operations.
  - Air movement of general supplies, equipment, and personnel, including logistics over-the-shore (LOTS) when applicable.
  - Aerial noncombatant evacuation operations when directed by applicable commander/ authority.
  - Air movement of nuclear and other special munitions.
  - Self-deployment of organic CH-47 helicopters to a theater of operations when the aircraft have been configured with extended range fuel systems.

        - Retail level aircraft refueling when aircraft are configured with extended range fuel system (commonly referred to as "Fat Cow").
- Air traffic services support.
- Aeromedical evacuation.
    - Patient evacuation.
    - Patient movement between MTFs (patient transfers).
    - Class VIII resupply.
    - Joint blood program support.
    - Medical C2.
    - Movement of medical personnel and medical equipment.
    - Air crash rescue support.

2-87. Due to the CH-47's capabilities, heavy helicopter units can perform two unique missions:

- High-altitude operations.
- Oversized, heavy, and special munitions movement.

2-88. This GSAB is dependent upon the following organizations:

- Appropriate elements of the division aviation brigade for legal, HSS, FHP, financial management, and personnel and administrative services.
- United States Air Force weather team in the headquarters and headquarters company of the combat aviation brigade (TOE 01302G200) for air weather service support.
- Aviation support company (TOEs 01918G200 or 01919G000 [ARNG]) for intermediate level field maintenance support, DS maintenance for assigned aviation ground power units (AGPU), line item number (LIN) P44627, and support for aviation and common Class IX repair parts.

MEDICAL COMPANY, AIR AMBULANCE (HH-60), GENERAL SUPPORT AVIATION BATTALION, COMBAT AVIATION BRIGADE

Operational Information

2-89. The medical company, air ambulance (Figure 2-6), provides aeromedical evacuation for all categories of patients consistent with evacuation precedence and other operational considerations within the division. In today's ongoing efforts in building a modular force, medical company, air ambulance will fall under the GSAB which will provide aircraft maintenance and logistics support, aviation communications, and real-time operational picture associated with today's combat environment.

Mission

2-90. The mission of the medical company, air ambulance (HH-60) (TOE 08443G00) is to provide air ambulance evacuation support within the division and theater.

Assignment

2-91. The medical company, air ambulance, (HH-60) is organic to the GSAB (TOEs 01125G000, 01225G100, and 01305G200) for C2.

Employment

2-92. This unit is employed in the division and corps area of responsibility. It is tactically located where it can best control its assets and execute its patient evacuation mission.

Basis Of Allocation

2-93. The BOA is one per GSAB and two per theater of operations (TOEs 01125G100, 01225G100, and 01305G100).

Capabilities

2-94. At Level I, this unit provides—

- Twelve helicopter ambulances to evacuate critically wounded or other patients consistent with evacuation priorities and operational considerations, from points as far forward as possible, to division, corps, and EAC hospitals. Single patient lift capability for the company is 72 litter patients or 84 ambulatory patients, or some combination thereof.
- Four forward support medical evacuation teams (three helicopters each) that can be deployed as a team or an independent aircraft.
- Air crash rescue support.
- Expeditious delivery of whole blood, biological, and medical supplies to meet critical requirements.
- Rapid movement of medical personnel and accompanying equipment/supplies to meet the requirements for mass casualty (MASCAL), reinforcement/reconstitution, or emergency situations.
- Movement of patients between hospitals, ASFs, hospital ships, casualty receiving and treatment ships, seaports and railheads in the theater of operations.
- Military working dog evacuation and aeromedical evacuation support to a combat search and rescue task force.

Dependency

2-95. This unit is dependent upon—

- Appropriate elements of the division or corps for religious, legal, financial management, personnel and administrative services, and CE and COMSEC maintenance.
- The headquarters and headquarters company of the GSAB for religious, legal, HSS, FHP, personnel and administrative service, and unit CBRN support.
- The forward support company of the GSAB for Class III, automotive and generator maintenance, and field feeding.
- The aviation support company of the GSAB for unit level field maintenance of organic aircraft, including unit level supply support for aircraft Class IX.
- United States Air Force weather team in the headquarters and headquarters company of the division aviation brigade for air weather service support.

Organization and Functions

2-96. The company headquarters provides C2, logistical support, and administration for the company.

2-97. The four forward support MEDEVAC teams (FSMTs) provide a task-organized means for medical evacuation at the DS and GS level. They also, provide emergency movement of medical personnel and emergency delivery of whole blood, biological, and medical supplies and equipment.

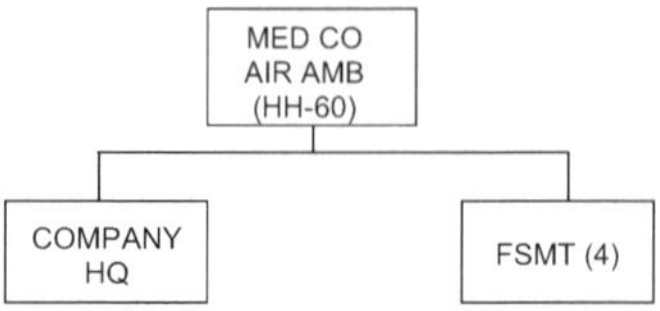

Figure 2-6. Medical company, air ambulance (HH-60)

AVIATION MAINTENANCE COMPANY, AVIATION UNIT MAINTENANCE, GENERAL SUPPORT AVIATION BATTALION, COMBAT AVIATION BRIGADE

MISSION

2-98. The mission of the aviation maintenance company is to perform aviation unit maintenance for the GSAB.

ASSIGNMENT

2-99. This unit is organic to a GSAB (TOE 01305G100).

EMPLOYMENT

2-100. The aviation maintenance company is employed as part of combined forces and provides unit level field maintenance support to the GSAB. The aviation maintenance company is located and its operations are normally conducted immediately behind the main battle area (MBA) in the EAB. The aviation maintenance company operations in forward areas are usually limited to daily inspections, minor adjustments, emergency and contact maintenance, and aircraft recovery. Contact teams and slice elements could be utilized as necessary.

BASIS OF ALLOCATION

2-101. The BOA is one per GSAB (TOE 01305G100).

CAPABILITIES

2-102. The aviation maintenance company provides unit level field maintenance support for 8 UH-60, 12 HH-60, and 12 CH-47 aircraft. Individuals of this organization can assist in the coordinated defense of the unit's area or installation. This unit performs unit level field maintenance on only the supported aircraft and some associated maintenance equipment.

DEPENDENCY

2-103. This unit depends on—

- Appropriate elements of the division or corps for religious, legal, HSS, FHP, financial management, personnel and administrative services, and logistical support (including supplemental transportation).
- Headquarters and headquarters company of the GSAB (TOE 01306G100) for religious, legal, HSS, FHP, financial management, personnel, and administrative services, unit-level logistical support, and communications maintenance.

- Forward support company (TOE 63317G200) for Class III, automotive and generator maintenance, and field feeding service support.
- Aviation maintenance battalion (TOE 01918G200), for unit level field maintenance support, backup unit level field maintenance support, DS maintenance for assigned AGPU, and support for aviation and common Class IX repair parts.

ORGANIZATION AND FUNCTIONS

2-104. This unit inspects aircraft systems, subsystems, components, and the maintenance performed by crew chiefs and repairers to determine airworthiness and compliance with applicable technical publications; performs operational checks and troubleshoots aircraft systems to isolate malfunctions; participates in maintenance test flights and maintains historical maintenance files.

2-105. This unit provides company maintenance operations. It maintains and regulates uniform flow of aircraft and associated components. Further, it maintains the company's aircraft logbooks, provides unit level Class IX support, and performs maintenance on aviation life support equipment (ALSE).

2-106. This unit provides unit level maintenance on UH/HH-60/CH-47 helicopters and associated equipment; services and lubricates helicopter systems and subsystems; and diagnoses and troubleshoots operational malfunctions.

MEDICAL COMMAND

2-107. The MEDCOM serves as the senior medical command within the theater, and works directly for the theater commander. With augmentation, MEDCOM is joint capable. In the absence of a theater commander the MEDCOM works for the regional combatant commander during a major combat operation (MCO). It possesses a centralized capability to effectively and efficiently task-organize medical elements based on theater specific medical requirements. The MEDCOM is depicted in Figure 2-7.

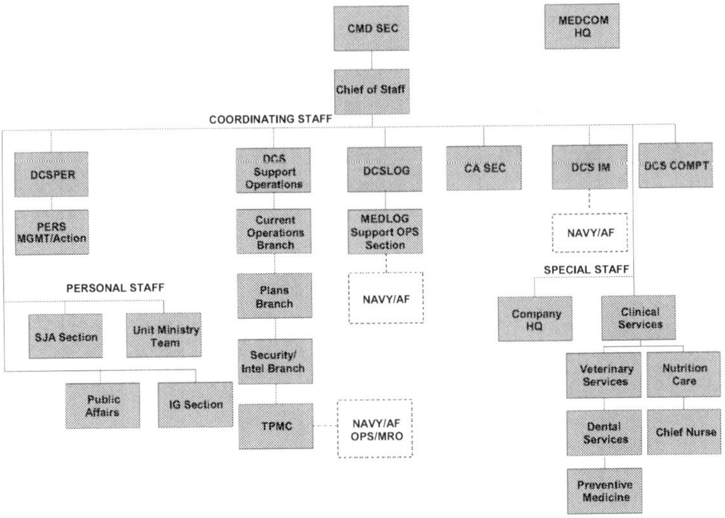

Figure 2-7. Medical command

2-108. The MEDCOM serves as the medical force provider (Figure 2-8) for the corps and EAC and provides the requisite medical C2 necessary to provide campaign-quality health care to deployed forces. The MEDCOM also focuses on theater medical operations plans and medical contingency plans.

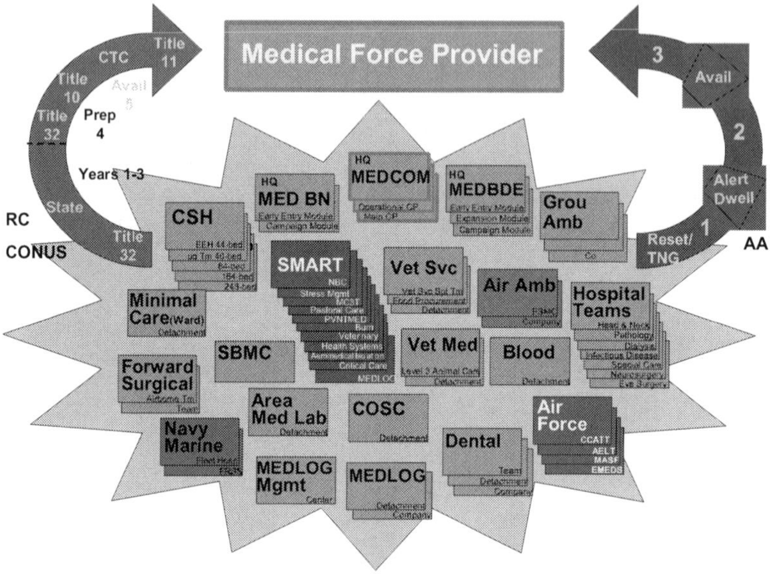

Figure 2-8. Medical force pool

2-109. The ability of the MEDCOM staff to develop accurate patient estimates based on real-time medical data ensures the right mix of medical specialties and resources are in the right place and at the right time to treat casualties on the battlefield. Care must be continuous and seamless from the POI or wounding through a phased system providing essential care in theater, with evacuation for definitive and rehabilitative care in CONUS.

2-110. The MEDCOM is a dedicated, regionally focused command with a basis of allocation of one per theater.

2-111. The MEDCOM is a versatile, modular medical C2 structure composed of a main command post (MCP) and an operational command post (OCP) which are 100 percent mobile. The MEDCOM is comprised of two standard requirements code (SRC) identified modules to provide a scalable medical C2 capability to the corps commander. Further, the MEDCOM is capable of—

- Providing subordinate medical organizations to operate under the subordinate MEDBDE and/or MMB and providing FSTs and/or other augmentation required by BCTs, and echelons above brigade (EAB) medical companies.
- Operating as the medical force provider for provisioning of required assets to the MEDBDEs.
- Advising the corps and other senior level commanders on the medical aspects of their operations.
- Providing staff planning, supervision of operations, and administration of assigned and attached medical units.
- Providing reach to the MHS.

- Providing medical Army support to other Services (ASOS)/Title 10 responsibilities of the ASCC commander.
- Coordinating with the USAF theater patient movement requirements center (TPMRC) for medical regulating and movement of patients from MEDCOM MTFs.
- Providing consultation services, technical advice, and clinical policy development in all aspects of medical and surgical services.
- Containing the functional staff to coordinate hospitalization, nursing, PVNTMED (entomology, epidemiology, OEH surveillance, potable water inspection, pest management, food facility inspection, and control of medical and nonmedical waste), medical evacuation, medical regulating, MEDLOG to include blood management, dental services, veterinary services (zoonotic disease control, investigation and inspection of subsistence, and animal medicine), pharmacy, optometry, nutrition care, and COSC, neuropsychiatric (NP), and BH services, medical laboratory services, and area medical support to supported units.
- Providing technical advice and consultation on medical automated information systems and programs such as the Theater Medical Information Program (TMIP) and Medical Communications for Combat Casualty Care (MC4).
- Functioning as the senior theater joint medical C2 element when augmented with joint augmentation package.
- Peacetime planning, training for and with the MEDBDEs, and participation in theater engagement plans.
- Providing advice and assistance in facility selection and preparation.
- Coordinating and orchestrating MEDLOG operations to include Class VIII supply, distribution, medical maintenance and repair support, optical fabrication, and blood management.
- Planning for the single integrated medical logistics manager (SIMLM), mission when designated.
- Planning and coordination of religious support for MEDCOM staff, assigned and attached medical units, and casualties in MEDCOM subordinate unit MTFs.

Operational Command Post

2-112. The MEDCOM OCP is a deployable, versatile module, which provides medical C2, policy development, and technical guidance to subordinate medical units and provides interface and liaison with supported theater forces in EAC, corps, and subordinate BCTs. The MEDCOM OCP has an assigned SRC to facilitate the placement of personnel in the CP and to integrate the CP into the time-phased force deployment list (TPFDL). The MEDCOM OCP is depicted in Figure 2-9.

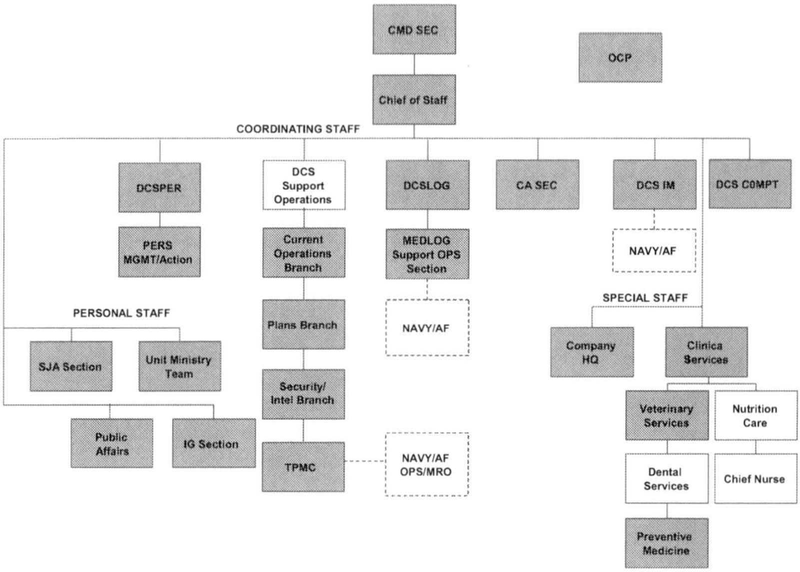

Figure 2-9. Operational command post, medical command

2-113. As the force continues to build within the theater, the MEDCOM's OCP can be incrementally expanded or augmented with additional planning and professional expertise. The deployment of personnel from the MCP model can then be built into the medical C2 required structure in order to provide campaign-quality health care to the deployed force.

Main Command Post

2-114. The MEDCOM MCP is a deployable augmentation module, which completes the staffing of the MEDCOM headquarters to enhance the delivery of campaign-quality health care to deployed forces in EAC, corps, and supported BCTs. It facilitates medical C2, policy development, and technical guidance to subordinate MEDBDEs and provides interface and liaison with supported theater forces in EAC, corps, and subordinate BCTs. If the OCP is deployed and needs additional personnel or clinical skills, the MEDCOM MCP can deploy personnel to the deployed OCP. This will provide a more robust medical C2 element and additional operational and planning capabilities; increased medical logistics staff; automated medical information systems; a more diverse clinical staff; and more depth in the command judge advocate (CJA), inspector general (IG), civil affairs (CA), and public affairs sections. The MEDCOM MCP has an assigned SRC to facilitate the placement on and the integration into the TPFDL. The MCP is depicted in Figure 2-10.

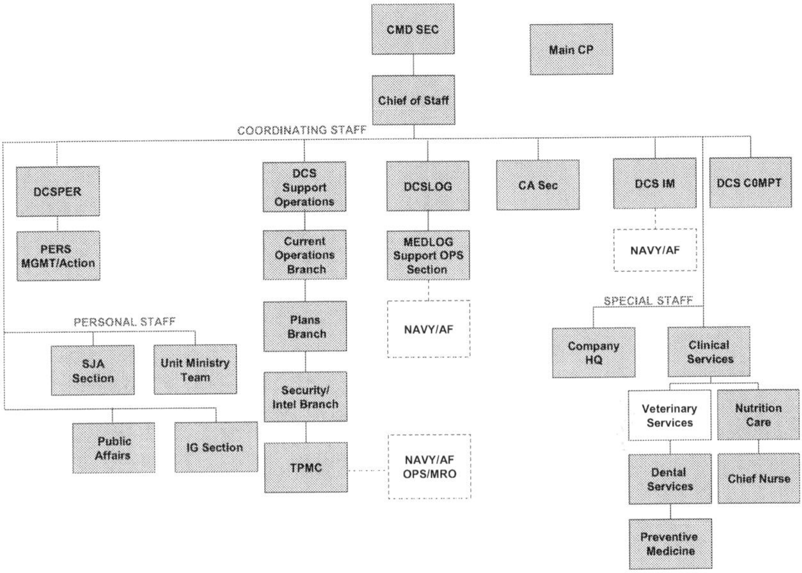

Figure 2-10. Main command post, medical command

MEDICAL BRIGADE

2-115. The MEDBDE provides a scalable expeditionary medical C2 capability for assigned and attached medical functional organizations task-organized in support of the deployed division/corps forces. The MEDBDE brings all requisite medical C2 and planning capabilities to provide responsive and effective HSS/FHP throughout the corps AO. The MEDBDE is depicted in Figure 2-11.

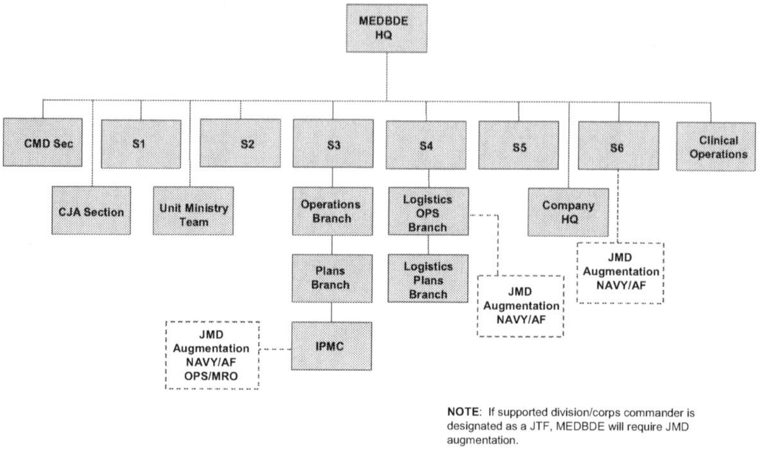

Figure 2-11. Medical brigade

2-116. The MEDBDE ensures the division/corps commander has the right mixture of medical professional (operational, technical, and clinical) expertise to synchronize the complex system of AHS functions required to maintain the health of his command by promoting fitness, preventing casualties from DNBI, and promptly treating and evacuating those injured on the battlefield.

2-117. The design and flexibility of the MEDBDE permits the division/corps commander the ability to meet expeditionary medical support requirements in support of early entry forces. As the supported corps commander's forces grow in both size and complexity, the MEDBDE can deploy additional modules that build upon one another to support full spectrum operations. The MEDBDE also provides the corps commander the appropriate medical C2 to continue to build his medical force capabilities through the integration of Army, joint, and multinational medical forces to ensure he can identify and counter any health threats in his supported areas. This will permit him to transition from expeditionary medical support operations to providing campaign-quality health care in his AO.

2-118. The MEDBDE has the capability to provide SRC elements for an early entry module (EEM), an expansion module, and the campaign module (CM), thus enabling its capability to be tailored to METT-TC factors of a specific division/corps operation.

2-119. The MEDBDE provides medical C2, administrative assistance, and technical supervision of assigned and attached medical units providing AHS in support of a division/corps. Further, it is capable of—

- Providing a rapidly responsive early entry medical C2 module that can quickly integrate into the early entry deployment sequence for crisis management.
- Providing full spectrum continuous medical C2 in support of all Army corps and joint (when augmented) forces.
- Providing operational medical augmentation to Role 2 BCT medical companies.
- Advising corps and BCT commanders on the medical aspects of their operations.
- Providing medical staff planning, operational and technical supervision, and administrative assistance for MMBs and CSHs task-organized under the MEDBDE.
- Coordinating with supporting TPMRC for medical regulating and medical evacuation from MMBs and hospitals to supporting EAC MTFs and CONUS.

- Providing medical consultation services and technical advice in—
  - Preventive medicine (medical and OEH surveillance, environmental health, sanitary engineering, and medical entomology).
  - Nursing services.
  - Dental services.
  - Behavioral health (to include COSC) and NP care.
  - Veterinary services (including food safety and inspection, animal medical care, and veterinary PVNTMED services).
  - Nutrition care and medical food service.
  - Medical laboratory support.
- Providing advice and recommendations for conduct of civil-military operations (CMO).
- Control and supervision of Class VIII supply and resupply movement to include blood management. When designated by the combatant command, serves as the SIMLM.
- Providing joint capable medical C2 capability when augmented with appropriate joint assets.
- Serves as the executive agent for veterinary services according to DODD 5200.31E.
- Coordinates ASOS for ship-to-shore/shore-to-ship medical evacuation mission.
- Plans and coordinates religious support for MEDBDE staff, assigned and attached medical units, and casualties in MEDBDE subordinate unit MTFs.

Chapter 3

Operational and Tactical Evacuation Planning

A comprehensive medical evacuation plan is essential to ensure effective, efficient, and responsive medical evacuation is provided to all wounded, injured, and ill Soldiers in the JOA. The Army medical evacuation plan flows from the GCC's guidance and intent and incorporates all missions and tasks directed by the GCC to be accomplished and is synchronized with supporting and supported units. In some scenarios, Army air and ground evacuation resources may be directed to provide support to sister Services, allied and coalition partners, and HN forces. Examples of a medical evacuation OPLAN and OPORD are contained in Appendix D.

SECTION I — THEATER MEDICAL EVACUATION PLANNING RESPONSIBILITIES

JOINT PLANNING

3-1. All medical evacuation planning within the theater reflects the GCC's guidance and intent. Planning at the joint level is governed by Joint Publications (JPs) 5-0, 5-00.1, and 5-00.2. Guidance on medical evacuation in joint operations is addressed in JP 4-02. When directed by the GCC, Army medical evacuation assets may be tasked to support other than Army forces engaged in the execution of the joint mission. These additional support missions will be clearly articulated in the GCC's OPLAN and OPORD. The ASCC surgeon, with the advice of the senior medical evacuation planner, will coordinate and synchronize these support operations with the combatant command surgeon, joint task force (JTF) surgeon, and the other Services and/or multinational partners as required to ensure that a comprehensive and effective, efficient, and responsive plan is developed and implemented.

MEDICAL COMMAND AND CONTROL ORGANIZATIONS

3-2. Army evacuation assets support the full spectrum of all Army operations and are capable of supporting Army forces engaged in the execution of the joint mission, as directed. Major combat operations supported by an ASCC will have a deployed MEDCOM. The C2 coordination and orders flow for AHS is illustrated in Figure 3-1. Evacuation planning should occur at each major level of command in order to effectively execute command, control, and influence over evacuation planning.

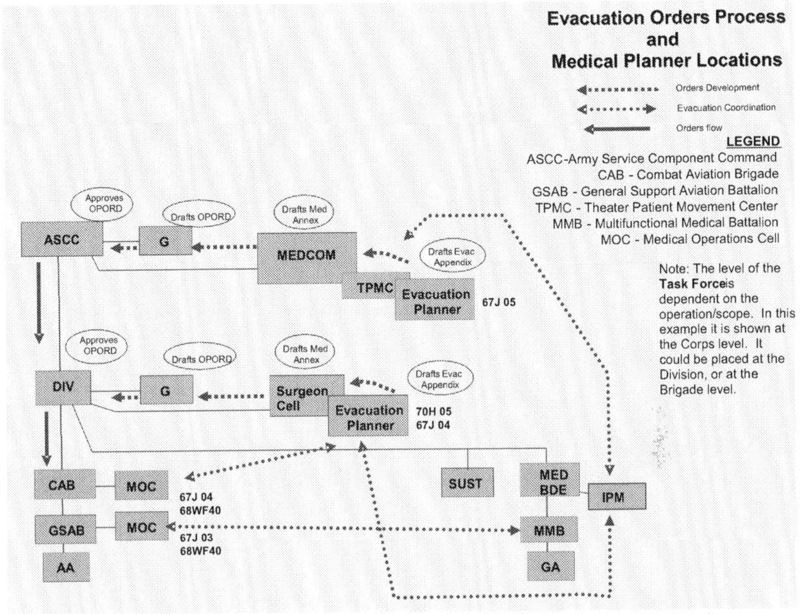

Figure 3-1. Command and control organizations for coordination and orders flow

3-3. Under the new Army modular design, some early entry operations may not require the deployment of a theater or a brigade C2 structure. In these cases, medical evacuation planning, coordination, and synchronization are initiated by the most senior medical command/staff level (such as the division surgeon, medical brigade, GSAB, and/or MMB). The senior leaders responsible for air and ground medical evacuation provide estimates and recommendations for the plan. This plan is normally provided as an appendix to the HSS annex and is disseminated down through the chain of command.

3-4. Dedicated Army rotary-wing aircraft (air ambulances) are under the C2 of the GSAB within the transformed Army aviation structure. Ground ambulances remain under the C2 of medical units/elements that are organic to movement and maneuver, fires, and sustainment units. The medical company, ground ambulance is under the C2 of the MMB.

3-5. To ensure an integrated medical evacuation system, synchronization of air with ground ambulance assets is essential. To affect this synchronization, the medical C2 headquarters must maintain influence over the execution of the medical evacuation plan. This influence is accomplished through the orders process.

THEATER PATIENT MOVEMENT CENTER

3-6. In the design of the modular MEDCOM, the medical reengineering initiative (MRI) MEDCOM's medical regulating office (MRO) staff was reevaluated, realigned, and augmented with medical evacuators to more effectively coordinate and synchronize the various ground and air evacuation aspects of the operation and to supervise and facilitate the medical regulating functions. This staff section was renamed as the theater patient movement center (TPMC). The addition of medical evacuators enables this section to develop, coordinate, and synchronize the medical evacuation portion of the MEDCOM AHS hour coordination and management responsibility for patient regulating and administration within the MEDCOM AO and directs patient evacuation between facilities within its AO as required to avoid patient overflow. This staff is responsible for preparing patient statistical reports for the command. This section also coordinates with the supporting TPMRC for the transfer of patients out of the MEDCOM/theater. The TPMRC regulates the flow of patients out of the theater to an MTF capable of providing the required specialty care and arranges for the transport of these patients by USAF aeromedical evacuation assets.

INTRATHEATER PATIENT MOVEMENT CENTER

3-7. As with the modular MEDCOM, the MRO of the modular MEDBDE was also redesigned in a similar fashion to enhance the capabilities for medical evacuation planning within the MEDBDE AO. This staff section was renamed the intratheater patient movement center (IPMC). This office is responsible for coordinating and synchronizing medical evacuation operations with the TPMC and the division surgeon cell. It regulates patients within its subordinate hospitals and arranges for transportation to transfer patients between facilities to relieve surgical backlogs, to obtain specialty care, and to ensure bed availability prior to start of operations, as appropriate. It also coordinates with the MEDCOM TPMC and the TPMRC for the evacuation of patients out of theater. This office maintains a 24-hour coordination and management responsibility for patient regulating and administration with the MEDBDE AO. It also prepares patient statistical reports for the MEDBDE commander. In the absence of the TPMC, IPMC assumes duties of the TPMC.

SECTION II — PLANNING PROCESS

EVACUATION PLANS AND ORDERS

CHARACTERISTICS

3-8. Operation plans and orders are the means by which commanders express their visualization, commander's intent, and decisions. They focus on results the commander expects to achieve. These OPLANs and OPORDs provide the basis for ensuring that medical evacuation operations are comprehensively planned, synchronized, responsive, and ensure a seamless continuum of care.

3-9. Evacuation plans and orders help form the basis commanders use to synchronize military operations. They encourage initiative by providing the *what* and *why* of a mission, and leave the *how to accomplish the mission* to subordinates. They give subordinates the operational and tactical freedom to accomplish the mission by providing the minimum restrictions and details necessary for synchronization and coordination. As medical evacuation operations are complex and may cross command and service boundaries, it is essential that the medical evacuation annex to the medical OPLAN/OPORD clearly delineates responsibilities, communications interoperability and procedures, and coordination requirements.

PLANS AND ORDERS

3-10. Publishing the commander's OPLAN/OPORD permits subordinate commanders to prepare supporting plans and orders. They can discern and implement instructions derived from a higher commander's plan and order. Appendix D provides an example of operational planning and a sample OPORD. Additionally, the commander's OPLAN/OPORD—

- Focuses subordinates' activities.
- Provides tasks and activities, constraints, and coordinating instructions necessary for mission accomplishment.
- Encourages agility, speed, and initiative during execution.
- Conveys instructions in a standard, recognizable, clear, and simple format.

3-11. The amount of detail provided in an evacuation plan or order depends on several factors, to include the experience and competence of subordinate commanders, cohesion and tactical experience of subordinate units, and the complexity of the operation.

3-12. Commanders balance these factors with their guidance and commander's intent, and determine the type of plan or order to issue. To maintain clarity and simplicity, plans and orders include annexes only when necessary and only when they pertain to the entire command. Annexes contain the details of support and synchronization necessary to accomplish the mission.

3-13. For a complete discussion of the military decision-making process and development of OPLANs/OPORDs refer to FM 5-0 and FMI 5-0.1.

Medical Evacuation Planning

3-14. Medical evacuation planning which supports the HSS/FHP plan is an ongoing process and is accomplished by medical operations element at battalion, brigade, division, corps, and theater levels. The HSS/FHP operations planning process must take under consideration all issues that could affect or influence HSS/FHP support for tactical operations. These issues should be considered in the initial developmental stages of theater, corps, division, or brigade plan as the supporting HSS/FHP annex is developed.

3-15. Planning is a dynamic and continuous process. The framework established in a robust initial plan can be built on in subsequent phases of an operation. The evacuation plan for a given AO will require adjustment and change during different phases of an operation. For example, transitioning from offensive operations to stability operations will generally dictate a significant change in evacuation coverage priorities.

3-16. Medical evacuation planning considerations, although essentially the same at all levels of command, may entail a broader scope, complexity, and detail at the higher levels of command due to the coordination and synchronization requirements which exist across command lines and service or multinational boundaries. Table 3-1 provides medical evacuation planning considerations at various levels of command based on METT-TC.

Table 3-1. Medical evacuation planning considerations

Mission	1. Define the protocol for a valid evacuation mission (mission statement). This protocol will serve as the authority to execute the mission by each medical evacuation unit/element/crew. 2. Use the HSS/FHP annex and mission analysis to: • Task evacuation units where needed. This may involve tasking part or all of DS units to perform a GS or area support mission. 3. Based on the type of operation: • Define trigger points for changes in the evacuation support plan. • Develop a concept of operations for all of the corps evacuation resources. 4. Define task organization of units assigned to the corps (and when directed to the JTF) and support relationships, if possible. Lines of coordination and mutual support relationships should be clearly delineated. 5. Area support and GS coverage areas will be

	defined with flexibility built into the plan. Coverage areas along division boundaries, for example, should not restrict or hinder evacuation of neighboring units that are geographically close. 6. Define pilot certifications for shore-to-ship medical air evacuation. 7. Develop standing operating procedure (SOP) requirements for single versus dual ship medical air evacuation. Address theater-specific procedures for in-flight link-up between medical evacuation aircraft and nonmedical evacuation aircraft for the purpose of dual ship requirements and/or security of pick-up zone. 8. Delineate any nonstandard air evacuation missions and allocate resources accordingly, such as support to Army special operations, joint and coalition forces. 9. Coordinate with any units that will have shared responsibilities for evacuation in a specific area and/or a specific mission. 10. Evaluate medical evacuation specific A2C2 considerations. Specifically, coordinating the crossing of brigade/division/corps boundaries/ area of operations. Medical evacuation must be afforded special expedited procedures for medical evacuation missions. Additionally, METT-TC dependent air corridors may have to be developed for the medical evacuation mission.
Enemy	1. Define the level or air-to-air threat and the threat faced by ground evacuation resources and the impact of this threat on the overall evacuation plan. This will be a major consideration in dictating which means of evacuation will be the primary means. 2. Based on the estimate of the number of enemy prisoners of war (EPW) or other categories of detainees involved in the operation— • Define procedures for evacuating wounded, injured, or ill persons through medical channels to include responsibilities of the echelon commander for designating guards for those evacuated directly from the point of capture to an MTF. • Allocate assets as necessary to provide this support and to support detainee collection points and holding areas.
Terrain and Weather	Define the types of terrain in the theater and use information as a major consideration in dictating the primary means of evacuation (ground or air in

	each sector) and any factors that may impair evacuation efforts.
Troops and Support Available	1. Disseminate locations of facilities used in evacuation, MTFs, MASFs, and aerial ports of debarkation (APOD). 2. Disseminate the capabilities and availability of joint evacuation assets and how to request their assistance. 3. Collect contact information from all evacuation units, assemble and publish for theater use. 4. Standardize a mandatory evacuation mission data collection format and collection method and schedule.
Time Available	1. Define the acceptable limits of evacuation time based on the distance from evacuation units. 2. Determine the proper allocation of resources to support the entire theater requirements. 3. Plan for trigger points (in time or distance) for changes in the evacuation plan to occur. 4. Define phases of the operation that reflect the commander's intent. Major changes in evacuation coverage usually occur when changes occur in the operational tempo (OPTEMPO). 5. Define briefing levels for launch authority. Consider developing a uniform theaterwide aviation risk assessment sheet.
Civil Considerations	1. Based on the HSS/FHP estimate and plan, determine requirements for the evacuation of HN civilians and others. 2. Define and disseminate protocols and procedures for evacuating civilians. 3. Allocate resources to support this mission, when directed.

3-17. For an effective HSS/FHP plan, medical staffs must coordinate and share operations information in order to synchronize support operations for mission accomplishment. See FMs 4-02, 4-02.1, 4-02.6, 4-02.12, 4-02.17, 4-02.25, 8-42 (4-02.42), and 8-55 (4-02.55), for doctrinal guidance on HSS/FHP operations. After the HSS/FHP plan is completed, it is incorporated into the C2 headquarters plan. After all annexes and appendixes of the plan are approved by the commander, it is incorporated into the OPORD.

SURGEON'S RESPONSIBILITIES

3-18. The surgeon is responsible for supervision and development of AHS input for the OPORD. The HSS plan serves as the base document for this input. The HSS plan is revised or updated based on mission analysis or changes in HSS requirements. The surgeon is tasked by the Assistant Chief of Staff (Operations and Plans) (G3)/S3 for HSS/FHP input to the OPORD for support of operations. The G3/S3 indicates timeline requirements. The surgeon and his staff are involved in all stages of the planning process, to identify all HSS requirements. Information on development of running estimates and the OPLAN/OPORD are discussed below. The medical plan/operations cell develops a HSS/FHP plan based on guidance received from the commander and the surgeon and provides HSS/FHP operational planning updates to the surgeon. The plan is briefed to the commander for approval, as required and provided to the G3/S3 according to the format of the in FM 8-55 (4-02.55).

MEDICAL EVACUATION TOOLS

3-19. It is essential that the evacuation plan for all combat operations be well planned, coordinated, and disseminated. In designing the medical evacuation plan, the medical planner should consider the following:

- Casualty collection points normally are predesignated along the axis of advance or evacuation routes.
- Forward of the BAS, the trauma specialist, CLS, and combat troops take casualties to the CCPs. These points facilitate acquisition by supporting ambulance teams and reduce evacuation time.
- When used by the BAS, CCPs help preserve BAS mobility, preclude carrying casualties forward, and reduce evacuation time to the sustainment area. When designating a CCP, the designating authority makes a decision whether or not to provide medical staff at the location. This decision is based upon the assessment of risk versus the availability of personnel.
- Normally, the role of care designating the point is responsible for staffing. Medical personnel may not be available to staff these points, and CLSs and ambulatory patients may be required to perform self-aid, buddy aid, or enhanced first aid. Casualty collecting points should be identified on operational overlays.

Note. A CCP staffed by a trauma treatment team is designated as a BAS (minus) rather than as a CCP.

3-20. Ambulance exchange points are a predetermined position where patients are exchanged from one evacuation platform to another.

3-21. These points are normally preplanned and are a part of the HSS/FHP annex to the OPLAN. In the forward area, the threat of enemy ground activities, large concentrations of lethal weapons systems, and effective use of antiaircraft weapons may dictate that the AXP be a predetermined rendezvous point for the rapid transfer of patients from one evacuation platform to another. The location of AXPs should be frequently changed to preclude attracting enemy fires.

3-22. Ambulance exchange points are established for many different reasons:

For example, the ambulance platoon of the heavy BSMC possesses a mixture of wheeled and tracked ambulances.

- The brigade tracked ambulances are provided so that they may keep up with maneuver elements. These vehicles carry the patients from the BAS to an AXP where the brigade wheeled ambulances take over for the relatively longer trip to the sustainment area.
- Ambulance exchange points are not limited to ground evacuation assets. Another example is a situation where the threat air defense artillery capability is such that air ambulances cannot fly as far forward as the BASs. However, an AXP could be established a few kilometers to the rear, still well forward of the BSA.
- The brigade combat team (BCT) tracked or wheeled ambulances could then transfer the patients to the air assets, thereby facilitating the rapid evacuation of patients and realizing a significant time savings.

3-23. By using AXPs evacuation assets are returned to their supporting positions faster. This facilitates evacuation as the returning crews are familiar with the road network and the supported unit's tactical situation. In the case of air ambulance (AA) assets, it is important because of the requirements for integration into the A2C2 system at each role and the enhancement to survivability provided by current threat and friendly air defense information.

3-24. The ambulance shuttle system is an effective and flexible method of employing ambulances during combat. It consists of one or more ambulance loading points, relay points, and when necessary, ambulance control points, all leveled forward from the principal group of ambulances, the company location, or basic relay points as tactically required. The ambulance—

- Loading point is a point in the shuttle system where one or more ambulances are stationed ready to receive patients for evacuation.
- Relay point is a point in the shuttle system where one or more empty ambulances are stationed. They are ready to advance to a loading point or to the next relay post to replace an ambulance that has moved from it. As a control measure, relay points are generally numbered from front to rear.
- Control points consists of a Soldier (from the ambulance company or platoon) stationed at a crossroad or road junction where ambulances may take one more directions to reach loading points. The Soldier, knowing from which location each loaded ambulance has come, directs empty ambulances returning from the sustainment area. The need for control points is dictated by the situation. Generally, they are more necessary in forward areas.

3-25. In the establishment of the ambulance shuttle system, once the relay points are designated, the required numbers of ambulances are stationed at each point. If the tactical situation permits, the ambulances may be delivered to the relay points by convoy.

3-26. Staffing of relay, loading, and ambulance control points. Important points may be manned to supervise the blanket, litter, and splint exchange and to ensure that messages and medical supplies to be forwarded are expedited.

3-27. Advantages of the ambulance shuttle system are that the system—

- Places ambulances at CCPs and BASs as needed.
- Permits a steady flow of patients through the system to MTFs.
- Avoids unnecessary massing of transport in forward areas.
- Minimizes the danger of damage to ambulances by the enemy.
- Permits the commander or platoon leader to control his elements and enables him to extend their activities without advancing the headquarters.
- Facilitates administration and maintenance.
- Maximizes the use of small battle command elements (sections or platoons) to operate the ambulance shuttle without employing the entire parent unit.
- Provides for flexible use of other ambulance assets for specific situations.

Obstacles Marking

3-28. Ambulance crews must know and recognize the standard land/gap marking patterns. Unit TSOPs on the types of materials used should be available to ambulance crews. Refer to FM 3-34.2 for additional information.

Chapter 4
Army Medical Evacuation

Medical evacuation operations are planned to provide comprehensive, responsive, flexible, and agile support to the tactical commander in conformity with the commander's intent and OPLANs. This chapter discusses the employment of medical evacuation resources and the coordination and synchronization required to effectively execute medical evacuation operations (to include the transfer of patients between MTFs and to staging facilities) by air and ground evacuation assets.

SECTION I — MEDICAL EVACUATION SUPPORT

EVACUATION PRECEDENCE

4-1. Casualties requiring evacuation are prioritized to ensure the most seriously injured or ill receive timely medical intervention consistent with their medical condition. As with medical treatment, medical urgency is the only factor used to determine the medical evacuation precedence. (Appendix A provides an in-depth discussion of the provisions of the Geneva Conventions.)

This paragraph implements STANAG 3204.

4-2. The determination to request medical evacuation and assignment of a precedence is made by the senior military person present or, if available, the senior medical person at the scene. This decision is based on the advice of the senior medical person at the scene (if available), the patient's condition, and the tactical situation. Assignment of a medical evacuation precedence is necessary. The precedence provides the supporting medical unit and controlling headquarters with information that is used in determining priorities for committing their evacuation assets. For this reason, correct assignment of precedence cannot be overemphasized; over classification may result in an increase in evacuation which could burden the HSS system.

4-3. The patient's medical condition is the overriding factor in determining the evacuation platform and destination facility. The AA operates wherever needed on the battlefield, dependent on risk and METT-TC factors. The crew of the AA, assisted by onboard patient monitoring and diagnostic equipment, is trained in aeromedical procedures to provide optimum en route patient care. It is the preferred method of evacuation for most categories of patients. Air ambulances are a low-density, high-demand resource and must be managed accordingly. To conserve these valuable resources, medical planners should plan to use AA to primarily move Priority I, URGENT and Priority IA, URGENT-SURG patients with the other categories on a space-available basis. Health service support planners must plan for a synchronized air and ground evacuation plan. Depending on the length of time required for an AA to be dispatched and arrive at the POI, it may be prudent to evacuate the casualty by ground evacuation assets to a BAS and/or Role 2 MTF for stabilization by a physician. Further evacuation could then be accomplished by AA.

4-4. Patients will be picked up as soon as possible, consistent with available resources and pending missions. Table 4-1 depicts the categories of evacuation precedence and the criteria used to determine the appropriate precedence.

Table 4-1. Categories of evacuation precedence

Priority I—URGENT	Is assigned to emergency cases that should be evacuated as soon as possible and within a maximum of 2 hours in order to save life, limb, or eyesight, to prevent complications of serious illness, or to avoid permanent disability.
Priority IA—URGENT-SURG	Is assigned to patients who must receive far forward surgical intervention to save life and to stabilize them for further evacuation.
Priority II—PRIORITY	Is assigned to sick and wounded personnel requiring prompt medical care. This precedence is used when the individual should be evacuated within 4 hours or his medical condition could deteriorate to such a degree that he will become an URGENT precedence, or whose requirements for special treatment are not available locally, or who will suffer unnecessary pain or disability.
Priority III—ROUTINE	Is assigned to sick and wounded personnel requiring evacuation but whose condition is not expected to deteriorate significantly. The sick and wounded in this category should be evacuated within 24 hours.
Priority IV—CONVENIENCE	Is assigned to patients for whom evacuation by medical platform is a matter of medical convenience rather than necessity.
The NATO STANAG 3204 has deleted the category of Priority IV—CONVENIENCE; however, it will still be included in the US Army evacuation priorities as there is a requirement for it on the battlefield.	

SECTION II — MEDICAL EVACUATION REQUESTS

4-5. This section discusses requests for medical evacuation. Specific procedures, frequencies, and security requirements for transmittal of medical evacuation requests are delineated through the orders process and are made a part of the unit/command SOPs. Each sector based on the METT-TC may be designated with a different method of evacuation as the primary means to effect evacuation. In sectors which have a high ground-to-air or air-to-air threat may rely on ground evacuation assets to move the majority of patients. In other sectors where the ground threat is high and comprised of small arms, improvised explosive devices, and bombs, medical evacuation operations may be more efficiently and effectively executed by AAs. An additional consideration in planning medical evacuation operations is to determine whether armed escorts are required for either the ground or AA mission. Those missions that require armed escort must be thoroughly coordinated and synchronized between the medical assets and force protection assets that will accompany them.

4-6. Medical evacuation requests often are sent from the POI, through intermediaries, such as higher headquarters, who then transmit the request up to the nearest medical evacuation unit. The unit relaying the request must ensure that it relays the exact information originally received. The radio call sign and frequency relayed (Line 2 of the request) should be that of the requesting unit and not that of the relaying unit. However, the intermediaries contact information can be given as additional information if a callback is necessary to clarify details of the mission. Figure 4-1 depicts the communication flow for evacuation requests.

4-7. The necessity for secure communication in transmitting a 9-line request is METT-TC dependent. A technically capable enemy may be able to intercept a nonsecure 9-line request and strike at the pickup zone causing additional harm to the patients and the medical evacuation crew. In combat and when directed by higher headquarters, always use secure communications when requesting medical evacuation.

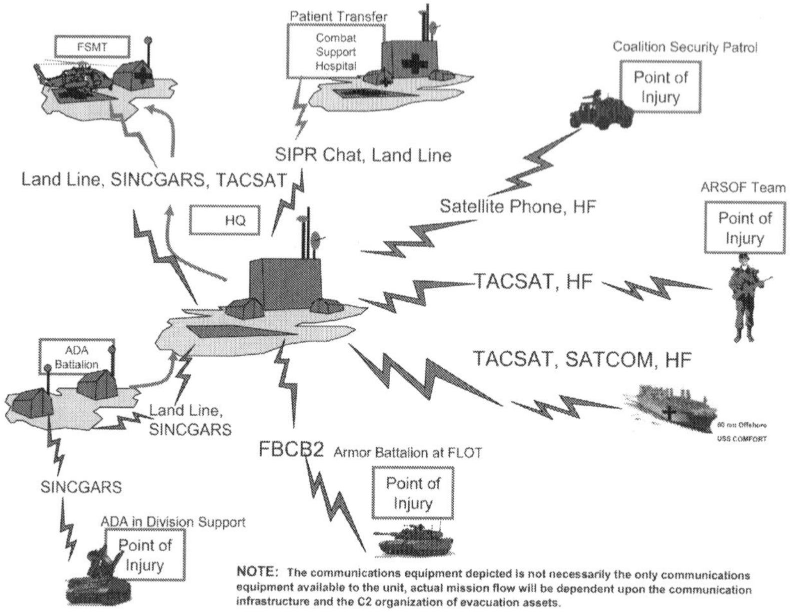

Figure 4-1. Flow of communication for evacuation request

GROUND EVACUATION REQUEST

4-8. Figure 4-2 provides an example of a request from the POI for ground evacuation. The patient is a ROUTINE precedence patient requiring evacuation from the POI to the supporting BAS (Role 1 MTF) with follow-on evacuation to an AXP for further evacuation to the rear. Should the patient require further treatment, either an air or ground evacuation could be used.

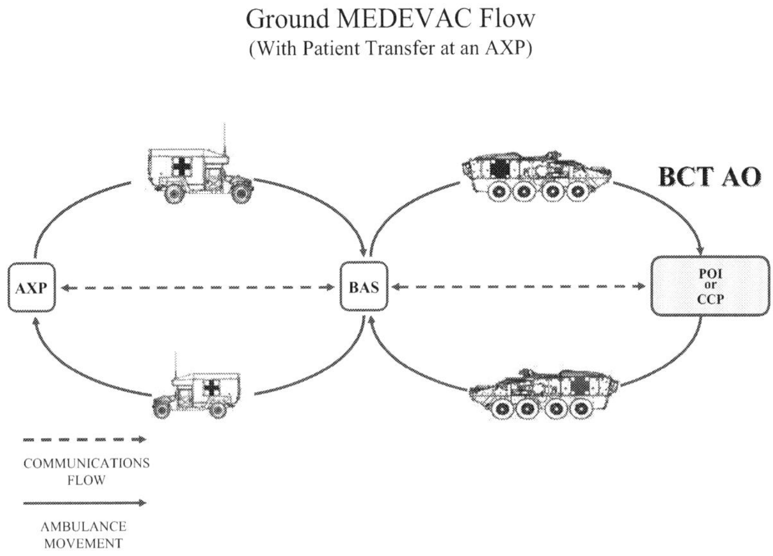

Figure 4-2. Ground evacuation request in a maneuver unit

4-9. Ground medical evacuation assets also provide DS and GS to supported units in the EAC. Figure 4-3 and Figure 4-4 depict the processes of providing DS to an ASMC and the GS support role of transferring a patient from a corps hospital to a MASF for further evacuation out of the theater.

4-10. In the corps area, the ASMC provides Roles 1 and 2 medical support on an area basis. The ASMC has organic evacuation assets and receives augmentation when required from the medical company ground ambulance assigned to the MMB.

4-11. Medical evacuation resources are also used to transfer a patient between hospitals within the theater and from a hospital to MASF for further evacuation out of theater. In Figure 4-3, this mission is being accomplished by a ground ambulance. However, depending upon the distance from the hospital to the MASF and/or APOD, this mission can also be accomplished by AA assets.

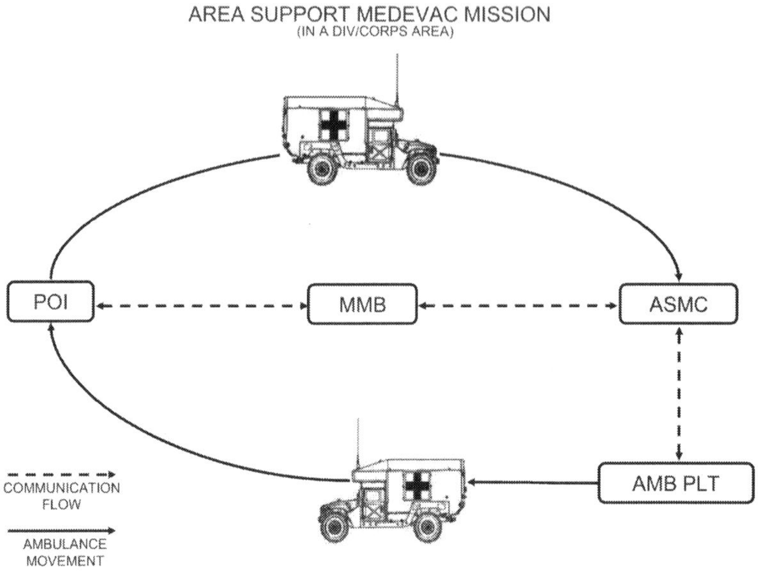

Figure 4-3. Ground evacuation request in the corps

AIR EVACUATION REQUEST

4-12. Air ambulances may fly as far forward as possible on the battlefield. Although evacuation by AA is the preferred means for all casualties, when high evacuation workloads exist, evacuation by AA should be the primary means used for URGENT and URGENT-SURG patients. Figure 4-5 depicts the zones of evacuation. The GSAB/CAB will place AA assets where they can best support the tactical commander's plan and facilitate the timely and responsive evacuation. To coordinate and synchronize the ground and air medical evacuation plans, a staff element was designed within the GSAB headquarters. This cell consists of the medical planners on the GSAB and CAB staff. The medical operations cell (MOC) will:

- Establish flight procedures specific to air evacuation missions within the CAB. This may include special routes or corridors as well as procedures for escort aircraft linkup.
- Ensure lines of communication (LOC) to supported units and higher levels of medical command are available. Further the MOC ensures supported units understand medical evacuation procedures and capability.
- Establish medical evacuation briefing and launch procedures. Ensure there is 24-hour access to those able to launch high and very high-risk missions.
- Maintain awareness of tactical and medical situation. Coordinate with MROs at higher levels to efficiently conduct the general support mission and to work in concert with adjacent units.
- Assist the AA CO and GSAB/CAB staff conduct medical evacuation operations.

- Coordinate missions with supported command surgeons. To ensure situational awareness and coordination of medical evacuation efforts, this staff keeps command surgeons informed on air evacuation missions being executed in their AO.
- Consults and coordinates with supported command surgeons when AAs cannot be launched to execute a requested mission. Ensures that the appropriate command surgeon is notified to ensure mission can be accomplished by ground evacuation assets.

4-13. Figure 4-4 depicts a request for air medical evacuation. The scenario shows the request being sent to the FSMT, the AA company, and the GSAB. The routing of the request would be dependant on how our forces are arrayed. In a similar situation where ground evacuation may not be readily available or the tactical situation precludes its use the requesting unit could transmit their medical evacuation request directly to a supporting FSMT as a primary means for a medical evacuation request or the AA company as a secondary option with a third being sent directly to the GSAB.

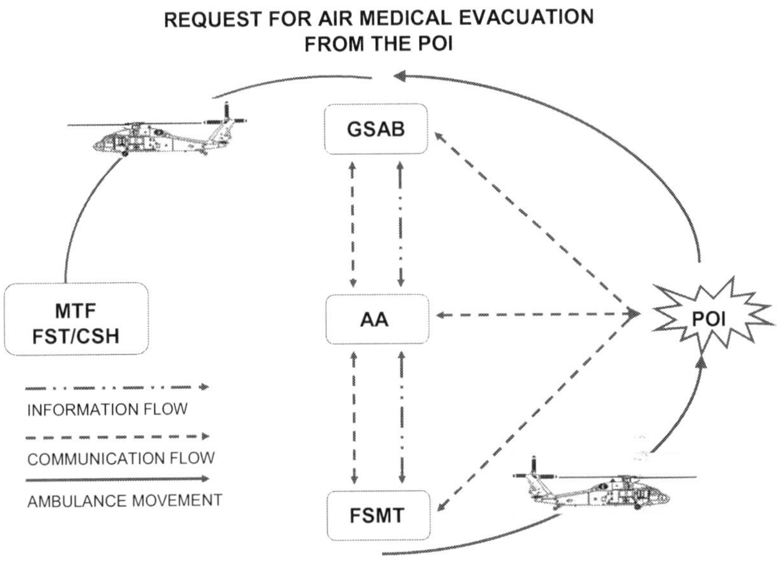

Figure 4-4. Air evacuation request in the corps

4-14. Figure 4-5 illustrates evacuation zones during a noncontiguous conflict. Here, no forward line of own troops (FLOT) is clearly visible.

- Forward support medical evacuation teams are placed where they are most needed. This can be with troops most often engaged in combat, high population density areas, area of famine or disease with high civilian casualties, refugee areas, geographically centralized location, and such.

- Area or general support FSMTs accomplish the patient transfer mission that develops between FSMTs and MTFs, between MTFs, and between MTFs and intertheater movement locations. These FSMT must also be responsible for the DS mission in their immediate vicinity.

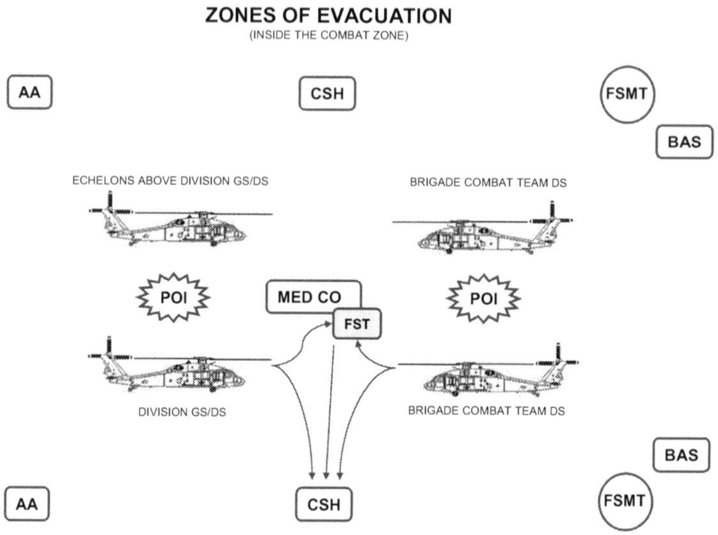

Figure 4-5. Air ambulance zones of evacuation

4-15. Figure 4-6 illustrates a comparison in aviation mission planning/execution cycle for medical evacuation and typical aviation.

- In contract to a typical aviation mission cycle, continuous aeromedical evacuation coverage results in extended operational duty days that often exceed 24 hours in length in order to provide seamless coverage to our forces. Medical evacuation units must plan and develop a detailed battle rhythm that addresses composite risk management, mission execution processes, and tactics, techniques, and procedures unique to 24-hour continuous evacuation operations.
- A medical evacuation crew cycle begins much like any other mission planning however they will not know the location or the time of execution for their mission. Since during these cycles it is critical the crew responds as rapidly as safely possible when the last information is received, it is imperative that crews are afforded the ability to manage rest while ensuring the procedures are in place to maintain battlefield situational awareness and facilitate a rapid, safe execution when a 9-line mission is received. This includes but is not limited to crew rest cycles, briefing procedures, and aviation and logistical support to remote/split-based crews to ensure the safe and timely execution of missions.

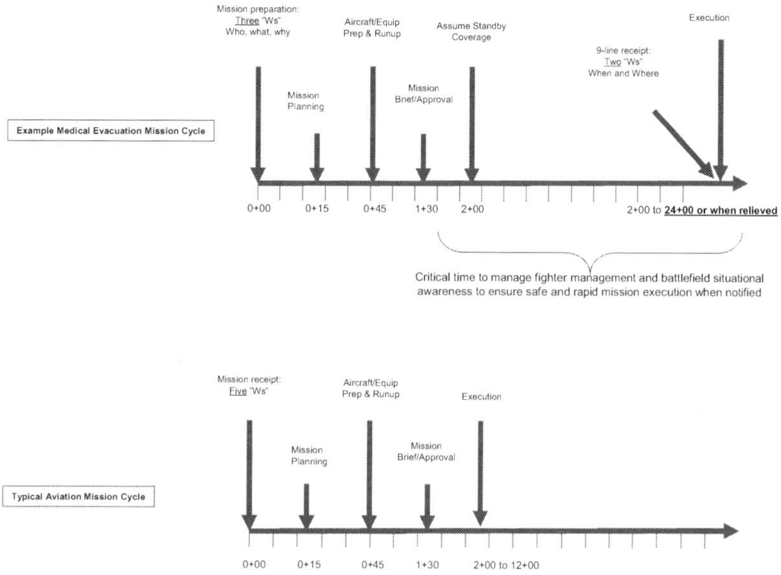

Figure 4-6. Aviation mission planning/execution cycle

JOINT INTERCONNECTIVITY

4-16. When directed by the GCC, Army evacuation assets may be used to evacuate patients from the other services or may be placed in direct support of other service, allied, or coalition units participating in the operation. Prior to the initiation of the operation, communications, and procedural issues should be synchronized to ensure the seamless execution of the operation. Interoperability issues can result in slower response times and may adversely effect medical evacuation operations. Figure 4-7 depicts the communications and patient flow in a joint scenario.

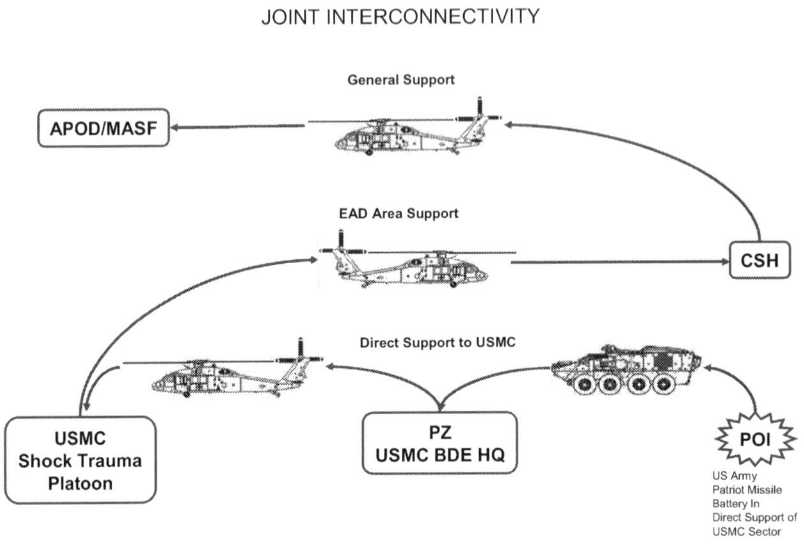

Figure 4-7. Air and ground ambulance evacuation in a joint environment

SECTION III — MEDICAL EVACUATION UNITS, ELEMENTS, AND PLATFORMS CONSIDERATIONS

MEDICAL EVACUATION SUPPORT PROTOCOL

4-17. Medical evacuation is performed by the higher role of medical care supporting forward and evacuating from the lower role. Medical evacuation support for theater, corps, and division is provided by organic ambulance teams, squads, and platoons. In addition, it is provided by the MMB medical company, ground ambulance medical company (GAMC), ASMC, and the GSAB medical company, AA.

4-18. The MMB maximizes the effective employment of ground ambulances resources whether in support of a division, corps, or at theater level. It coordinates through the combat aviation brigade's GSAB for effective use of AA support of its ASMCs. The medical brigade, MMBs, GSAB, sustainment brigade, BCTs and division surgeon section (DSS) coordinate and synchronize medical evacuation operations for the division AOs. The DSS coordinates with the MEDBDE for the evacuation of patients from division MTFs to supporting CSHs. The MEDBDE coordinates and arranges the required intratheater medical evacuation of patients to and from supporting CSHs. The MEDCOM coordinates for evacuation of patients out of the theater. To ensure that patients are evacuated to the appropriate MTFs, MROs are organic to the MEDCOM, and MEDBDE.

MEDICAL EVACUATION PLATFORMS FACTORS

4-19. Evacuation platforms must be capable of keeping pace with the troops supported. The patient's medical condition is the overriding factor in determining the evacuation platform and destination MTF. The AA operates wherever needed on the battlefield, dependent on risk and METT-TC factors. Use of hardened armored MEV may be the vehicle of choice for some missions for short evacuation to an MTF or to a secure AXP for transfer to an air or wheeled ground ambulance. The criteria for an efficient and effective evacuation system were identified in Chapter 1.

4-20. The crew of the AA, assisted by onboard patient monitoring and diagnostic equipment, is trained in aeromedical procedures to provide optimum en route patient care. It is the platform of choice for most categories of patients. However, insufficient numbers of AAs are available to evacuate all patients expected in a battle zone. To conserve these valuable resources, medical planners should plan to use AAs to primarily move Priority I, URGENT, Priority IA, URGENT-SURG, and Priority II patients with other categories on a space-available basis.

SECTION IV — MEDICAL EVACUATION AT UNIT LEVEL

CONSIDERATIONS FOR MEDICAL EVACUATION MISSIONS

4-21. As METT-TC factors affect all medical evacuation missions and the employment of ambulance assets. The medical commander must consider the basic tenets that influence the employment of medical evacuation assets. These factors include the patient's medical condition and the:

- Tactical commander's plan for employment of combat forces.
- Enemy's most likely course of action.
- Anticipated patient load.
- Expected areas of patient density.
- Availability of medical evacuation resources to include ground and air crews.
- Availability, location, and type of supporting MTFs.
- Protection afforded medical personnel, patients, and medical units, vehicles, and aircraft under the provisions of the Geneva Conventions.
- Army airspace command and control plan.
- Engineer obstacle plans.
- Fire support plan (to ensure medical evacuation assets are not dispatched onto routes and at the times affected by the fire support mission).
- Road network/dedicated medical evacuation routes (contaminated and clean).
- Weather conditions.

PATIENT ACQUISITION

4-22. Units with organic medical evacuation assets have the primary responsibility for patient acquisition. Methods of employment and evacuation techniques differ depending upon the nature of the operation.

4-23. Units without organic ambulance assets are provided medical evacuation support on an area support basis. Units must develop techniques which facilitate the effective employment of their combat medics/company/platoon medics, enhance the ability to acquire patients in forward areas, and rapidly request medical evacuation support. The techniques developed should be included in the unit TSOP. As a minimum, the TSOP should include the—

- Vehicle assignment for the organic medical personnel.
- Vehicles designated to be used for casualty transport and/or patient evacuation.
- Procedures for requesting medical evacuation support (during routine operations or during mass casualty situations).
- Role of the first sergeant (1SG), platoon sergeants, and CLSs in medical evacuation.

MANEUVER BATTALION MEDICAL PLATOONS

4-24. The medical platoon leader (a physician) along with the medical platoon field medical assistant should be included in all battalion tactical planning. This officer—

- Maintains knowledge of the concept of operations, commander's intent, and the anticipated medical requirements.
- Develops the HSS/FHP plan (FM 8-55 [FM 4-02.55] and FM 4-02.4) and provides HSS/FHP overlays with preplanned evacuation routes and CCPs, to ambulance squads or teams for inclusion in the battalion OPLAN.
- Ensures that the platoon sergeant provides strip maps or other navigational tools to the ambulance drivers, if needed.
- Requests augmentation support from the supporting medical company in advance of the operation, if required. When elements of other maneuver battalions are attached as part of a task force, then ensures that adequate medical elements are included in the support package.
- Ensures that orientation and support are provided for the medical personnel. This precludes taxing the medical elements of the receiving unit. These responsibilities are normally delegated to the medical operations officer (field medical assistant).

4-25. The ambulance/evacuation section squad leader/NCO ensures that the ambulance teams have a working knowledge of the terrain features in the AO. Whenever possible, he familiarizes with primary and secondary medical evacuation routes through route reconnaissance conducted by the squad leaders. This platoon sergeant and squad leader manages the employment of the ambulance teams and monitors the communications net to remain abreast of the tactical situation.

4-26. The following factors should be considered when selecting ambulance routes:

- Tactical mission.
- Coordinating evacuation plans and operations with the unit movement officer.
- Security of routes and security escort.
- Availability of routes.
- Physical characteristics of roads and cross-country routes (to include natural obstacles).
- Requirements to traverse roads in urban areas and potential obstructions from rubble and debris.
- Traffic density.
- Time and distance factors.
- Proximity of possible routes to areas that may be subject to enemy fire.
- Lines of patient drift.
- Cover, concealment, and available defilade for moving and stationary vehicles.
- Engineer obstacle plans.
- Fire support plan (to ensure medical evacuation assets are not dispatched onto routes and at the times affected by the fire support mission).

4-27. Depending upon the combat situation, the modes of evacuation may include walking soldiers who are wounded, manual and litter carries, nonmedical transportation assets, or dedicated medical evacuation platforms. Evacuation in the battalion area normally depends on the organic ambulances assigned. Evacuation by AA is dependent upon the patient's medical condition, availability of air assets, tactical situation, air superiority, and weather conditions.

4-28. The ambulance team is normally deployed to each maneuver company trains and one remains with the Role 1 MTF/BAS in the combat trains. Ambulances operate as far forward as the tactical situation permits. An ambulance team operating in a maneuver company AO, is normally under the tactical control of the maneuver company executive officer or 1SG. This ambulance team, however, remains under the OPCON of the medical platoon.

4-29. The medical operations officer ensures that the ambulances are located close to the anticipated patient workload. An ambulance team consists of one ambulance and two combat medics (on track vehicles, a third emergency care specialist is required to provide en route medical care). One or two of these teams serve in DS of a maneuver company. To become familiar with the specific terrain and battlefield situation, the team maintains contact with the supported maneuver company during most combat operations. Maneuver medical platoon ambulances not in DS are positioned strategically throughout the battalion area or are sited at the BAS to—

- Evacuate patients from the company aid posts, CCPs to the BAS.
- Reinforce the forward positioned ambulance teams.
- Support the combat forces held in reserve and/or scout, mortar and forward observer platoons.

4-30. Another employment option is to forward site the additional ambulance teams at company CCPs, as well as at the BASs. Many times the ambulance team finds battlefield casualties who have not been seen by a trauma specialist. In these cases, the team members dismount and then find, treat, and evacuate the patients. Ambulance teams not specifically dedicated to support combat elements can be used as messengers in medical channels and to provide transport of emergency medical personnel, equipment, and supplies.

4-31. During static situations where the maneuver company is not in enemy contact or is in reserve, the ambulance team returns to the BAS to serve as reinforcement to other elements in contact. However, during movement to contact, the ambulance team immediately deploys with its supported unit. In moving patients back to the CCP, the team may be assisted by nonmedical personnel. Specific duties of the ambulance team are to—

- Maintain contact with supported elements.
- Find and collect the wounded.
- Administer EMT.
- Initiate or complete the FMC.
- Evacuate patients to the BAS.
- Direct or guide ambulatory patients to the BAS.
- Resupply company/platoon medics.
- Serve as messengers in medical channels.
- Route recons from Role 1 MTF to Role 2 MTF or the higher role of medical care.

Casualty Collection Points

4-32. During the offense, CCPs may be used to avoid hampering the movement of the maneuver elements. In fast-moving situations, preplanned CCPs are included in the HSS plan and activated based on the crossing of phase lines, upon the occurrence of predetermined events, or on the execution of other control measures. It may be necessary to set up multiple CCPs for each phase of an operation. Rotating the use of these points precludes the enemy from using them to pinpoint maneuver elements or from attracting enemy fires. When the situation permits, medical evacuation of patients from CCPs, or AXPs may be accomplished by AAs.

4-33. Ambulance teams move using available terrain features for cover and concealment. They avoid prominent terrain features and likely targets. When stationary, the ambulance crew should conceal the vehicle as much as possible. When a casualty occurs in a tank, Bradley infantry fighting vehicle (BIFV), a Stryker wheeled armored vehicles, the ambulance team moves as close to the armored vehicle as possible. Assisted by the armored crew, if possible, the casualty is extracted from the vehicle and then administered EMT. The ambulance team moves the patient to the Role 1 MTF/BAS or to a CCP point to await further evacuation.

4-34. The senior company medic/trauma specialist normally remains with the company combat trains, but may be used anywhere in the company area, even assisting the ambulance teams in some situations. He may be used to direct ambulance teams to locations where vehicle crews need assistance or where injured or wounded crew members have been left. In some situations, crew members may have to rely on self-aid or buddy aid until the CLS or the platoon medic/trauma specialist arrives.

Area Medical Evacuation Support

4-35. Medical evacuation on an area basis is required at all roles in the AHS system. The division units without organic medical evacuation resources, such as combat engineers, will require evacuation support on an area basis. To ensure that these elements receive adequate support, the medical planner must include their requirements into the OPLAN. Prior coordination is essential to ensure that the locations of CCPs, AXPs, and BASs are disseminated to these elements and that any unique support requirements are included in the OPLAN. Further, the triggers for relocating these assets and their next locations should also be included.

MEDICAL EVACUATION SUPPORT FROM ROLE 1 AND ROLE 2 MEDICAL TREATMENT FACILITIES

4-36. Evacuation from Role 1 MTFs operated by treatment squad or team is normally provided by the BSMC/FSMC ambulance platoon and the GSAB AA company. Further, these ambulance assets provide medical evacuation support on an area basis to other units in the brigade area, as required.

Ground Ambulances

4-37. The ambulance teams from the ambulance platoon are normally collocated with the BSMC/FSMC treatment platoon for mutual support. They establish contact and locate one ambulance team with the medical platoon of each maneuver battalion. The remaining ambulances are used for BCT operations and area support. The ambulances are positioned at AXPs or CCPs, or are field-sited with Role 1 MTFs operated by the BSMC/FSMC treatment teams.

4-38. A FSMT from the supporting GSAB medical company, AAs provides DS AA support for the BSMC/FSMC Role 2 MTF. The teams normally remain under the operational control of the GSAB. The DS relationship provides authority to the BSMC/FSMC in coordination with the brigade support battalion/forward support battalion (BSB/FSB) distribution support section, brigade surgeons section to direct the integrated air and ground medical evacuation support for the BCT. The section leader of the FSMT should be included in the brigade tactical planning process.

Air Ambulances

4-39. The AA team evacuates Priority I, URGENT patients from as far forward as possible to the BSMC/FSMC. Further, when a FST is colocated with a BSMC/FSMC, AAs evacuate Priority IA, URGENT-SURG to this facility if the patient can not be flown to a Role 3 MTF condition due to distance or patient condition. External lift capabilities of aeromedical evacuation helicopters add an important dimension to its role on the battlefield in moving medical supplies and equipment.

4-40. The GSAB AAs may operate from the sustainment area and BSAs providing around-the-clock, immediate response, evacuation aircraft, as well as provide the BSMC/FSMC commander flexibility and agility in the movement of treatment teams and equipment to the forward battle area. It also provides the capability to rapidly resupply Class VIII supplies to combat units.

4-41. The medical elements within the BSB/FSB will request AA support through the brigade aviation element (BAE). The FSMT leader will provide coordination between the BSB/FSB, BAE, or the BCT S3.

4-42. In the BSMC/FSMC, the executive officer (XO) is the principal assistant to the commander for the tactical employment of the company assets. The XO should be included in all brigade tactical planning. Officer preparation will be required to reinforce or reconstitute forward HSS elements and to request augmentation through the brigade surgeon section (BSS), if required. The FSMT leader keeps the BSMC/FSMC XO apprised of the operational capability. This enables the XO to effect timely reinforcement or augmentation. The BSMC/FSMC XO must be familiar with the specific terrain and battlefield situation. Further, the XO should have a thorough understanding of the division and brigade commanders' ground tactical plan.

Brigade Medical Evacuation Plan

4-43. The BSS is responsible for developing the BCT medical evacuation plan. The BSMC/FSMC commander is responsible for the execution of brigade medical evacuation plan, to include the use of both ground and air assets. The BSS medical planner should include the medical company commander, XO, brigade S1, brigade XO, medical platoon leaders, FSMT leader, the BSB/FSB medical plans and operations officer, and the BAE company commander (if available) in the planning process.

Ground Ambulance Evacuation Brigade Role 2 Medical Treatment Facilities

4-44. Evacuation from the BSMC/FSMC is provided by ground ambulances from the GAMC or from AAs from the GSAB AA company. The GAMC provides support to the BSMCs and medical elements in the division AO. They usually evacuate only those patients who will not RTD within 72 hours.

Ground Ambulance Support for Echelons Above Brigade/Division Units

4-45. The ASMC ambulance platoon normally collocates with the treatment platoon for mutual support and area taskings. It performs ground evacuation and en route patient care for supported units in the sustainment area. The ASMC may also evacuate patients from the supported Role 2 MTFs in the sustainment area to the ASMC Role 2 MTF. The ambulance platoon is mobile in operations as its assets may be totally deployed at one time. The platoon normally forward stations a portion of its teams in support of those units in the sustainment area.

4-46. The remaining teams are used for GS support to division sustainment, movement and maneuver and fires units, and for reinforcing support or ambulance shuttles. Platoons or squads from the GAMC will be in DS, or OPCON to, and collocated with supported division units.

SECTION V — EXCHANGE OF PROPERTY

PROPERTY EXCHANGE AND PATIENT MOVEMENT ITEMS

4-47. United States Army medical evacuation operations require that whenever a patient is evacuated from one MTF to another, or is transferred from one ambulance to another, medical items of equipment (casualty evacuation bags [cold weather-type bags], blankets, litters, and splints) remain with the patient. To prevent rapid and unnecessary depletion of supplies and equipment, the receiving Army element exchanges like property with the transferring element. This reciprocal procedure will be practiced to the fullest extent possible through all phases of evacuation from the most forward element through the most rearward hospital.

4-48. With USAF aeromedical evacuation operations, a major factor in the evacuation of patients is that specific medical equipment and durable supplies designated as patient movement items (PMI) must be available to support the patient during the evacuation. Examples of PMI include—

- Ventilators.
- Litters.
- Patient monitors.
- Pulse oximeters.

These items will be available for exchange at the supporting ASFs and MASFs. Refer to JP 4-02 and FM 4-02.1 for additional information on PMI.

This subparagraph implements STANAG 2128 and QSTAG 436.

4-49. Medical property of allied nations (NATO and ABCA armies) accompanying patients of allied nations will be returned to the parent nation at once, if possible. If it is not possible, like items will be exchanged as in the paragraph above.

4-50. When medical property of coalition forces without ratified standardization agreements accompanies patients of coalition forces, it is returned as soon as practical. Commanders should consult with their staff judge advocate early in the planning process to ensure appropriate policy and procedures are developed and disseminated.

SECTION VI — MEDICAL EVACUATION SUPPORT FOR THE OFFENSE, THE DEFENSE, AND STABILITY OPERATIONS

MEDICAL EVACUATION SUPPORT FOR THE OFFENSE

4-51. The offense is the decisive form of war, the commander's only means of attaining a positive goal or of completely destroying an enemy force (FM 3-0). The offense is characterized by rapid movement, deep penetrations, aggressive action, and the ability to sustain momentum regardless of counterfires and countermeasures.

4-52. When considering the evacuation plans to support an offensive action, the medical planner must consider many factors (FM 8-55 [FM 4-02.55]). The forms of maneuver, as well as the enemy's capabilities, influence the character of the patient workload and its time and space distribution. The analysis of this workload determines the allocation of AHS resources and the location or relocation of MTFs. Evacuation support of offensive operations must be responsive to several essential characteristics. As operations achieve success, the areas of casualty density move away from the supporting facilities. This causes the routes of medical evacuation to lengthen. Heaviest patient workloads occur during disruption of enemy main defenses, at terrain or tactical barriers, during the assault on final objectives, and during enemy counterattacks. The accurate prediction of these workload points by the medical planner is essential if medical evacuation operations are to be successful.

4-53. In traditional combat operations, the major casualty area of the division BCTs is normally the zone of the main attack. As the main attack accomplishes the primary task of the division, it receives first priority in the allocation of combat power. The allocation of combat forces dictates roughly the areas which are likely to have the greatest casualty density. As a general rule, all BCT MTFs are located initially as far forward as combat operations permit. This allows the maximum use of these facilities before lengthening evacuation lines force their displacement forward. In operations that feature deep battles with CBRN weapons targeted at supporting logistical bases, MASCAL operations may be conducted in sustainment area.

4-54. As advancing combat formations extend control of the battle area forward, supporting medical elements overtake patients. This facilitates the acquisition of the battle wounded and reduces the vital time elapsed between wounding and treatment. In offensive operations, two basic problems confront the supporting evacuation units.

- First, contact with the supported unit must be maintained. Responsibility for the contact follows the normal HSS pattern of rear to front. The contact is maintained by forward deployed air and ground evacuation resources.
- Secondly, the mobility of the MTFs supporting the combat formations must be maintained. Periodically, BCT medical companies and collocated FSTs, are cleared of their patients so that they may move forward. In some scenarios, CSHs may also have to evacuate patients to reduce patient overflow, surgical backlogs, or to increase bed availability.

4-55. This requirement for prompt evacuation of patients from forward MTFs requires available ambulances to be leveled well forward from the outset. The requirement for periodic movement of large numbers of patients from divisional and corps facilities further stresses the evacuation system.

TYPES OF OFFENSIVE OPERATIONS

Movement to Contact/Attack

4-56. Medical evacuation support in movement to contact/attack is key to the tactical plan. Prior deployment of evacuation resources with parent and supported units permits uninterrupted and effective evacuation support.

Exploitation and Pursuit

4-57. Evacuation support of exploitation and pursuit operations resembles those discussed for the envelopment (paragraph 4-59). Since exploitation and pursuit operations can rarely be planned in detail, evacuation operations must adhere to TSOPs and innovative C2. These actions are often characterized by—

- Decentralized operations.
- Unsecured ground evacuation routes.
- Exceptionally long distances for evacuation.
- Increased reliance on convoys and AAs.

MEDICAL EVACUATION SUPPORT FOR DIFFERENT FORMS OF OFFENSIVE MANEUVER

Penetration and Attack

4-58. In this tactic, the attack passes through the enemy's principal defensive position, ruptures it, and neutralizes or destroys the enemy forces. Of all forms of offensive maneuver, the penetration of main enemy defenses normally produces the heaviest medical evacuation workload. Patient acquisition starts slowly, but becomes more rapid as the attack progresses. The evacuation routes lengthen as the operation progresses. The penetration maneuver is often preceded by heavy preparatory fires which may evoke heavy return fire. These enemy fires may modify the decision to place evacuation assets as far forward as possible. Patient evacuation may be slow and difficult due to damage to roads or the inaccessibility of patients. Evacuation support problems multiply when some combat units remain near the point of original penetration. This is done to hold or widen the gap in enemy defenses while the bulk of division forces exploit or pursue the enemy. Treatment elements are placed near each shoulder of the penetration; ground evacuation cannot take place across an avenue of heavy combat traffic. Besides the heavy traffic, the area of the penetration is normally a target for both conventional and nonconventional CBRN weapons.

Envelopment

4-59. In the envelopment, the main or enveloping attack passes around or over the enemy's principal defensive positions. The purpose is to seize objectives which cut the enemy's escape routes and subject him to destruction in place from flank to rear. Since the envelopment maneuver involves no direct breach of the enemy's principal defensive positions, the medical evacuation system may not be confronted with a heavy workload in the opening phase—

- Ambulances are echeloned well forward in all levels of AHS to quickly evacuate the patients generated by suddenly occurring contact.
- Medical treatment facilities moving with their respective formations overtake patients during evacuation and reduce delays in treatment. After triage and treatment, the patients are evacuated to division/corps-level facilities by accompanying division/corps ground ambulance assets or AAs.

- When the isolated nature of the envelopment maneuver precludes prompt evacuation, the patients are carried forward with the treatment element. Again, nonmedical vehicles may be pressed into emergency use for this purpose. When possible, nonmedical vehicles should be staffed by a combat medic or CLS.
- When patients must be carried forward with the enveloping forces, AHS commanders use halts at assembly areas and phase lines to arrange combat protection for ground ambulance convoys through unsecured areas.
- Further, the commander may take advantage of friendly fires and suppression of enemy air defenses to call for prearranged AA support missions, or emergency use of medium-lift helicopter backhaul capabilities.

Infiltration

4-60. Infiltration is a form of maneuver used during offensive operations.

- The division can attack after infiltration or use it as a means of obtaining intelligence and harassing the enemy.
- Though it is not restricted to small units or dismounted actions, the division employs these techniques with a portion of its units, in conjunction with offensive operations conducted by the remainder of its units.

Army Health System of Infiltration/Maneuver

4-61. The AHS of infiltration/maneuver is restricted by the amount of medical equipment, supplies, and transportation assets that can be introduced into the attack area. No deployment of division-level medical units without their organic transportation should be attempted.

4-62. Elements of unit-level AHS should be accompanied by their organic vehicles, and ambulances should receive priority for deployment. It may be necessary to man-carry enough BAS equipment into the attack area to provide EMT and ATM; however, this results in degrading mobility. When the element is committed without its ambulances, patients are evacuated to the BAS by litter bearer teams. This requires reinforcement of the medical platoon by division or corps medical personnel or improvisation of litter teams using combat troops (if available and approved by the tactical commander). Patient evacuation from the BAS and medical resupply of the force may be provided by litter bearers, depending upon distances and degree of secrecy required.

4-63. When airborne and air assault forces are used, infiltrating elements may land at various points within the enemy's rear area and proceed on foot to designated attack positions. As in surface movement, the amount of medical equipment taken may be limited. In airborne operations, the evacuation of patients will be by litter bearers or frontline ambulances to CCPs or the BAS and then by division-level ambulances to the Role 2 MTF. In air assault operations, the evacuation is by litter bearers to CCPs or to the BAS and then by AAs to a Role 2 MTF. Once the combat element begins the assault on the objective, secrecy is no longer important and its isolated location requires HSS characteristic to airborne and air assault operations until ground linkup.

Turning Movement

4-64. The turning movement is a variant to the envelopment in which the attacker attempts to avoid the defense entirely; rather, the attacker seeks to secure key terrain deep in the enemy's rear and along his LOC. Faced with a major threat to his rear, the enemy is thus "turned" out of his defensive positions and forced to attack rearward at a disadvantage.

Note. General MacArthur's invasion at Inchon during the Korean War is an example of a classic turning movement. Casualties were initially light as the main defenses were avoided; however, as the invasion developed, resistance stiffened and higher casualty rates were experienced. Further, as fighting occurred in a populated area (Seoul), significant civilian casualties resulted. The lack of Korean health care providers caused many of these civilians to seek medical aid from US field medical units.

Medical Evacuation for Turning Movement

4-65. Medical evacuation support to the turning movement is provided basically in the same manner as to the envelopment. As the operation is conducted in the enemy's rear area, LOC and evacuation routes may be unsecured resulting in delays in resupply and evacuation. In the Inchon example, a hospital ship was located off the coast to accept patients evacuated from the fighting. However, due to the precarious tides, evacuation and resupply were often delayed for hours and sometimes days since the harbor could not be navigated by small vessels. It was not until Kimpo Airfield fell that timely evacuation could occur. The deployed HSS units must be able to quickly clear the battlefield of patients, evacuate them from the forward areas, and sustain the patients in rear areas until evacuation routes are established.

MEDICAL EVACUATION SUPPORT FOR THE DEFENSE

TYPES OF DEFENSE

4-66. There are three forms of the defense:

- **Area defense** concentrates on denying enemy access to designated terrain for a specific period of time, rather than on the outright destruction of the enemy.
- **Mobile defense** focuses on denying the enemy force by allowing him to advance to a point where he is exposed to a decisive counterattack by the striking force. The primary defeat mechanism, the counterattack, is supplemented by the fires of the fixing force.
- **Retrograde** is an organized movement to the rear and away from the enemy. The enemy may force these operations or a commander may execute them voluntarily. Within the retrograde operation there are three forms: delay, withdrawal, and retirement.

4-67. Support is generally more difficult to provide in the defense. The patient load reflects lower casualty rates, but forward area patient acquisition is complicated by enemy actions and the maneuver of combat forces.

4-68. Medical personnel generally are permitted much less time to reach the patient, complete vital EMT, and remove him from the battle site. Increased casualties among exposed medical personnel further reduce the medical treatment and evacuation capabilities. Heaviest patient workloads, including those produced by enemy artillery and CBRN weapons, may be expected during the preparation or initial phase of the enemy attack and in the counterattack phase.

4-69. The enemy attack may disrupt ground and air routes and delay evacuation of patients to and from treatment elements. The depth and dispersion of the defense create significant time and distance problems for evacuation assets. Combat elements may be forced to withdraw while carrying their remaining patients to the rear. The enemy exercises the initiative early in the operation which may preclude accurate prediction of initial areas of casualty density. This makes the effective integration of air assets into the evacuation plan essential. The use of AAs must be coordinated and synchronized with the supporting GSAB to ensure the synchronized execution of evacuation operations occurs.

4-70. The support requirements for retrogrades may vary widely depending upon the tactical plan, the enemy reaction, and the METT-TC factors. Firm rules that apply equally to all types of retrograde operations are not feasible, but considerations include:

- Requirement for maximum security and secrecy in movement.

- Influence of refugee movement that may impede medical evacuation missions conducted in friendly territory.
- Integration of evacuation routes and obstacle plans should be accomplished.
- Difficulties in controlling and coordinating movements of the force which may produce lucrative targets for the enemy.
- Movements at night or during periods of limited visibility.
- Time and means available to remove patients from the battlefield. In stable situations and in the advance, time is important only as it affects the physical well-being of the wounded. In retrograde operations, time is more important. As available time decreases, medical managers at all levels closely evaluate the capability to collect, treat, and evacuate all patients.
- Medical evacuation routes will also be required for the movement of troops and materiel. This causes patient evacuation in retrograde movements to be more difficult than in any other type of operation.
  - Command, control, and communications may be disrupted by the enemy.
  - Successful medical evacuation requires including ambulances on the priority list for movement.
  - Providing for the transportation of the slightly wounded in cargo vehicles and providing guidance to subordinate commanders defining their responsibilities in collecting and evacuating patients.
  - Special emphasis must be placed on the triage of patients and consideration given to the type of transportation assets available for evacuation.
- When the patient load exceeds the means to move them, the tactical commander must make the decision as to whether patients are to be left behind. The medical staff officer keeps the tactical commander informed in order that he may make a timely decision. Medical personnel and supplies must be left with patients who cannot be evacuated (refer to FM 4-02 for additional information).

Air Ambulance Evacuation Support to Covering Forces in Defensive Operations

4-71. The division covering force uses combat power to screen the FLOT and force the early deployment of the enemy's main force. Organic AHS elements of the covering forces establish aid stations of minimal size or operate in the split-team mode (FM 4-02.4). Ambulances are deployed well forward to promptly clear patients from combat units. Medical care is limited to EMT and ATM, followed by rapid evacuation whenever possible.

4-72. The covering force surgeon maintains communications with attached aviation elements and uses these assets (augmented by medical personnel to provide en route medical care when feasible) to provide backhaul of casualties, rather than use ground ambulances. The early evacuation of patients from BASs ensures their mobility for rearward displacement. The BCT or armor cavalry regiment (ACR) medical company/troop is equipped to receive patients generated in the covering force area. Depending upon the fluidity of the defensive situation, the FSMT (if available) may have to rapidly evacuate patient's further rearward to a BSMC/FSMC and collocated FST or a CSH.

4-73. The wide dispersion of units and the manner in which they withdraw make patient acquisition difficult. When covering forces withdraw, patients are transported to the rear by the medical element that acquired them. Seriously wounded or injured patients are given priority for evacuation by air. There are usually few COSRs during covering force operations, although delayed symptoms may occur once the element is withdrawn to a safe area. Ground ambulances augmented by nonmedical transportation assets evacuate the remaining patients.

MEDICAL EVACUATION SUPPORT FOR STABILITY OPERATIONS

4-74. Stability operations promote and protect US national interests by influencing the threat, political, and information dimensions of the operational environment. They include developmental, cooperative activities during peacetime and coercive actions in response to crisis.

4-75. Some of the purposes for which the US Forces are employed to conduct stability operations are as follows:

- Civil control.
- Civil security.
- Provide essential services.
- Support economic and infrastructure development.
- Governance.

4-76. Stability operations may complement the main effort or they may themselves constitute the main effort. Table 4-2 identifies the types of stability operations and the types of missions the BCT may conduct.

Table 4-2. Types of stability operations

Types Of Operations	*Types Of Missions*
Civil Control	Establish police forces training.
	Integrate trained police into operations.
	Counter organized crime.
	Establish judicial system.
	Transition to indigenous police forces.
Civil Security	Initiate security forces training.
	Establish National Security Forces.
	Integrate security forces into operations.
	Isolate population from insurgency.
	Defeat insurgency.
	Transition to National Security Forces.
Essential Services	Sewage treatment plants operating.
	Establish trash disposal.
	Water treatment plants operating.

Table 4-2. Types of stability operations

	Restore electrical power. Reopen hospitals and clinics.
	Reopen schools and universities.
Economic/Infrastructure Development	Implement Employment Programs.
	Secure vital natural resources.
	Repair/rebuild distribution infrastructure.
	Prioritize reconstruction projects.
	Essential banking services available.
	Implement public works projects.
Governance	Identify and recruit local leaders.
	Facilitate establishment of sector representation.
	Facilitate establishment of neighborhood councils.
	Facilitate establishment of district councils.
	Support and secure elections.

4-77. Medical evacuation support to forces deployed in stability operations is dependent upon the specific type of operation, anticipated duration of the operation, number of forces deployed, theater evacuation policy, medical troop ceiling, and anticipated level of violence. In most situations, medical evacuation support follows the traditional support provided to combat forces. If there is a shortened theater evacuation policy, a limited medical troop ceiling, and limited hospitalization assets within the AO, organic and DS ambulance support is provided from the POI to the supporting Roles 1 or 2 MTF and, once the patient is stabilized for further evacuation, from the treatment element to an airfield for evacuation out of the theater.

4-78. Reconstruction operations use Army forces to assist civil authorities, foreign or domestic, as they prepare for or respond to crises and relieve suffering. In reconstruction operations, Army forces provide essential support, services, assets, or specialized resources to help civil authorities deal with situations beyond their capabilities. The purpose of reconstruction operations is to meet the immediate needs of designated groups for a limited time, until civil authorities can do without Army assistance.

4-79. Humanitarian and civic assistance operations can include a number of activities such as disaster relief, civil support, refugee assistance, the provision of medical care to isolated populations, and refeeding programs resulting from famines or natural disasters. Medical evacuation assets may be used to evacuate the injured from disaster sites, to provide the emergency transport of critically needed medical supplies and personnel to remote locations, or to perform emergency rescues during times of flooding, wildfires, or other natural disasters.

SECTION VII — MEDICAL EVACUATION SUPPORT FOR ENABLING/SHAPING OPERATIONS

PASSAGE OF LINES

4-80. Passage of lines presents a challenge for the medical planner. There will be a number of medical evacuation units using the same air and road networks. Coordination and synchronization are essential if confusion and over evacuation are to be avoided. The information required to operate in the BCT or division AO includes:

- Radio frequencies and call signs.
- Operations plans and TSOPs.
- Location of MTFs.
- Location of CCPs and AXPs.
- Main supply route, forward arming and refueling points (FARPs), and A2C2 data.

SECURITY OPERATIONS

4-81. In security operations, the covering forces are dependent upon organic resources found in the maneuver battalion medical platoon for initial support. The level of command for the covering force (division or corps) determines the responsibility for the subsequent evacuation plan. In a corps covering force, for example, the corps AHS structure has the responsibility for establishing and operating the medical evacuation system to support the forward deployed corps forces. This is done to prevent the divisions following the covering forces from becoming overloaded with patients prior to the hand off and passage of lines. The use of CCPs, AXPs, and nonmedical transportation assets (CASEVAC) to move the wounded is essential. The covering force battle may be extremely violent. Patient loads will be high and the distance to MTFs may be much longer than usual. The effectiveness of the medical evacuation system depends upon the forward positioning of a number of ground ambulances and the effective integration of AAs into the evacuation plan.

ADVANCE, FLANK, AND REAR GUARDS

4-82. In advance, flank, and rear guards, these forces normally receive medical evacuation support through the attachment of evacuation teams. The teams evacuate patients to predesignated CCPs along a main axis of advance or to the nearest treatment element providing area support. Employment of AAs provides a measure of agility and flexibility.

RIVER CROSSING OPERATIONS

4-83. The river barrier itself exerts decisive influence on the use of division and BCT medical units. Attack across a river line creates an AHS delivery problem comparable to that of the amphibious assault. Army health system protection elements cross as soon as combat operations permit.

4-84. Early crossing of treatment elements reduces turnaround time for all crossing equipment that is used to load patients on the far shore. Maximum use of AA assets is made to prevent excessive patient buildup in far shore treatment facilities. Near shore MTFs are placed as far forward as assault operations and protective considerations permit to reduce ambulance shuttle distances from off-loading points. For detailed information on river crossing operations, refer to FM 90-13.

RECONNAISSANCE OPERATIONS

4-85. The reconnaissance in force is an attack to discover and test the enemy's position and strength or to develop other intelligence. The division usually probes with multiple combat units of limited size, retaining sufficient reserves to quickly exploit known enemy weaknesses.

4-86. Ambulances are positioned well forward at both unit and division levels. Ambulances are moved at night to enhance secrecy. The echeloning of ambulances is an indication to the enemy that an attack is imminent due to the forward placement of HSS. Role 2 MTFs are not established until a significant patient workload develops.

4-87. Patients received at BASs of reconnoitering units are evacuated to Role 2 MTFs as early as practical or are carried forward with the force until a suitable opportunity for evacuation presents itself. Maximum possible use of AA assets is made to cover extended distances and to overcome potentially unsecured ground evacuation routes.

RETROGRADE OPERATIONS

4-88. A retrograde operation is a movement to the rear or away from the enemy. This type of operation may be forced by enemy action or may be executed voluntarily.

AEROMEDICAL EVACUATION SUPPORT IN RETROGRADE OPERATIONS

4-89. Aeromedical evacuation support in retrograde operations may vary widely depending upon the enemy reaction and the situation. Firm rules that apply equally to all types of retrograde operations are impossible to establish. Factors to consider in planning aeromedical evacuation support for retrograde operations include—

- Patient's condition/status.
- Mission, enemy, terrain and weather, troops and support available, time available and civil considerations.
- Requirement for maximum efforts in secrecy and employment of aircraft survivability equipment.
- Influence of refugee movement (which may impede military medical movements conducted in friendly territory).
- Integration of evacuation routes and other Army aviation routes.
- Difficulty in controlling and coordinating movements of the force which may produce lucrative targets for the enemy.

4-90. During retrograde operations, battalion, brigade, and division surgeons must evaluate their capability to collect, treat, and evacuate patients. Three factors influence the number of patients evacuated from the battlefield and the time and place of patient treatment:

- Time available prior to MTF displacement.
- Location of the enemy advance.
- Means of evacuation.

4-91. Due to congestion on main supply routes and general evacuation routes, the AAs may provide the best means of evacuation during retrograde operations. Seriously wounded patients should be evacuated by the fastest and most comfortable means and should receive medical care en route. Proper sorting (triage) and rapid evacuation of patients lessen the need for establishing a complete Role 2 MTF operation. Therefore, the FSMT (if available) leader should be prepared to surge his efforts in the retrograde. Combat and operational stress reaction casualty rates are usually low as compared to wounded in action (WIA) rates during controlled retrograde operations, but increase after safety is reached.

4-92. Commanders must decide to displace medical evacuation units before they hamper the maneuver forces conducting the retrograde movement. During mission planning, the medical evacuation units/elements integrate with the supported unit's displacement plan. The medical evacuation element leaders must coordinate with the supported unit commander and the BSMC for displacement decision points and locations.

WITHDRAWAL OPERATIONS

4-93. A withdrawal operation is one in which a deployed force disengages from an enemy force. It may be forced by enemy pressure or conducted voluntarily.

4-94. Although the deployed force disengages from the enemy, contact is maintained by security elements while the main force moves to the sustainment area, forms march columns, and moves to a predesignated location. Aeromedical evacuation elements must be prepared to support the withdrawing force as well as any forces left in contact. Time available to acquire, treat, and evacuate patients from the line of contact is limited.

AEROMEDICAL EVACUATION SUPPORT IN WITHDRAWAL OPERATIONS

4-95. The AA medical company element supporting the withdrawing force must ensure it maintains sufficient station capability at its present location to support the patients while taking action to move to a new location.

4-96. In addition to the evacuation support provided to the force left in contact, there is a requirement to support the force moving to the sustainment area. If the main force infiltrates to the sustainment area, patients are carried by their parent units to the assembly area or MTFs in the sustainment areas. Medical assets are positioned in the assembly area to receive patients.

4-97. The division establishes Role 2 MTFs at a minimal size and well to the rear of the first line of alternate positions. If combat and environmental conditions indicate a light patient workload and the road network permits rapid ambulance movement, the division may use another Role 2 MTF to provide GS to all withdrawing brigades.

4-98. Role 2 MTFs conduct leapfrog rearward, occupying successive positions placed along the withdrawal route to minimize the requirement for multiple displacements by any one MTF. This avoids unnecessary interference with combat operations while providing continuous medical support. With the exception of an ambulance squad to support the covering force the division normally does not employ ambulances further forward than the BAS. AA elements will normally evacuate to sustainment area and corps positions to avoid impeding the withdrawal of brigades.

4-99. Preparation for the withdrawal operation includes distribution of extra medical consumable supplies and nonexpendable exchange items to each medical element. This is done to overcome the potential effects of isolated MTFs or the possible intermittent operation of the medical evacuation system. The AA medical company may receive the mission to assist with this distribution.

4-100. If withdrawal is made under enemy pressure, the provision of aeromedical evacuation is modified. Since the time available is critical, aeromedical resources cannot remain in the forward areas. The assets are usually withdrawn as a unit. Patients occurring during the withdrawal are carried to the sustainment areas by the parent unit, normally using nonmedical vehicles.

DELAYING OPERATIONS

4-101. Delaying operations occur when forces are insufficient to attack or to defend and when the defensive plan calls for drawing the attacker into an unfavorable situation. The delay maneuver may include friendly maneuver to successive positions across a broad front.

4-102. Delaying brigades split their combat power, moving their less mobile forces directly to the next defensive position while the elements remaining in contact fight to the sustainment area.

4-103. Aeromedical support for a delaying operation is conducted like withdrawal operations.

RETIREMENT OPERATIONS

4-104. A retirement operation is a rearward movement of a force not in contact with the enemy. It is conducted according to the force's OPLAN and without pressure by enemy forces.

4-105. Because the division is no longer in contact with the enemy, it can march (in multiple columns) directly to the rear. The AHS requirements for this type of operation are similar to those in a movement to contact. The patient workload is light. The FSMT (if available) leader will coordinate with the BSMC commander, and the AA medical company CP will coordinate with the DSS for guidance in the positioning of company assets to support this operation.

SECTION VIII — SPECIAL FORCES OPERATIONS

4-106. Army special operations forces (ARSOF) have limited organic AHS resources. They are, therefore, dependent upon the theater AHS for the majority of their health care requirements. To ensure that ARSOF receive comprehensive and timely AHS and logistics, medical planners must consider each medical function and determine the support organic and theater AHS assets will provide. The significant challenge for ARSOF medical planners is to ensure the efficient delivery and use of the conventional Class VIII (medical materiel) system for support of ARSOF missions.

4-107. Army special operations forces consist of special forces (SF), rangers, special operations aviation (SOA), psychological operations (PSYOP), and CA. The SOA, CA, and PSYOP have limited medical capabilities and are very dependent upon the theater HSS assets and the organizations they are supporting. Special forces and rangers often operate far removed from conventional AHS and must be more self-reliant and sustaining than conventional forces. Accordingly, SF and ranger medical personnel receive enhanced medical training above that provided for a trauma specialist. The Army educates the SOF medic to be an independent care practitioner qualified to provide ATM to combat casualties. When deployed on independent operations, the two SF medics are the sole source of medical care for their operational detachment, the indigenous forces, and their families that the detachment supports. They can train the indigenous populace in basic medical skills and establish an austere AHS system. Nonmedical ARSOF personnel receive medical training at the combat lifesaver level. Special operation combat medics (SOCMs) are trained in emergency medical treatment and in combat trauma management, but they do not possess the invasive surgical skills of the SF medical sergeant. The SOCMs provide medical care in combat environments where resuscitative surgery is more rapidly accessible than in the small unit missions of SF. Physicians and physician assistants assigned to units with SOCMs can provide ATM. Army health system planning for missions where the risk of penetrating trauma is high and further complicated by time and extended distance requires extensive planning and coordination. The capability for medical personnel and units that can perform resuscitative surgery and invasive stabilization must be provided regardless of the ARSOF unit involved.

4-108. Although the ARSOF health care provider receives enhanced medical training exceeding the level and scope found in conventional forces, he depends heavily on the conventional AHS system to conserve the combat strength of the ARSOF (particularly in the area of medical evacuation where the ARSOF does not have a dedicated system).

4-109. Ideally, medical evacuation for ARSOF personnel should follow the doctrinal flow sequence. The ARSOF medical planner must be innovative and follow the tenets of immediate far forward stabilization. He directs evacuation to the appropriate MTF when the condition of the patient warrants it, with whatever means of transportation are available.

4-110. Medical evacuation of ARSOF casualties is an operational matter. That is, it must reflect the commander's concept of the operation. It can only succeed when the medical planner integrates the medical evacuation plan with the tactical plan and logistics airflow.

4-111. Conventional medical evacuation assets are not normally used when operations occur in hostile and denied areas. Use the conventional medical evacuation system once the casualty is extracted from the hostile or denied area. For additional information on AHS to ARSOF, refer to FM 3-05, FM 4-02.6. FM 8-42 (FM 4-02.42), and FM 4-02.43.

SECTION IX — URBAN OPERATIONS

TERRAIN AND ENVIRONMENT

4-112. Throughout history, battles have been fought on urbanized terrain. Some recent examples urban operations (UO) include Hue, Beirut, Panama City, and Baghdad. Military operations on urbanized terrain are those military actions planned and conducted on a terrain where man-made structures impact on the tactical options available to the commander. This terrain is characterized by a three-dimensional battlefield, having considerable rubble, ready-made fortified fighting positions, and an isolating effect on all movement and maneuvers and fires, and sustainment units. In this environment, the requirement for a sound and understandable evacuation plan cannot be overstated. Of concern to medical and tactical planners is the need to plan, train, prepare, and equip for evacuation from under, above, and at ground level.

TERRAIN CONSIDERATIONS

4-113. Medical evacuation in the urban environment is a labor-intensive effort. Due to rubble, debris, barricades, and destroyed roadways, much of the evacuation effort will be by manual litter teams. When this occurs, establish a litter or ambulance shuttle system. The shuttle system reduces the distance that litter teams have to carry the wounded or injured Soldiers. This enhances the litter teams' effectiveness by providing brief respites, reducing fatigue. Further, the litter teams stay in the forward areas. They are familiar with the geography of the area and what areas have or have not been searched for casualties. In moving patients by litter, you should:

- Use covered evacuation routes such as subways, whenever possible.
- Use easily identifiable points for navigation CCPs.
- Rest frequently by alternating litter teams.

EVACUATION PLANNING CONSIDERATIONS FOR URBAN OPERATIONS

4-114. Conducting medical evacuation operations in the urban environment challenges the medical planner. He must ensure that the HSS/FHP plan includes special or unique materiel requirements or improvised use of standard equipment. The plan must be sufficiently flexible to support unanticipated situations.

4-115. Special equipment requirements include, but are not be limited to:

- Axes, crowbars, and other tools used to break through barriers.
- Special harnesses; portable block and tackle equipment; grappling hooks; collapsible litters; lightweight collapsible ladders; heavy gloves; and casualty blankets with shielding. This equipment, using pulleys, is for lowering casualties from buildings or moving them from one building to another at some distance above the ground.
- Equipment for the safe and quick retrieval from craters, basements, sewers, and subways. Casualties may have to be extracted from beneath rubble and debris.

4-116. Air ambulances equipped with a rescue hoist may be able to evacuate casualties from the roofs of buildings or may be able to insert needed medical personnel and supplies.

COMMUNICATIONS IN THE URBAN ENVIRONMENT

4-117. The urban environment degrades communications. Line of sight radios will be ineffective and individual Soldiers will not have access to radio equipment. The medical evacuation teams will have difficulty in locating injured or wounded Soldiers due to their isolation within buildings, or by being hidden in rubble and debris. Injured Soldiers can display alternate forms of communications, such as markers, panels, or field expedients (field/uniform jackets or T-shirts) indicating where they may be found.

ESTABLISHING CASUALTY COLLECTION POINTS

4-118. Preplan and establish CCPs at relatively secure areas accessible to both ground and AAs. The location of these points are depicted on the HSS/FHP overlay to the OPLAN. Casualty collection points should—

- Offer cover from enemy fires.
- Be as far forward as the tactical situation permits.
- Be identifiable by an unmistakable feature (natural or man-made).
- Allow rapid turnaround of ambulances.

4-119. The tactical commander must approve route markings to the MTF and if to display of the Geneva Red Cross at the facility. (Camouflaging or not displaying the Geneva Red Cross can forfeit the protections, for both medical personnel and their patients, afforded under the Geneva Conventions. Refer to Appendix A and FM 4-02 for additional information.) The location of the MTF must be as accessible as possible, but well separated from fuel and ammunition depots, motor pools, reserve forces, or other lucrative enemy targets, as well as civilian hazards such as gas stations or chemical factories.

GROUND EVACUATION FOR URBAN OPERATIONS

4-120. When using ground evacuation assets in support of UO, the medical planner must be aware that built-up areas may have significant obstacles to vehicular movement. Factors requiring consideration include the following:

- Rubble and other battle damage complicate and canalize transportation operations within the urban terrain.
- Bypassed pockets of resistance and ambushes pose a constant threat along evacuation routes.
- Land navigation with most tactical maps proves to be difficult. Using commercial city maps when available can aid in establishing evacuation routes.
- Ambulance teams must dismount from the ambulance, search for, and rescue casualties.
- Movement of patients becomes a personnel intensive effort. There are insufficient medical personnel to search for, collect, and treat the wounded. Medical units may require assistance in the form of litter bearers and search teams from supported units.
- Refugees may hamper movement into and around urban areas.
- Dislocated civilians, EPW, and detainees receive medical treatment according to the command policy and the Geneva Conventions.

AIR EVACUATION FOR URBAN OPERATIONS

4-121. When using AA assets in support of UO, the medical planner must consider enemy air defense capabilities and terrain features, both natural and man-made, within and adjacent to the built-up areas. Air ambulances are the preferred means of evacuation in UO. Considerations in the use of AAs include the following:

- Movement is highly restricted and is canalized over secured areas, down wide roads, and open areas.
- Telephone and electrical wire and communication antennas hinder aircraft movement.
- Secure landing zones (LZs) must be available.
- Landing zones may include buildings with helipads on their roofs or sturdy buildings, such as parking garages.
- Snipers with air defense capabilities may occupy upper stories of the urban area's taller buildings.

Specialty Training

4-122. Medical personnel require special training in the tactics, techniques, and procedures required to operate in an UO environment. If they are to survive in this environment, they must know how to:

- Cross open areas safely.
- Avoid barricades and mines.
- Enter and depart safely from buildings.
- Recognize situations where booby traps or ambushes are likely and would be advantageous to the enemy. Detailed information on the conduct of combat operations in the urban environment is contained in FM 3-06 and FM 3-06.11.

Note. Medical personnel do not engage in offensive-type actions. They must rely on the supported unit to provide covering fires and to clear rooms and buildings prior to entry.

4-123. Units can modify and apply many of the extraction and patient evacuation techniques used in a mountainous terrain for evacuation in UO.

SECTION X — HOMELAND SECURITY

4-124. To preserve the freedoms guaranteed by the Constitution of the US, the Nation must have a homeland that is secure from threats and violence, including terrorism. Homeland security (HS) is the Nation's first priority, and it requires a national effort; in which the DOD has a key role in that effort. The National Strategy for Homeland Security (NSHS) complements the National Security Strategy of the US by providing a comprehensive framework for organizing the efforts of federal, state, local, and private organizations whose primary functions are often unrelated to national security. Critical to understanding the overall relationship is an understanding of the distinction between the role that DOD plays with respect to securing the Nation and HS, and the policy in the NSHS, which has the Department of Homeland Security (DHS) as the lead. Homeland security at the national level has a specific focus on terrorist threats. The DOD focus in supporting HS is broader. Military application of the NSHS calls for preparation, detection, deterrence, prevention, defending, and responding to threats and aggression aimed at the homeland. The DOD also provides military assistance to civil authorities (MACA), including consequence management activities. The Armed Forces of the US support the NSHS through two distinct but interrelated mission areas–homeland defense (HD) and civil support. The United States Northern Command (USNORTHCOM) is the unified command that has the DOD responsibility for HD and civil support.

HOMELAND DEFENSE

4-125. Homeland defense is the protection of US territory, sovereignty, domestic population, and critical infrastructure against external threats and aggression. The purpose of HD is to protect against and mitigate the impact of incursions or attacks on sovereign territory, the domestic population, and defense critical infrastructure. For HD missions, as directed by the President of the US and/or the Secretary of Defense, DOD serves as the lead federal agency.

Homeland Defense Characteristics

4-126. The DOD is the lead, supported by other agencies, in defending against traditional external threats/aggression (such as air and missile attack). However, against internal asymmetric, nontraditional threats (such as terrorism), DOD may be in support of DHS. When ordered to conduct HD operations within US territory, DOD will coordinate closely with other federal agencies or departments.

TYPES OF HOMELAND DEFENSE OPERATIONS

4-127. The types of homeland defense operations that the future brigade combat teams is expected to perform include—

- Offensive operations.
- Defensive operations.
- Humanitarian and civic assistance.
- Support to counterdrug operations.
- Show of force.
- Domestic support operations.
- Security operations (screen, guard, area security, and local security).
- Reconnaissance operations (zone, route, and area reconnaissance).

CIVIL SUPPORT

4-128. Civil support missions consist of military support to US civil authorities for domestic emergencies, and for designated law enforcement within the scope of restrictions required by the Posse Comitatus Act and other support approved by the Secretary of Defense (see FM 3-07).

4-129. Employment of military forces within the US, its territories, and possessions, under the auspices of civil support, typically falls under the broad mission of MACA. The MACA missions consist of three mission subsets. These mission subsets consist of military support to civil authorities, military support to civilian law enforcement agencies, and military assistance for civil disturbances.

MILITARY SUPPORT TO CIVIL AUTHORITIES

4-130. Military support to civil authorities (MSCA) is the most widely recognized form of DOD civil support because it usually consists of support for high-profile emergencies such as natural or man-made disasters that often invoke Presidential or state emergency/disaster declarations:

- **Natural Disasters.** In the event of a natural disaster or emergency, there will be a concerted US Government effort to support the affected areas. The DOD may be asked to provide assistance to DHS/Federal Emergency Management Agency (FEMA) in an attempt to save lives, protect property, and lessen the threat of a catastrophe in the US. Examples of natural disasters include, but are not limited to severe weather, wildland fire fighting, and animal disease eradication. When natural disasters occur and military assistance is anticipated, DHS/FEMA will request a defense coordinating officer (DCO) who serves as the single DOD point of contact within the disaster area. The DCO will be under the OPCON of the designated supported combatant command. Units supporting the event are normally OPCON to the supported combatant command.
- **Special Events.** The DOD provides support to a special event, a planned program of athletic competition or related activities involving participants from the US and/or other nations. Historic examples of these events are the Olympic Games and the Pan American Games.
- **Man-made Disasters.** Manmade disasters can be accidental or intentional. An example of an accidental event is an oil spill. The implications of a deliberate or unintentional large release of a CBRNE are severe. A catastrophic CBRNE event or attack may occur with little or no warning, resulting (either immediately or over time) in mass casualties and producing a mass exodus of evacuees. The response capabilities and resources of the local jurisdiction (to include mutual aid from surrounding jurisdictions and response support from the state) may be insufficient and quickly, if not immediately, overwhelmed.

Domestic Support Operations

4-131. Domestic disasters occur due to various causes and many may require the assistance of DOD and more specifically, the Army medical evacuation assets to fill response shortfalls. In the post-9/11 environment, terrorism or other intentional acts are recognized as a more likely cause of contingencies within the US than many had acknowledged since the civil defense era of the forties and fifties. This threat increases the likelihood that a large-scale disaster will occur that requires the augmentation of civil authorities' resources by DOD. The US DHS and USNORTHCOM have been established to coordinate the preparedness for and response to such events for civil authorities and DOD respectively.

4-132. Domestic support operations are typically conducted to prevent loss of life and property. Certain operations may be conducted in the form of immediate and automatic response by US military commanders as defined in DOD Directive 3025.15, to save lives, prevent human suffering, or to mitigate great property damage under imminently serious conditions. Other requests for support for civil disasters are made by civil authorities or a DCO (in a Stafford Act declared emergency) through the Joint Directorate of Military Support per the directive and work down through command channels from the Secretary of Defense.

4-133. Homeland Security Presidential Directive 5 recognized that there currently exists an artificial separation of local, state, and federal response to an incident and the role of DOD is not well streamlined or defined. This recognition led the White House and the DHS to initiate the transition from the Federal Response Plan to the National Response Plan, which serves as the doctrine for how the US responds to disasters no matter the cause.

4-134. Medical evacuation support plans to augment civil authorities are necessary for a variety of contingencies. Domestic support planning must address a range of problems such as—

- Preidentifying medical evacuation capabilities, units and personnel available to support various contingencies large enough to require civil support.
- Command and control relationships between civil authorities and DOD forces especially when DOD units are divided.
- Support for deployed DOD forces when no DOD logistics operations are deployed, including medical support.
- Cost capture and reimbursement from civil authorities to DOD in non-Stafford Act emergencies.

SECTION XI — OTHER TYPES OF MEDICAL EVACUATION SUPPORT MISSIONS

EVACUATION OF MILITARY WORKING DOGS

4-135. Military working dogs (MWDs) when injured or ill may be evacuated on any transportation means available. The using unit is responsible for the evacuation of the animal. Use of dedicated medical evacuation assets (air or ground ambulances) is authorized based on mission priority and availability. When possible, the handler should accompany the animal during the evacuation. Using units should include the location of veterinary treatment facility/support units on operational overlays. See FM 4-02.18 for information on veterinary support for MWDs.

PERSONNEL RECOVERY OPERATIONS

4-136. Air ambulances do participate in PR operations in a support role and are not protected from attack while engaged in a PR mission over contested or denied territory. If the involvement in these operations consists solely of evacuating wounded crewmembers from a crash site in friendly territory, AAs retain the protection accorded to them under the provisions of the Geneva Conventions. Air ambulances flying in contested or denied areas are not protected from attack unless flying at times and routes agreed to by the enemy and may be summoned to land. Air ambulances must obey a summons to land. Personnel retain their protections under the Conventions. However, if AAs participate in the actual search and rescue phases of the operation, they are not solely engaged in the provision of AHS and are, therefore, not afforded the protections. Further, because the mission does not fall within the protected activities, the AAs participating in the PR operation must remove the Geneva Conventions emblem from the aircraft.

SHORE-TO-SHIP EVACUATION OPERATIONS

4-137. It is imperative that Army AA units be able to interface on the first day of battle with US Navy (USN) air-capable ships. Lessons learned from past operations, have shown that US Army helicopters should be able to operate to and from USN air-capable ships. An interservice agreement between the Army and the Navy allows for deck-landing qualification of Army pilots (refer to FM 1-564 for additional information). It is important that units having contingency missions requiring Navy support establish training requirements to obtain naval-operations orientation, water egress training, water survival, and deck-landing qualification. This enhances the successful accomplishment of the aeromedical evacuation mission to naval vessels.

COMMUNICATIONS

4-138. In past joint operations, communications have been burdensome for both Army and Navy elements. It is essential to establish commonality of communication requirements during training exercises and to establish communication equipment and frequencies for medical evacuation to Navy vessels. This will provide smooth integration of Army helicopters into the Navy airspace management system during actual operations.

NAVIGATION

4-139. As Navy vessels may operate relatively long distances from the ground combat operations, Army AA units need to be proficient in over water navigation. The use of navigational aids (NAVAIDS) from the Navy element in support of the operation is the first priority for over water navigation. Basic dead-reckoning remains a secondary measure.

MEDICAL EVACUATION OF ENEMY PRISONERS OF WAR AND DETAINEES

4-140. Sick, injured, and wounded EPW and detainees are treated and evacuated in military police (MP) channels when possible. They must be physically segregated from US, allied, and coalition patients. Guards for these prisoners are provided according to the BCT, division or corps TSOP and are from other than medical resources. The echelon commander is normally responsible for this support. The US provides the same standard of medical care for wounded, sick, and injured EPW and detainees as that given to US, allied, and coalition Soldiers. Wounded, sick, or injured EPW and detainees in the CZ may be treated and returned to MP channels for evacuation; or the EPW and detainees may be stabilized and moved through medical channels to theater hospitals for treatment. Enemy prisoners of war and detainees are not evacuated from the theater for medical treatment.

4-141. When EPW are evacuated through medical channels, medical personnel—

- Report this action through medical channels to the next higher headquarters.
- Request disposition instructions from the MEDBDE IPMC.
- The MEDBDE IPMC is responsible for—
  - Coordinating the transportation means.
  - Identifying the MTF to which the EPW/detainees will be taken.
  - Coordinating, in conjunction with the hospital commander, with the Detainee Reporting System to account for EPW within medical channels.

SECTION XII — MEDICAL EVACUATION REQUEST

4-142. Procedures for requesting medical evacuation support must be institutionalized down to the unit level. Procedural guidance and standardization of request procedures are provided below. The same format used to request aeromedical evacuation is also used for requesting ground evacuation.

Table 4-3. Procedures for information collection and medical evacuation request preparation

Line	*Item*	*Explanation*	*Where/How Obtained*	*Who Normally Provides*	*Reason*
1	Location of pickup site	Encrypt the grid coordinates of the pickup site. When using the DRYAD Numeral Cipher, the same "SET" line will be used to encrypt the grid zone letters and the coordinates. To preclude misunderstanding, a statement is made that the grid zone letters are included in the message (unless unit SOP specifies it's use at all times)	From map	Unit leader(s)	Required so evacuation vehicle knows where to pickup patient. Also, so that the unit coordinating the evacuation mission can plan the route for the evacuation vehicle (if the evacuation vehicle must pick up from more than one location).
2	Radio frequency, call sign and suffix	Encrypt the frequency of the radio at the pickup site, not a relay frequency. The call sign (and suffix if used) of person to be contacted at the pickup site may be transmitted in the clear.	From SOI	RTO	Required so that evacuation vehicle can contact requesting unit while enroute (obtain additional information or change in situation or directions).
3	Number of patients by precedence	Report only applicable information and encrypt the brevity codes. A—URGENT B—URGENT-SURG C—PRIORITY D—ROUTINE E—CONVENIENCE If two or more categories must be reported in the same request, insert the word "BREAK" between each category.	From evaluation of patient(s)	Medic or senior person present	Required by unit controlling vehicles to assist in prioritizing missions.

4	Special equipment required	Encrypt the application brevity codes. A—None B—Hoist C—Extraction equipment D—Ventilator	From evaluation of patient/situation	Medic or senior person present	Required so that the equipment can be placed on board the evacuation vehicle prior to the start of the mission.
5	Number of patients by type	Report only applicable information and encrypt the brevity code. If requesting medical evacuation for both types, insert the word "BREAK" between the litter entry and ambulatory entry. L+# of patients–Litter A+# of patients–Ambulatory (sitting)	From evaluation of patient(s)	Medic or senior person present	Required so that the appropriate number of evacuation vehicles may be dispatched to the pickup site. They should be configured to carry the patients requiring evacuation.
6	Security of pickup site (wartime)	N—No enemy troops in area P—Possibly enemy troops in area (approach with caution). E—Enemy troops in area (approach with caution). X—Enemy troops in area (armed escort required).	From evaluation of situation	Unit leader	Required to assist the evacuation crew in assessing the situation and determining if assistance is required. More definitive guidance can be furnished to the evacuation vehicle while it is enroute (specific location of enemy to assist an aircraft in planning its approach).
6	Number and type of wound, injury or illness (peacetime)	Specific information regarding patient wounds by type (gunshot or shrapnel). Report serious bleeding, along with patient's blood type if known.	From evaluation of patient(s)	Medic or senior person present	Required to assist evacuation personnel in determining treatment and special equipment needed.
7	Method of marking pickup site	Encrypt the brevity codes. A—Panels B—Pyrotechnic signal C—Smoke signal D—None E—Other	Based on situation and availability of materials	Medic or senior person present	Required to assist the evacuation crew in identifying the specific location of the pick up. Note that the color of the panel or smoke should not be transmitted until the evacuation vehicle contacts the unit (just prior to its arrival). For security, the crew should identify the color and the unit verifies it.
8	Patient	The number of patients in each category need not	From evacuation	Medic or	Required to assist in

	nationality and status	be transmitted. Encrypt only the applicable brevity codes. A—US military B—US citizen C—Non-US military D—Non-US citizen E—enemy prisoner of war (EPW)	platform	senior person present	planning for destination facilities and need for guards. Unit requesting support should ensure that there is an English speaking representative at the pickup site.
8	CBRN contamination (wartime)	Include this line only when applicable. Encrypt the applicable brevity codes. C—Chemical B—Biological R—Radiological N—Nuclear	From situation	Medic or senior person present	Required to assist in planning for the mission. (determine which evacuation vehicle will accomplish the mission and when it will be accomplished)
9	Terrain description (peacetime)	Includes details of terrain features in and around proposed landing site. If possible, describe relationship of site to prominent terrain feature (lake, mountain, tower).	From area survey	Personnel present	Required to allow evacuation personnel to assess route/avenue of approach into area. Of particular importance if hoist operation is required.

Chapter 5

Medical Evacuation in Specific Environments

This chapter addresses medical evacuation in specific environments or under special circumstances. The medical evacuation effort must be well-planned and its execution synchronized to be effective. Further, medical evacuation personnel must be flexible and ready to improvise, if needed, to meet the demands of unique situations.

SECTION I — MOUNTAIN OPERATIONS

5-1. In the past, armies have experienced great difficulty in evacuating patients from mountainous areas because they are extremely diverse in nature. Some mountains are dry and barren with temperatures ranging from extreme heat in the summer to extreme cold in the winter. In tropical regions, mountains are frequently covered by lush jungles and heavy seasonal rains occur. Many areas display high rocky crags with glaciated peaks and year-round snow cover. Elevations can also vary from as little as 1,000 feet to over 16,000 feet with drastic and rapidly occurring weather changes.

5-2. Operations in mountainous terrain require some procedure modifications. This is due to the environmental impact on personnel and equipment. Important physical characteristics and considerations that influence medical evacuation are—

- Rugged peaks, steep ridges, and deep valleys.
- Limited number of trafficable roads.
- Reduced communications ranges.
- Unpredictability of and severe changes in weather.
- Decreased partial pressure of oxygen.
- Limited availability of LZs.

5-3. In order to effectively support the tactical plan, the HSS/FHP plan must provide maximum flexibility. The HSS planner should consider using all means of evacuation. Due to the length of evacuation times and the limited means of ground evacuation, it is important to triage and prioritize patients prior to movement.

- The availability of improved, hard-surfaced roads is extremely limited, if they exist at all. Usually, improved roads are only found in valley corridors. Such roads are often dependent upon a system of narrow bridges spanning mountain streams and ravines. They may also twist along ridgelines and cling to steep shoulders.
- Secondary roads and trails may be primitive and scarce. However, they may provide the only routes capable of vehicular traffic. Cross-compartment travel between adjacent valleys may be impossible by ground vehicle. Off-road travel requires detailed planning, even for short distances.
- Because of rough terrain, the Role 2 MTF may not be able to reach the BAS by ground vehicle. An ambulance shuttle system is established with an AXP for air and ground evacuation vehicles to meet litter bearers. Litter bearers and beasts of burden may be the only means of evacuation available. Any available personnel may be used as litter bearers (nonmedical personnel from supported units may be required to augment the litter bearer teams). Close coordination between Role 2 MTFs and BASs in establishing CCPs or AXPs is necessary to—
  - Reduce distance traveled by litter bearers.
  - Reduce evacuation time.

  - Conserve personnel.
  - Locate the best potential LZs for AAs.
- In mountainous areas, evacuation of patients by air is the preferred means. Air ambulances permit the rapid movement of patients over rugged terrain. For example, to travel a distance of only 6 kilometers on foot could take up to 2 hours, while flying time could be less than 2 minutes.
- Frequency-modulated radios are the principal means of communication in this environment. The ability to transmit is hampered by the limitations of line of sight transmissions.
- The briefing of ambulance drivers needs to be extensive, including detailed strip maps and overlays. Further, specific instructions on what to do in various situations should be covered (such as if the vehicle breaks down or the unit moves).

5-4. The mountain environment, with its severe and rapidly changing weather, impacts on aircraft performance capabilities; accelerates crew fatigue; and requires special flying techniques. Having to rely on continuous aviation support for a successful mountain operation is risky.

- Flying in mountainous areas requires special training. Both the terrain and the weather influence basic flying techniques and operational planning. Rugged, mountainous terrain complicates flight route selection. Direct routes can seldom be flown without exposing the aircraft to detection and destruction by the enemy.
- Important considerations for AA operations in mountainous areas are—
  - Density altitude which is the most important factor affecting aircraft performance. Density altitude combines temperature, humidity, and pressure altitude, and provides the basis for lift capability. Density altitude can vary significantly between the pickup point and the LZ because of the time of day and changes in elevation. Frequent performance planning updates are essential.
  - Unpredictable winds which can produce significant turbulence, wind sheers, updrafts, and downdrafts. This further increases the risk of a catastrophe in a seemingly routine mission. Adverse winds along with high density altitude demand current and accurate performance planning. Pilots must plan for greater margins of safety.
  - Ice can clog intake ports, thus starving the engine of air, or it can collect on rotor blades resulting in a significant loss of lift. Asymmetrical shedding can cause severe out-of-balance rotor conditions.
  - Visibility can be reduced by low clouds or fog greatly decreasing the ability to navigate or to avoid obstacles.
  - Lack of LZs is critical because the characteristics of mountain terrain do not usually afford adequate LZs. The terrain may only allow the aircraft to hover while loading patients on board.
  - Hoist operations can be expected in mountainous terrain requiring the use of the internal or external rescue hoist. Mounting the rescue hoist on the aircraft as standard equipment in mountain operations may be required. When possible, orientation and training sessions with supported troops should be conducted to help minimize the difficulty of such missions. Depending on the terrain, the forest penetrator may also be needed to accomplish the mission.
  - Enemy air defenses must be considered because when enemy air defense capabilities preclude using AAs in forward areas, they should be used to evacuate patients from AXPs or from Role 2 MTFs.
  - Ambulatory patients may be reported as litter patients in mountainous terrain because these patients may be unable to move unassisted over the rugged terrain. Once placed on the AA, their status may be upgraded.
  - Additional crew training for ground and air evacuation crews should include training and orientation in mountaineering skills, handling patients, and survival skills. Such training might include cold weather survival training (cold injury prevention, mountain [rock] climbing, and the use of ropes and vertical rescue techniques). Training may also include individual and unit

movement at high altitudes and techniques of patient evacuation by litter, emphasizing the use of pack animals (if available from the host country), and the improvised travois litter.

- Patient loading can be hazardous and care must be taken when loading patients where there is a great deal of slope to the LZ. Emphasis on approaching and loading the aircraft from the down-slope side of the aircraft must be reinforced.

DANGER

Approaching the aircraft from the up-slope side is hazardous.

5-5. Troops operating in mountainous areas are exposed to other injuries and illnesses that frequently occur in this environment. These conditions include—

- An increased rate of fracture, sprain, and dislocation injuries.
- Incidents of acute mountain sickness, high-altitude pulmonary edema, and cerebral edema caused by rapid ascent to heights over 7,500 feet.
- Cold weather injuries.
- Dehydration and heat exhaustion.
- Sunburns and snow blindness.
- Aggravated sickle cell anemia. Although this condition is not considered a mountain illness, personnel with the sickle cell trait can be seriously affected by the decrease in barometric pressure and lower oxygen levels found at higher altitudes.

5-6. The proportion of litter cases to ambulatory cases is increased in mountainous terrain, for even the slightly wounded may be unable to move unassisted over rough terrain. Litter relay stations may be required along the evacuation route to conserve the energy of litter bearers and to speed evacuation. It is important to be able to predict the number of patients that can be evacuated with available personnel. When the average terrain grade exceeds 20 degrees, the four-man litter team is no longer efficient and should be replaced by a six-man team. The average mountain litter team should be capable of climbing 120 to 150 vertical meters of average mountain terrain and return with a patient in approximately 1 hour.

5-7. Mountain operations may require medical personnel to carry additional equipment. Items such as ropes, pulleys, pitons, piton hammers, and snap links are all necessary for evacuating patients and establishing BASs. All unnecessary items of equipment including those for which substitutes or improvisations can be made should be left behind. Heavy tentage, bulky chests, extra splint sets, excess litters, and nice-to-have medical supplies should be stored. Such medical supplies, if stored, should be readily available for airdrop or other means of transport. Medical items that are subject to freezing should be safeguarded; they should not be exposed to the low temperatures experienced in mountainous areas.

5-8. For further information on mountain operations, refer to FM 3-97.6.

SECTION II — JUNGLE OPERATIONS

5-9. Health service support elements in a jungle environment retain the same basic capabilities as in other environments. Jungle operations, however, subject personnel and equipment to effects not found in other environments. The jungle environment degrades the ability to maneuver. Security problems are also increased and affect medical evacuation operations as much as they do the combat forces.

5-10. In jungle operations, combinations of air and ground evacuation units are used to maximize the patient evacuation potential. Using this dual system of evacuation ensures that the inherent limitations of one system can be compensated for by the other. Jungle variations affect the organizing, positioning, and securing of HSS/FHP. Due to the terrain, aerial resupply is usually a common practice. The responsiveness provided by aerial resupply requires fewer supplies to be stockpiled in the combat trains.

5-11. Jungle combat operations are characterized by ambushes and other unconventional warfare operations. The security threat caused by infiltrators requires that LOC be patrolled often and that convoys be escorted. It is, therefore, essential that HSS be performed as far forward as the tactical situation permits. Positioning assets forward—

- Improves response time.
- Reduces road movement.
- Allows the HSS elements to take advantage of the security offered by combat units.

5-12. The thick foliage often makes evacuation by ground more difficult than in other types of terrain. Factors such as the threat, limited road network, and reliance on nonmedical personnel for convoy security make air evacuation the preferred means. By using the ambulance shuttle system, patients can be transferred from forward operating ground ambulances to either ground or AAs operating further to the rear. In situations where evacuation assets are delayed by various factors (weather or terrain), patients are held for longer periods of time at forward locations. This will dictate the need for additional medical supplies. Health service support planners must try to anticipate these delays whenever possible. The increased disease and infection incidences associated with the jungle environment may worsen the patient's condition; therefore, timely evacuation is essential.

5-13. In some remote and densely foliaged jungles, the only means of evacuation may be by litter. Ambulances may not be practical on trails, unimproved muddy roads, or in swamps. As in mountain operations, there is a higher proportion of litter cases than usual. In the jungle even a slightly wounded Soldier may find it impossible to walk through dense undergrowth. At best, litter teams can carry patients only a few hundred meters over rough jungle terrain before needing rest or relief. Litter carries should be kept as short as possible and medical elements prepositioned and retained forward.

5-14. Other special planning considerations in jungle operations include—

- Water is vital in the jungle and is plentiful. Water from natural sources, however, should be considered contaminated. Water purification procedures must be taught to all Soldiers. (Refer to FMs 21-10 and 4-25.12 for additional information.)
- Due to the tropical climate, units should pack hot weather clothing when deploying to jungle areas. Insect (mosquito) nets, insect repellent, permethrin treated uniforms and sunscreen should be issued to all Soldiers operating in this environment.
- The jungle environment is ideal for the transmission of a large number of diseases. The rate of DNBI casualties is potentially the highest in this climate. The heat, humidity, and terrain places the troops at high risk for dehydration, heat injury, skin diseases, endemic diseases, and immersion foot. Small wounds can rapidly become infected and lead to loss of effectiveness and possibly require evacuation. High standards of personal hygiene must be taught, encouraged, and maintained by the command. Mosquitoes and other arthropods that carry disease flourish under jungle conditions. Use of all possible personal protective measures must be ensured. Food- and waterborne diseases leading to diarrhea or other symptoms will abound. Food service sanitation measures must be strictly followed. The potential for contamination of food and water increases with each time they are handled, stored, or transported. Soldiers must be encouraged to consume adequate amounts of water that has been purified and to eat only approved foods. In the jungle it is necessary for the commander to pay meticulous attention to the details of PVNTMED measures to maintain an effective fighting force. For additional information on PVNTMED measures, refer to FM 21-10 and FM 4-25.12.
- Health service support personnel should be trained in survival and support techniques in jungle environments. For example, training should be conducted in—
  - Hot weather acclimatization and survival.
  - Prevention, early detection, and treatment of arthropod-, food-, and waterborne diseases.
  - Land navigation in a jungle environment.
  - Field sanitation and other PVNTMED measures.
  - Care and maintenance of equipment and supplies.

- Due to the increased heat and humidity, vehicles and equipment require additional maintenance. Equipment tends to rust quickly and must be cleaned and oiled more frequently. Canvas items rot and rubber deteriorates much faster than in more temperate climates.
- The range of FM communications in the jungle is significantly reduced due to the dense undergrowth, heavy rains, and hilly terrain. The range of a radio set operated in the jungle may be reduced by 10 to 25 percent. The heavy rain and high humidity of the tropics also reduce the range (about 20 percent) and reliability of wire communications. The transmission range can be extended by using additional radio relays and field expedient antennas.
- Utility helicopters are not able to lift the same size loads that can be lifted in more temperate areas. This results in a reduced patient load in some evacuation aircraft. Again, frequent and accurate performance planning is essential for mission accomplishment.
- There may be few suitable LZs. Many LZs will only be large enough to support one or two helicopters at a time.
- Hoist operations may be required more frequently in the thick jungle vegetation where LZs are not available. The forest penetrator should be carried on all operations.
- For aviation-specific information, refer to training circular (TC) 1-400.

SECTION III — DESERT OPERATIONS

5-15. Deserts are arid, barren regions of the earth incapable of supporting normal life due to a lack of fresh water. Although deserts are often thought of as hot climates, it is important to note that temperatures range from over 136 degrees (°) Fahrenheit (F) in some deserts, to bitter cold in others. Day to night fluctuations in temperature can exceed 70° F. Desert terrain can have mountains, rocky plateaus, or sandy dunes; some desert areas may contain all of these characteristics. Rain, when it falls, often causes flooding in low-lying areas. Winds can have a devastating effect upon AHS operations by destroying equipment and supplies and causing dust storms. Dust storms make navigation and patient treatment difficult. Since deserts vary considerably in the type of terrain and temperature, and in their cultural makeup, current medical intelligence should be obtained prior to deployment on operations conducted on desert terrain.

5-16. People have lived and fought in desert areas for thousands of years. However, the environmental effects on personnel can be extreme, especially for Soldiers not prepared for these operations.

- To be effective, Soldiers must be properly acclimatized to the desert. Two weeks are usually required to satisfactorily acclimatize troops to hot environments, using progressive degrees of heat exposure and physical exertion. Other potential acclimatization problems that may be encountered are the effects of dry air and altitude on the respiratory system. Since many desert areas are located in mountainous terrain, Soldiers must be acclimatized for both the altitude and the temperature. In some areas of the world, such as the Gobi Desert in East Asia, people must be acclimatized to the cold in addition to the dryness. (For additional information, refer to FM 21-10, FM 8-250, and TB MED 507.)
- Units deployed in desert areas typically have long LOC and are widely dispersed. As unit elements become more dispersed, commanders must rely more heavily on junior officers and NCOs to ensure that C2 is maintained and that discipline and PVNTMED programs are enforced. For a unit to be effective, a high level of discipline must exist at all levels of the organization.
- Water is the most basic need in a desert. Without it, Soldiers cannot function effectively for more than a few hours. Thirst is not an adequate indicator of the need for water. It is necessary for each commander to establish and enforce a supervised drinking program. Experience has shown many times that Soldiers do not drink enough fluids unless forced to do so. It is important to cool the water, if at all possible, to make it more appealing. Water supplies should be carefully guarded against accidental loss, sabotage, or contamination. Extra water must be carried by AHS vehicles for patients to drink and to cool heat casualties.
- Soldiers deployed in the desert are susceptible to endemic diseases and environmental injuries. Proper water discipline, vaccines, prophylactic measures, field sanitation measures, personal

hygiene, and other PVNTMED measures can reduce these risks. Cold weather injuries, heat injuries, and respiratory disease can also be prevalent. Proper clothing, equipment, and a water discipline program must have command emphasis in desert operations.

- Winds may very easily damage materiel such as aircraft, antennas, and tents. Equipment is protected by using covers, tie-downs, and shelters. Terrain helps shield equipment from the wind if site selection is done carefully. In some cases, special tools, such as extra long metal tent stakes, are necessary.
- The effects of wind and sand are interrelated. Desert sand starts to become airborne when the wind reaches about 20 knots. Sandstorms—
  - Restrict visibility.
  - Pose a hazard to eyes (especially for Soldiers wearing contact lenses).
  - Can contaminate water supplies (if they are not protected).
  - Make navigation difficult.
- The sun may cause sunburn of the skin and eyes (if protection is not used).

5-17. Seven characteristics of the desert environment that may adversely effect equipment are—

- Terrain. Trafficability varies with the type of terrain covered. Open, flat, and rocky terrain affords higher trafficability than do mountainous areas, lava beds, or salt marshes. Drivers must be well-trained in judging the terrain over which they are driving to select the best alternative routes.
  - Tracked vehicles are best suited for desert operations. However, they can throw tracks when traversing a rocky area. Their use is also limited in rough terrain with steep slopes.
  - Wheeled vehicles may be used in desert operations; however, they normally have a lower average speed than tracked vehicles and a higher incidence of damage and malfunction. Wheeled vehicles often bog down in sandy areas and cannot traverse many of the rougher areas.
  - In planning for desert operations, vehicles should carry extra repair parts (fan belts, tires, and other items apt to malfunction).
- Excessive heat causes vehicles to overheat, leading to greater than normal wear. The frequency of leaks on vehicles and aircraft is greater than in some other environments. Engine and transmission seals tend to dry out and crack; fuel lines wear out quickly; and water requirements for cooling vehicle engines are greater. Loss of water, through evaporation, must be included in logistical planning. Using vehicle and aircraft covers reduce the effects of heat while vehicles and aircraft are not in use. Refer to FM 1-202.
  - Batteries do not hold their charge efficiently in intense heat. Dry battery supplies should be increased to compensate for a higher usage rate.
  - Communications equipment must be protected from the heat in the desert. Dust covers are used on this type of equipment. If the equipment has ventilating ports, these should be cleaned regularly to avoid clogging.
  - Medical supplies must be protected from the heat to prevent deterioration. The shelf life of some medical supplies decreases when stored in hot climates.
  - Medical equipment must be protected from the effects of heat. It may be protected using the same techniques as those used to protect communications equipment.
- The sun burns unprotected skin and it may damage unprotected eyes. Soldiers should dress in loosely fitting clothing, use sunburn cream or oils to protect exposed skin, and wear sunglasses or goggles to protect their eyes. Soldiers should remain fully clothed. Removing clothing increases direct exposure of the skin to the sun and eliminates the beneficial cooling effects of the moisture trapped in clothing. Radiant light or its heat effects may be detrimental to plastics, lubricants, pressurized gasses, rubber, and some fluids. All vehicles and aircraft should be kept well-ventilated, and windshields should be covered to reduce heat buildup inside. Supplies of all types should be stored in a well-ventilated, shady area. Placing supplies in covered holes in the ground may reduce adverse heat effects.

- Dust and sand present one of the greatest dangers to the proper functioning of equipment. Sand mixed with lubricants forms an abrasive paste. Lubrication fittings, bearings, and filters should be inspected frequently and changed when required.
  - Aircraft should not be exposed to dust and sand any more than is absolutely necessary. Ground handling instead of hovering reduces sand ingestion. Dust and sand increase failure of microphone switches, signal distribution panels, and circuit breakers, and cause electrical motors and generators to burn out. Wheel and flight control bearings require more frequent cleaning; engines should be flushed frequently.
  - Medical and communications equipment may be adversely affected by dust and sand. Over a period of time, electrical insulation is damaged by windblown sand. When combined with the effects of lubricants on the insulation, dust and sand can become a major communications problem. Special care should be taken to brush dust off radio equipment and to keep ventilating ports and channels clear.
  - Sand can accumulate in airframes, on the bottom of armored vehicles, and in bearings on all types of equipment. This accumulation, combined with oil and condensation, adds extra weight to aircraft and vehicles as well as jamming their control linkages. Sand and grease buildups must be removed from bearings to ensure safe operation and control of aircraft and vehicles.
  - Dust trails created by hovering aircraft or ground vehicles can be seen in excess of 10 miles on a relatively flat desert. This exposes these assets to direct and indirect enemy fires. Ground vehicles should reduce their speed to the point that they do not create a dust signature.
- Humidity is a factor in some desert areas of the world, especially in the Middle East. Humidity can become a problem for short periods of time in other desert areas. Light coats of lubrication can help prevent rust; however, these benefits should be weighed against the dust-gathering qualities of oil. Demisting equipment is used on optics and night vision equipment to combat the effects of humidity.
- Temperature variation can cause condensation in humid desert areas affecting optics, fuel lines, air tanks, and weapons. Expansion and contraction of air and fluids cause tires to over inflate during the day and under inflate at night. Fuel tanks may overflow during the day causing a fire hazard. Oil fluid levels become overfull and cause leaks during the day, or insufficient lubrication occurs when the oil cools. Vehicle operators and crew chiefs must ensure that the effects of temperature variations do not become a significant problem.
- Static electricity is prevalent in the desert. This is important to remember during refueling operations and when oxygen is being used onboard vehicles or aircraft. Proper refueling procedures must be followed. Static electricity also causes severe shock to ground personnel in sling load and hoist operations.

CAUTION

During hoist and sling load operations, the load must touch the ground before the ground crew can handle it.

5-18. To ensure success in desert operations, detailed planning is required. Factors to consider include—

- Additional quantities of water are required for HSS/FHP operations for the survival of both medical personnel and their patients. Load plans for all vehicles and aircraft must include water. Water is as mission essential as any piece of unit equipment. It should be a priority item when load plans are developed.
- Prescribed load lists are expanded to carry sufficient quantities of repair parts easily degraded by the environmental factors. For example, rubber and plastic fittings and tubes, or spare parts for communications equipment.

- Plan for the effects of wind, sand, and sun on all equipment. All plastic and glass surfaces on vehicles, aircraft, and other equipment should be covered when not in use. Covers should be ordered or made prior to deployment.
- Fuel planning is critical due to power limitations, extended range requirements, and increased vulnerability of refueling sites in the relatively open desert terrain. Careful planning of FARPs is essential for mission accomplishment.
- Units should plan to pack both hot and cold weather clothing when deploying due to varying temperatures between day and night.
- Petroleum, oils, and lubricants (POL) should be of the proper viscosity for desert operations. Maintenance services are also performed more frequently on ground vehicles and aircraft, thus requiring a larger amount of POL than normal.
- Extra filters of all types are planned for due to a higher consumption rate.

5-19. Training for desert operations is not significantly different than training for operations in other areas except for the following:

- Because many desert areas are in mountainous terrain and because high temperatures increase density altitude, aeromedical evacuation units should conduct mountain training to prepare for contingencies in desert areas. Further, procedures and techniques for evacuation in mountainous terrain must be practiced by all AHS personnel. Special equipment requirements (paragraph 5-7) must also be considered.
- Navigation in desert terrain varies from simple to extremely difficult. Factors affecting navigation are the type of desert and the scale and quality of the available navigational charts. At times, aircraft may have to use dead-reckoning navigational techniques (time, distance, and heading). Ground vehicles must have compasses available, as they have to rely on compass headings and odometer readings to navigate. Ground and AA crews should be able to interpret navigational charts and maps of all types and scales. Use of convoys is a viable technique to improve security and to ensure that ground vehicles do not get lost. Aircraft may be used to assist in navigation by convoys in those areas in which there are poor road networks and the terrain offers no distinctive features by which to navigate.

5-20. In principle, medical evacuation operations in the desert do not differ greatly from these operations in other environments.

5-21. Helicopter landing sites should be chosen with care. Common mistakes made by many units when establishing the LZ are—

- Locating the pad relative to the patient and tents, vehicles, and other obstacles. A common tendency is to locate the helipad downwind of MTFs so that approaches may be made into the wind towards the facility. In mountainous deserts, winds normally channel down the valleys and are more predictable along valley floors. A better site selection for an LZ is with the MTF, alongside the approach and takeoff zone. Thus, the landing direction is up or down the valley, depending on the airflow, and the MTF is not over flown.
- Situating landing sites in washes, small confined areas between large rocks, or close to moving tracked vehicles. Map coordinates are rarely accurate unless the site is beside a major terrain feature. Therefore, LZs should be located next to major terrain features or on higher ground where they can be seen from the air at a distance of 2 to 3 kilometers, if possible. Lack of distinctive features in the open desert and on large-scale maps makes pinpoint navigation difficult, especially at night.
- Marking of helicopter LZs is done so that the pad can be seen from the air, but the markings should not be a hazard in themselves. If engineer tape is used, it should be firmly secured to prevent it from blowing loose. Panel markers are not a good tool to use as they are difficult to see. If panel markers are used, they need to be secured. If used, flares or marker smoke should not be deployed on or directly upwind from the pad. Smoke grenades or flares should not be thrown under the aircraft as it lands. Avoid using white smoke to mark the LZ. Colored smoke is probably the best daylight marking method. It is difficult to detect a smoke grenade more

than 2 to 3 kilometers away, but an aircraft in the general vicinity can normally see it. Radios are used to guide aircraft to the LZs, but this creates an electronic signature. Units requesting medical evacuation must be prepared to signal the evacuation aircraft upon its arrival. Normally, map coordinates will guide the aircraft to within 2 to 3 kilometers of the LZ. Even from nap-of-the-earth (NOE) altitudes, the aircrew may be able to see several units in the area. The requesting unit must signal the aircraft to ensure the designated LZ is used.

5-22. When planning for night flights, considerations include the following—

- Moonlight aids the medical evacuation pilot by providing him with the light to see with either unaided vision or night vision goggles (NVGs). When adequate ambient light exists, medical evacuation crews function almost as effectively at night as they do during daylight. The small arms threat is somewhat reduced at night, although it still exists from radar-guided weapons, infrared-sited weapons, and passive night vision device-equipped weapons systems. Flying into a bright moon with NVGs on can be compared to flying into the sun during the day. The goggles darken and visibility becomes extremely poor. Flight routes should not be planned to fly directly into a bright moon if NVGs are to be used.
- The lack of visual cues over sand is similar to that over water. It is very easy for pilots to become disoriented and fly into the ground. Reliance on radar altimeters is a must over flat sandy areas of the desert.
- Frequently, desert areas do not have sufficient ambient light to allow adequate night vision, even with the aid of NVGs. A pilot wearing NVGs is often unable to see the ground at an altitude of 100 feet using a landing light equipped with a pink light filter. Under these conditions, dead reckoning is the only effective navigation method unless global positioning satellites, Doppler equipment, or NAVAIDS are available. Unfiltered light can be used with or without NVGs; however, this increases the risk of exposing the aircraft's position to the enemy.
- Desert warfare is usually characterized by extended battle zones which increase evacuation distance and time. Combat health support units are located further to the sustainment area in the desert. Establishing an ambulance shuttle system or CCPs is useful. Combat health support units require a greater number of vehicles for operating in deserts than in other environments. Air evacuation by fixed and rotary-wing aircraft is the preferred method due to their speed and range. Further, using aircraft reduces the load on ground vehicles. Augmentation from higher echelon HSS may also be required to meet the extended evacuation needs.
- Smoke is used extensively on the modern battlefield by both sides. It can be effectively used to mask friendly actions to include medical evacuation. (Refer to Appendix E for further information.)
  - Smoke can be a major hazard, especially to medical evacuation helicopters. Smoke reduces visibility and forces an aircraft higher where it can be acquired by threat weapons systems. The phenomenon of inversion occurs often in the desert. When this happens, medical evacuation vehicles and aircraft may be able to work underneath the smoke using the smoke layer for overhead concealment.
  - Medical units must coordinate closely with supported organizations on smoke operations. Smoke can either help or hinder the evacuation mission, depending upon how it is used.
- Communications in the desert are affected by a number of factors. Atmospheric interference and the skip of signals occur frequently. Mineral deposits in the desert may unexpectedly disrupt communications. Many of these problems can be overcome by using additional radio relays, pre-established control measures, and visual signals.
- Artificial lights may be used at times in the desert. They are very easily detected. Even with blackout lights, vehicles using lights can be detected for miles with NVG. Serious consideration should be given to driving without using lights when the tactical situation dictates. Ground guides are used to help vehicles navigate through areas that are not clearly marked or through area where troops are present.
- Wind is one of the most significant environmental factors affecting medical evacuation in the desert. Wind can be destructive to both structures and equipment; tents, antennas, and aircraft

can be easily damaged. Wind direction and speed vary greatly within the space of a few miles. Velocity is substantially increased when wind channels between hills and direction changes due to interference of terrain features. The wind frequently makes aeromedical evacuation impossible by exceeding the operating limitations of the aircraft. At other times, it may limit the use of some potential LZs. Blowing sand and dust can slow down the evacuation system by making navigation by either ground or AA difficult, if not impossible. High winds are predictable to a certain extent. For example, at certain times of the year in the Mojave Desert high winds occur every day at dusk and last for 3 to 4 hours. At other times, high winds, based on frontal weather patterns, can remain for several days at a time. These factors should be considered by HSS planners, and medical evacuation assets should be massed or relocated accordingly.

- The desert provides little or no protection from enemy air defenses except in mountainous terrain. Aircraft may have to be flown above NOE altitudes to prevent a dust signature. These factors cause increased exposure and vulnerability of AAs to enemy air defenses and may limit their employment.

5-23. Further Information. Refer to FM 90-3 for additional information on desert operations.

SECTION IV — EXTREME COLD WEATHER OPERATIONS

5-24. Operations in the extreme cold have many of the limiting factors found in desert operations. The tundra and glacial areas are harsh, arid, and barren. Temperatures may reach lows of -80° F to -100° F which, combined with gale force winds, make exposure unsurvivable.

- The greatest environmental detriment to operations is blowing snow. This results in a loss of depth perception from total white conditions. Blowing snow is caused by the wind or by the rotor wash of helicopters; its effect reduces visibility to zero.
- Other environmental considerations are as extreme but easier to circumvent. Solid footing is suspect in both the dead of winter and in the summer. Snow and ice cover crevasses, holes, and otherwise unstable ground. In traversing suspect ground situations, consider linking Soldiers by rope. During the summer, ground transportation is more restricted than in any other environment due to the marsh and muskeg composition of the arctic tundra. Patients must be sustained for a longer duration due to terrain delays and the lack of direct lines of evacuation.
- Greater responsibility has to be placed on each Soldier, especially for maintenance of nutrition and water consumption. It is imperative to stress that leadership and training are important in the prevention of cold weather injury. Strict adherence to the guidelines found in FMs 21-10 and 31-70 assures an effective fighting force. Water conservation is essential; however, adequate consumption by the individual should be enforced.
- Factors to consider for conducting evacuation in arctic operations include the following:
  - Arctic warfare is usually characterized by extended battle zones that increase evacuation distance and time. Establishing an ambulance shuttle system or CCPs is useful. Augmentation from higher echelon AHS may also be required to meet the extended evacuation needs.
  - Additional supplies of water should be carried by ambulances and maintained at CCPs, if possible.
  - Due to the decreased temperature and frozen environment, ambulance maintenance requirements are increased. Lubricants must be of the correct viscosity for the temperature. In extreme cold, batteries perform less efficiently. Consult the specific aircraft maintenance manual for the recommended battery and procedures for extreme cold weather operations.

Note. All ambulances are considered deadlined without a functional heater for the patient compartment.

  - The proper storage of medical supplies is essential to prevent loss from freezing.

- There are few terrain features or road networks; therefore, evacuation routes must be surveyed and marked over open terrain. At extreme latitudes, operations during the winter months are conducted in extended hours of darkness. The use of NVG may be required. Compass accuracy is inconsistent due to a geomagnetic phenomenon. Beacons and homing devices are essential for air navigation.
- Weather is extremely unpredictable. There are too few observers to allow for accurate assessment of weather patterns. Unfavorable weather conditions cause unexpected delays; therefore, medical personnel must be prepared to provide survival measures for their patients and themselves.
- Landing zones must be chosen with extreme care in both winter and summer. Blowing snow mandates instrument-assisted takeoffs and running landings. Landing areas must be correspondingly larger. The full weight of the aircraft cannot be allowed to settle on the skis until after firm ground conditions are established. Movement of patients to and from the aircraft is difficult. Where an aircraft lands is where it stays. A rocking motion, to free the skis prior to lift off, is performed using the cyclic and antitorque controls.

- Thorough planning and strict preparation are the keys to survival. Factors to consider include the following:
  - Mud obstacles at noon may become an avenue of approach at midnight.
  - Snow complicates all work. Snow-covered terrain hampers reinforcements, muffles noise, makes cross-country driving hazardous, and creates different camouflage requirements.
  - Because of thermal sights, a complete reappraisal of concealment is required.
  - Tracks in the snow destroy concealment.
  - No Soldier is assigned to any job alone. The buddy system is used at all times.
  - Anticipate that all maintenance tasks will take twice as long.
  - Bare metal can stick to skin or wet garments in subfreezing temperatures.
  - Fuel spilled on skin or garments increases the freezing factor; it is one of the greatest causes of injury in winter operations.
  - When operating in the cold, anticipate increased POL needs. Fuel consumption can rise as much as 25 percent for vehicles operating in deep snow, slush, or mud.
  - The recommended fuel for Yukon stoves is diesel.
  - Make every effort to warm gearboxes and engines before starting.
  - The first consideration in the AO is heat; followed by shelter for sustained work.
  - Soldiers need to stand clear of taut cables; steel tends to be brittle and breaks in extremely cold temperature.
  - Fire extinguishers are winterized by adding 15 percent nitrogen to the carbon dioxide.
  - Degradation of battery life requires changes as much as six times more frequently than in a more temperate environment.
  - Radio sets are warmed up prior to transmission. The sets may be turned on but should not transmit for at least one-half hour.

Note. Single-channel ground and airborne radio system (SINCGARS) radios do not require a warm up period.

  - Frost shields (such as using the plastic bag in which the batteries are packed) should be placed over microphones.
  - Grounding rods have to be buried horizontally instead of pounded in vertically. Recovery of stakes and rods placed in the ground is significantly more difficult.
  - Flooring is needed in heated areas because of the thawing of the tundra.
  - Soldiers must take breaks for water and warmth.
  - Static electricity presents a serious safety hazard especially around flammable materials.

5-25. For additional information, refer to FM 31-70.

SECTION V — CHEMICAL, BIOLOGICAL, RADIOLOGICAL, AND NUCLEAR ENVIRONMENTS

5-26. Medical evacuation and treatment are conducted continuously throughout operations conducted in a CBRN contaminated environment. The AHS commander must have a comprehensive plan which is rehearsed on a periodic basis to ensure the timely evacuation and treatment of casualties in a CBRN environment. Techniques and procedures which are essential for operating in a contaminated environment should be contained in the unit TSOP. The number of casualties and their medical condition, type of contaminant, the size of the land area contaminated, the expected duration of operation, risk assessment and acceptable level of risk, and the number of AHS assets (medical personnel, medical units, and evacuation vehicles and aircraft) initially contaminated will determine the quantity and type of uncontaminated AHS resources, if any, which will be introduced into the contaminated environment to ensure timely medical treatment and evacuation occur.

5-27. The Commander must take into consideration the number of assets he is willing to commit during evacuation operations in a CBRN contaminated environment. Since the combinations of evacuation methods are nearly endless, the commander has greater flexibility in tailoring an evacuation plan to meet his particular tactical situation in a CBRN contaminated environment.

5-28. On the modern battlefield there are three basic modes of evacuating patients (personnel, ground vehicles, and aircraft).

- In using personnel to physically carry the casualties, the commander must realize the inherent stress involved. Cumbersome mission-oriented protective posture (MOPP) gear needed in a contaminated environment (added to climate, increased workloads, and the fatigue of battle) greatly reduces the effectiveness of unit personnel.
- If the commander must send evacuation personnel into a radiological contaminated area, he must establish operational exposure guidance (OEG) for the medical evacuation operation. Radiation exposure records are maintained by the unit CBRN NCO and are made available to the commander, staff, and surgeon. Based on OEG, the commander decides which medical evacuation assets to send into the contaminated environment.

5-29. Commanders should make every effort to limit the number of evacuation assets which are contaminated while still maintaining a timely and effective medical treatment and evacuation operation.

- It is expected that a certain number of both ground and AAs will become contaminated in the course of battle. The commander can, therefore, segregate the contaminated ones. This results in the smallest impact on his available assets and the greatest possibility for continuing the patient evacuation mission. Optimize the use of resources, medical or nonmedical, which are already contaminated before employing uncontaminated resources.
- Once a vehicle or aircraft has entered a contaminated area, it is highly unlikely that it will be able to be spared long enough to undergo a complete decontamination. This depends upon the contaminant, the tempo of the battle, and the resources available. Normally, contaminated vehicles (air and ground) have restricted use and are confined to dirty environments.
- Introducing uncontaminated aircraft into a contaminated area should be avoided, whenever possible. Ground ambulances should be used instead of AAs as long as their use does not adversely affect the patient's medical condition. Ground ambulances are more plentiful and are easier to decontaminate. This does not, however, preclude using aircraft in a contaminated environment or in the evacuation of contaminated patients.
- The relative positions of the contaminated area, FLOT, and threat air defense systems determine if and where helicopters are to be used. The commander may choose to restrict one or more helicopters to the contaminated areas and use ground vehicles to cross the line separating contaminated and clean areas. The ground ambulance can proceed to the receiving MTF with a

patient decontamination station. The patient can then be transferred to a clean ground or AA if further evacuation is required. The routes used by ground vehicles to cross between contaminated and clean areas are considered dirty routes and should not be crossed by clean vehicles. The effects of wind and time upon the contaminants must also be considered.

- The rotor wash of the helicopters must always be kept in mind when evacuating contaminated casualties. The intense winds disturb the contaminants in the area and further aggravate the condition by additionally spreading the contaminants. Ideally, the aircraft must be allowed to land and reduce to a flat pitch prior to bringing any patients near. This will be dictated by the tactical situation, but allows some reduction in the effects of the downwash. A helicopter must not land too close to a decontamination station (especially upwind) because any trace of contaminants in the rotor wash will compromise the decontamination procedure.
- Immediate decontamination of aircraft and ground vehicles should be accomplished to minimize crew exposure. Units should develop their own procedures for thorough decontamination and document them in their TSOPs. A sample aircraft decontamination station that may be tailored to a particular unit's needs is provided in FM 3-11.5.
- Evacuation of patients must continue even in a contaminated environment. The commander must recognize the constraints placed upon him by resources and plan and train to overcome deficiencies.
- Refer to FM 4-02.7 for additional information on AHS operations in a CBRN environment.

SECTION VI — SHORE-TO-SHIP EVACUATION OPERATIONS

5-30. It is imperative that Army aeromedical evacuation units be able to interface with USN air-capable ships. Lessons learned from past operations, such as Vietnam and Grenada, have shown that US Army helicopters should be able to operate to and from USN air-capable ships. An interservice agreement between the Army and the Navy allows for deck-landing qualification of Army pilots. (Refer to FM 1-564 for additional information.)

- It is important that units having contingency missions requiring Navy support establish training requirements to obtain naval-operations orientation, water egress training, water survival, and deck-landing qualification. This enhances the successful accomplishment of the aeromedical evacuation mission to naval vessels.
- In past joint operations, communications have been burdensome for both Army and Navy elements. Commonality of communication requirements should be established during training exercises. Communication equipment and frequencies for medical evacuation to Navy vessels must be established. This will provide smooth integration of Army helicopters into the Navy airspace management system during actual operations.
- As the Navy vessels may operate relatively long distances from the ground combat operations, Army aeromedical evacuation units need to be proficient in over water navigation. The use of NAVAIDS from the Navy element in support of the operation is the first priority for over water navigation. Basic dead-reckoning remains a secondary measure.

5-31. An important aspect of joint operations is the medical capabilities of Navy vessels servicing the CZ. Knowledge of ship's medical capabilities assists the MRO to direct patients to proper treatment sites. There are many classes of ships which can meet the medical needs of ground forces. Destroyer tenders and aircraft carriers have helicopter landing areas, one operating room (OR) and, at a minimum, one medical officer. Amphibious ships have the most extensive medical facilities of any Navy combat ship. The Navy has 59 amphibious ships in active commission plus two tank landing ships, which are operated by the Navy Reserve Force. The primary mission of the amphibious ships is to transport and support the Fleet Marine Force. The ships have the additional duty of casualty receiving and treatment ships (CRTS). During normal operations, the medical staff is kept to a minimum. The medical staff is augmented when expanded capabilities are needed. Current information regarding landing requirements and medical capabilities should be obtained during training periods with the Navy. Casualty care is secondary to the combat mission of all US combat ships.

- Amphibious assault ships "WASP" class are designated by the Navy as amphibious helicopter assault carrier dock (LHD) (followed by a number) and have the largest patient care facilities on any US combat ship. The WASP class ships have six main ORs, four dental ORs, bed capacity that can be expanded to 600, and it carries 1,500 pints of frozen blood. This ship can receive casualties from helicopters or landing craft.
- Amphibious assault ships "TARAWA" class are designated by the Navy as amphibious helicopter assault carrier (LHA) (followed by a number). The TARAWA class ships have three main ORs, two dental ORs, an overflow bed capacity of 300, and carries 1,500 pints of frozen blood. The ship can receive casualties from helicopters or landing craft.
- Amphibious assault ships "IWO JIMA" class are designated by the Navy as amphibious assault helicopter carrier—aviation (LPH) (followed by a number). These ships were specifically designed to operate helicopters. The IWO JIMA class ships have two ORs and an overflow bed capacity of 200.
- The amphibious transport dock (LPD) (followed by a number) has less medical capabilities than the LHD, LHA, or LPH ships. It can be designated as secondary casualty receiving ship.
- The older dock landing ship (LSD) (followed by a number) can be used as a secondary CRTS when augmented. The newer class of LSD currently under construction can be used as a casualty receiving ship with a capacity for 50 wounded.
- The tank landing ship (LST) (followed by a number) is another type of ship used in amphibious operations. It is designed with a helicopter platform and a stern ramp. Patients can be delivered by air or boat when required by tactical or mass casualty situations. When the LST is augmented with medical personnel and materiel, it can be used for the emergency treatment and evacuation of patients.
- The troop transport (AP) (followed by a number) is not in active service. When available, the troop transport can be outfitted with special medical facilities and carry sick, injured, and wounded personnel.

5-32. The Military Sealift Command operates two hospital ships. The United States Navy Ship (USNS) MERCY T-AH 19 and the USNS COMFORT T-AH 20. One ship is based on each coast and, when needed, will be assigned medical staffs from military hospitals, getting underway within 5 days. The hospital ships MTFs were designed for a total capacity of 1,000 casualties, including 500 acute care beds and 500 recuperation beds. The hospital ships have 50 trauma stations in the casualty receiving area; 12 operating rooms; a 20-bed recovery room; 80 intensive care beds; and 16 intermediate, light, limited care wards. The maximum patient flow rate, for which the helicopter facility and the casualty reception area were designed, is 300 patients per 24 hours. There is a limited capability to receive casualties from boats.

5-33. The US Army has the shore-to-ship medical evacuation mission on an area support basis for Marine forces deployed on land.

SECTION VII — AIRBORNE AND AIR ASSAULT OPERATIONS

5-34. Airborne and air assault operations are some of the most complicated military operations to undertake, not only from an operational perspective, but also from a sustainment and conservation of the fighting strength perspective. A number of personnel within the medical chain that need to coordinate, not only amongst the AMEDD, but also include the aviation community as what proponent own the aircraft (for example, BAE, BSS, medical platoon, and such). Manual evacuation techniques may have to be used and planners should plan for redundant communications, only ground medical evacuation may or may not be available.

5-35. The airborne and air assault operational forces are specialized forces employed to maximize their design characteristics. Airborne units are a flexible force that can be strategically or tactically deployed. They can be inserted rapidly anywhere in the world as either a deterrent or strike force. Air assault units are flexible and lethal fighting organizations. They are ideally suited for rapid employment to critical areas beyond the reach of ground forces.

5-36. After airborne forces have landed in the objective area, they reorganize and maneuver to seize objectives. When it is necessary for assault aircraft to land in the drop zone, they are parked and unloaded rapidly. Then, they may be used to transport Soldiers injured during the parachute assault. It must be understood that organic medical units may experience an overload of patients during the early phases of an airborne assault. These units have to hold the patients until either ground link-up is made or evacuation can be established at airheads. Aeromedical evacuation from the airhead is accomplished using tactical and strategic USAF aircraft.

5-37. The air assault division's organic aircraft have the ability to attack from any direction, over fly obstacles, and bypass enemy positions. Evacuation of patients in the assault phase is accomplished by AAs. Air ambulances may accompany the air assault task force (AATF) or respond from laager sites once the initial assault has taken place. If AAs are providing on-call support, it will be necessary to fly secure air avenues of approach.

5-38. When both airborne and air assault divisions have been employed and become a part of other conventional forces, their operations are similar to that of light infantry forces. During initial deployment, division medical evacuation assets may be used to evacuate patients to the airhead for air evacuation directly to division/corps hospitals.

Chapter 6
Medical Regulating

GENERAL

6-1. Medical regulating is the coordination and control of moving patients to MTFs which are best able to provide the required specialty care. This system is designed to ensure the efficient and safe movement of patients.

PURPOSE OF MEDICAL REGULATING

6-2. Medical regulating entails identifying the patients awaiting evacuation, locating the available beds, and coordinating the transportation means for movement. Careful control of patient evacuation to appropriate hospitals is necessary to—

- Effect an even distribution of cases.
- Ensure adequate beds are available for current and anticipated needs.
- Route patients requiring specialized treatment to the appropriate MTF.

6-3. The factors that influence the scheduling of patient movement include the following:

- Patient's medical condition (stabilized to withstand evacuation).
- Tactical situation.
- Availability of evacuation means.
- Locations of MTFs with special capabilities or resources.
- Current bed status of MTFs.
- Surgical backlogs.
- Number and location of patients by diagnostic category.
- Location of airfields, seaports, and other transportation hubs.
- Communications capabilities (to include radio silence procedures).

MEDICAL REGULATING TERMINOLOGY

6-4. As medical regulating may include coordination with other services, it is necessary to use the correct terminology. These terms include—

- Intracorps Medical Regulating—This is the system by which patients are transferred or evacuated from an FSB or main support battalion (MSB) to a combat support hospital (CSH).
- Intratheater Medical Regulating—This is the system by which patients are transferred or evacuated from one hospital to another within the AO. This includes evacuations between CZ hospitals, between EAC hospitals, or from CZ hospitals to EAC hospitals.
- Intertheater Medical Regulating—Moving patients between, into, and out of the different theaters of the geographic combatant commands and into the CONUS or another supporting theater. (JP 4-02.)
- Patient Administrator—The patient administrator (PAD) accomplishes the medical regulating function at the hospital level in addition to his normal duties. His medical regulating functions include consolidating all evacuation requests within the hospital and forwarding an evacuation request to his next higher headquarters for action. The PAD is also responsible for keeping his next higher MRO apprised of the current beds available and the OR status.

- Medical Regulating Officer—The MRO functions as the responsible individual at C2 headquarters for receiving and consolidating evacuation requests. These requests are initiated by the DSS or subordinate hospitals. The MRO also maintains the current patient status, bed status, and the surgical backlog at subordinate hospitals. His duties include—
  - Managing what patient classes are regulated into his facility.
  - Determining what resources are available to move the patients and coordinating for the use of these assets.
  - Maintaining accountability of patients within the MTFs.
  - Preparing reports as required.
- Theater Patient Movement Requirements Center—The TPMRC is a joint agency normally located at or near the unified theater headquarters. The theater surgeon supervises the functions of this office. These functions include—
  - Maintaining direct liaison with the Global Patient Movement Requirements Center (GPMRC), the MROs of component services, and the transportation agencies which furnish the means for evacuation.
  - Obtaining periodic reports of available beds from the services MROs providing hospitalization.
  - Selecting hospitals based on the reported bed availability to receive patients within EAC.
- Global Patient Movement Requirements Center—The GPMRC is a joint agency located in CONUS and established by the USTRANSCOM. The GPMRC receives requests from the TPMRCs. The primary role of the GPMRC is to apportion intertheater assets to the TPMRCs, collaborate and integrate proposed TPMRC intertheater plans and schedules, and communicate lift and bed requirements. The destination hospital is determined based on the patient's medical needs, the available transportation resources, and MTF capabilities.
- Joint Patient Movement Requirements Center (JPMRC)—Joint patient movement requirements center provides theater patient movement requirements center-type domain, automatic information system support and operations for a joint task force operating within a unified command area of responsibility. The theater patient movement requirements center maintains overall responsibility for theater patient movement operations, but the joint patient movement requirements center is responsible for patient movement operations within its area scope of responsibility and coordinates with the theater patient movement requirements center for intra-theater patient movement and the GPMRC for inter-theater patient movement. Also called JPMRC.
- Theater Aeromedical Evacuation System (TAES)—The TAES is a functional organization which is provided by the USAF and performs the mission of theater AE. Significant components of the TAES include the following:
  - Aeromedical evacuation command squadron.
  - Aeromedical evacuation control team (AECT).
  - Mobile aeromedical staging facility.
  - Contingency aeromedical staging facility (CASF) (under the C2 of the medical group commander).
  - Aeromedical evacuation crews.
  - Critical care air transport teams (CCATTs) (not necessarily Air Force).
- Aeromedical Evacuation Command Squadron—The AE command squadron provides C2 of all assigned TAES forces and can deploy in advance of other AE components to arrange support requirements for AE forces. The command squadron advises commanders and other personnel/agencies on AE operations, capabilities, and requirements and provides procedural and technical guidance for attached and transiting AE elements.
- Aeromedical Evacuation Control Team—The AECT is a CASF (under the C2 of the medical group commander) and is an USAF element responsible for linking validated patient movement requests with the appropriate airlift platform. The AECT is normally located within the Air

Mobility Division of the theater air operations center (AOC) for intratheater and the tanker airlift control center (TACC) for intertheater coordination of AE operations using USAF aircraft. The CONUS and strategic AOC for the USAF is the TACC located at Scott Air Force Base, Illinois. The AECT coordinates with and receives requirements from the PMRC for execution.

- Aeromedical Evacuation Liaison Team (AELT)–The AELT provides a direct communications link and immediate coordination between the user service and the TAES. The AELT is located at any level where Air Force fixed-wing requests are initiated, verifies and coordinates with the AECT/patient movement requirements center (PMRC) the physiology of flight issues and patient flight/movement requirements, assists with patient preparation for flight, and directs patient on-load activities.
- Mobile Aeromedical Staging Facility—The MASF is a USAF staging facility employed at forward airfields in the CZ to provide a temporary staging capability for preparation of patients being evacuated from corps to EAC hospitals. The MASF is employed to ensure patients are prepared for aircraft loading with the main focus of reducing AE aircraft ground time.
- Contingency Aeromedical Staging Facility—The CASF, similar to the MASF, is a USAF staging facility. The CASF, which provides greater capability and longer holding periods than the MASF, is a fixed facility located at theater AE strategic hubs and interfaces with the TAES. As opposed to the TAES elements, which are under the C2 of the AE command squadron, the CASF reports to the medical group commander.
- Military Sealift Command (MSC)—The MSC is the USN element responsible for coordinating movement of supplies, equipment, and personnel into the JOA by Navy ships. Further, it coordinates, through the TPMRC, the medical evacuation of patients by ship from the JOA to the support base or hospital ship, as required.
- Corps Movement Control Center (CMCC)—The CMCC is the corps movement control organization. It provides centralized movement control and highway regulation for movement of personnel and material into, within, and out of the corps area. When USAF capabilities are exceeded, the CMCC coordinates requests for additional air and ground resources. It also obtains the necessary clearances to support the medical evacuation mission from the CZ.
- Movement Control Battalion (MCB)—The MCB mission is to provide movement management services and highway traffic regulations and to coordinate for personnel and material movements into, within, and out of the theater. The MCB coordinates with allied and HN movement control agencies. It also coordinates with the USTRANSCOM and its subordinate units (such as the Air Mobility Command and MSC) and prepares movement and port clearance plans and programs.
- Joint Military Transportation Board (JMTB)—The JMTB is a joint staff composed of members of the Army, USAF, and USN that coordinates transportation requirements for patients requiring intertheater evacuation.
- Transportation Command Regulating and Command and Control Evacuation System—The Transportation Command Regulating and Command and Control Evacuation System is the single DOD automated information system that supports medical regulating, patient movement, and patient in-transit visibility requirements.
- Automated Patient Evacuation System—The Automated Patient Evacuation System is the system that automates the patient movement portion of medical evacuation.

MEDICAL REGULATING FOR THE DIVISION

6-5. Medical regulating in and from the division is the responsibility of the DSS. Medical regulating in the division is not as formalized as the rest of the medical regulating system. It is usually operated procedurally so as not to depend solely on communications to effect rapid evacuation. The medical regulating function in the DSS is concerned primarily with.

- Tracking the movement of patients throughout the division MTFs and into the corps facilities.
- Monitoring the use of ambulance assets.

6-6. Air and ground ambulances placed in GS of the BCTs can be field sited in the sustainment area when the mission directs. When these assets go forward to the BSMC or the ASMC to evacuate patients to corps MTFs, they have corps MTF destinations predetermined (blocks of beds). The DSS, in coordination with the medical group/brigade MRO, establishes the number of patients a supporting corps hospital can accept during a particular period of time. These blocks of available beds are then provided to the GS ambulances prior to the call for missions.

Note. Under the MRI force structure, the medical group functions were absorbed by the medical brigade.

- Once an evacuation mission is completed, the originating division MTF contacts the patient disposition section of the BSS and provides—
  - Patient numbers by category and precedence.
  - Departure times.
  - Modes of transportation.
  - Destination MTFs.
  - Any other information required by TSOP.
- The DSS, in turn, notifies the medical group/brigade MRO via the patient administration net, which is monitored by the corps MTFs. Since corps ground ambulances have no onboard communications ability and AAs have no amplitude-modulated high-frequency capability at present, all patient information is passed to the gaining MTFs via the patient administration nets. To reduce the turnaround time for ground ambulances and to move more serious patients to the CSHs in the sustainment area—
  - Air ambulances are given blocks of beds in the corps hospitals farther to the rear.
  - Ground ambulances are normally given blocks of beds in the more forward deployed CSHs.

6-7. Medical evacuation can be effected immediately, procedurally, and under conditions of communications silence without interrupting the continuum of care by—

- Preparing patient estimates.
- Prioritizing and task-organizing ambulance support.
- Assigning blocks of hospital bed designations prior to the start of the mission.
- The assistant chief of staff (personnel) (G1) is also involved in medical regulating as the G1 is responsible for personnel accountability and casualty reporting. Casualty liaison teams (CLTs) will normally be colocated at CSH to assist and coordinate patient tracking and casualty reporting.

MEDICAL REGULATING WITHIN THE COMBAT ZONE

6-8. The requirement to transfer patients from one hospital to another within the CZ occurs. This results from—

- Surgical backlogs.
- Mass casualty situations.
- Specialty care requirements.
- Planned movement of an MTF.

6-9. When it is necessary to transfer a patient, the attending physician notifies the hospital PAD. The PAD consolidates all such requests from the hospital and requests movement authority from the IPMC.

6-10. If the medical IPMC can transfer the patient or patients to its subordinate hospitals, he designates the hospitals to receive the patients and notifies both the requesting and receiving hospitals of the transfer. The IPMC also tasks subordinate medical evacuation units for the assets to transfer the patients.

6-11. If the MEDBDE cannot provide the needed hospitalization within its own resources, the MRO forwards the request to the medical TPMC for action. The TPMC then designates the receiving hospitals and notifies the subordinate IPMCs. The medical IPMCs disseminate the information to the hospital PAD and coordinate the evacuation resources for the transfer. The MRO also coordinates the regulation of patients to—

- Other US military service hospitals and naval hospital ships.
- Allied nations' military hospitals.
- Other authorized supporting facilities.

MEDICAL REGULATING FROM THE COMBAT ZONE TO ECHELONS ABOVE CORPS

6-12. Hospital attending physicians and oral and maxillofacial surgeons submit daily reports to the hospital PAD listing the patients requiring evacuation. The PAD assembles this information and transmits the report to the medical brigade headquarters. This report is a request for transportation, as well as a notification of the number of patients requiring evacuation. The report classifies the patients according to—

- Diagnostic category.
- Desired on-load points.
- When the patients will be available for evacuation.

6-13. The medical brigade TPMC consolidates these reports from each hospital attached to the medical brigade and forwards his report to the corps TPMC. The medical brigade MRO consolidates the reports and transmits the data to the theater TPMC.

6-14. If a JPMRC has been activated within the theater, the MEDCOM MRO consolidates all reports from the CZ medical brigades and forwards them to the JPMRC. The JPMRC designates hospitals in the EAC to receive the patients. The designation is based on the previously received bed status reports from all service components and available means of evacuation. The JPMRC then notifies the MEDCOM MRO of designated hospitals. The TPMC accomplishes this task if the JPMRC is not activated.

6-15. The primary means of moving patients from the CZ out of theater is USAF aircraft. With the elements of the TAES deployed, it is possible to find AELTs at each echelon and as far forward as Air Force fixed-wing requests are initiated. The PMRC monitors the MRO patient evacuation requests and passes requirements to the AELT. At the same time they pass airlift requirements to the AECT, seeking an aircraft to perform the evacuation mission. The AELT, through the MRO patient movement request, requests the PMRC/AECT to move patients. Included in the request are the originating medical facility (OMF) and the destination airfields. The airfields selected are those serving the hospitals designated to receive patients.

6-16. The AECT is a component of the TAES and performs the mission of coordinating the movement of and providing in-flight medical care to patients while under the USAF control. The AECT receives patient movement requirements from the TPMRC, then works with the airlift control team (ALCT) in the AOC to meet the evacuation requirements.

6-17. The AOC coordinates the forward movement of cargo and personnel aboard USAF aircraft with other USAF units, Army transportation representatives, and USN agencies. Certain of these aircraft are scheduled to evacuate patients on their return trips. These aircraft seldom go forward solely to evacuate patients; however, these missions may be used for retrograde airlift and mixed cargo/AE missions is the primary means of AE due to limited aircraft and the need to maximize the use of this asset.

6-18. After the schedules have been arranged, the AECT returns the detailed flight schedule to the JPMRC who passes the information to the TPMC/AELT and the parent AE element.

6-19. The MEDCOM MRO, in coordination with the AELT, issues these instructions to the IPMCs (with the authority to move patients in Army CZ facilities) and the receiving hospitals. The hospitals must prepare to receive the patients at the destination airfields. (This may be accomplished by collocating an ASMC at the airfield to receive the incoming patients.) The patients are sorted by destination hospital and moved by Army medical evacuation means. The instructions mentioned above include, as a minimum, the—

- Number of patients to be moved.
- On-load airfield.
- Destination airfield.
- United States Air Force aircraft mission number.
- Estimated time of arrival at the destination airfield.

6-20. The TPMC issues the flight and movement instructions to its subordinate medical IPMCs. The IPMCs then direct the evacuation units and hospitals within their AOs to move the patients to the on-load airfield according to the arrival time of the aircraft. This movement must be closely controlled, as a MASF can accommodate up to 50 patients. The patients cannot be delivered to the MASF earlier than 6 hours prior to arrival of the aircraft and no later than 1 hour prior to arrival.

MEDICAL REGULATING WITHIN ECHELONS ABOVE CORPS

6-21. Medical regulating within the EAC is similar to the system used within the CZ. Attending physicians within the Role 3 hospitals notify the hospital PAD of patients requiring evacuation to CSHs. The PAD then consolidates the requests from the hospital and forwards the consolidated requests to the IPMC. The IPMC, in turn, consolidates the requests and forwards them to the TPMC.

6-22. The TPMC, based on periodic bed status and availability reports from subordinate hospitals, designates specific hospitals to receive the patients. The hospitals are designated based on bed availability, to include specialty beds, to support the specific patient. The TPMC then notifies the requesting medical IPMC of the designated hospitals and, in turn, notifies the designated hospitals.

INTERTHEATER MEDICAL REGULATING

6-23. The patients who are evacuated to EAC are treated there and then further evacuated to CONUS. The attending physicians at the hospital notify the PAD. The PAD then consolidates these requests and forwards them to the IPMC. This IPMC forwards the consolidated request to the TPMC who, in turn, consolidates and forwards a request to the JPMRC (if established) or the TPMRC.

6-24. Upon request of the J/TPMRC for authority to evacuate patients to CONUS, the GPMRC directs the distribution of these patients into hospitals throughout the CONUS; advises the J/TPMRC of the destination hospital; and provides the authority for such movement. As a rule, the destination hospitals are military facilities. Civilian national disaster medical system member hospitals and other federal hospitals may also receive patients. The Department of Veterans Affairs hospitals, for example, may receive patients who are expected to be discharged from service. The GPMRC validates the TPMRC patient movement requirements and, if moving by air, tasks the TACC to plan, schedule, and execute the intertheater evacuation.

6-25. When the J/TPMRC receives the authorization to move patients, it notifies the TPMC of destination hospitals in CONUS. The TPMC coordinates with the JMTB to arrange movement of CONUS-bound patients. The TPMC then authorizes the movement to ASFs/aeromedical staging squadrons (ASTSs) that are located on or near air bases or airstrips capable of handling long-range aircraft. Transportation is arranged, within Army channels, to move patients from the hospitals to the staging facilities. The medical brigade, in coordination with the AELT, then notifies the subordinate CSHs of the flight schedule and the evacuation arrangements for movement to strategic airheads. At strategic airheads, there is an established ASF/ASTS. When the patients are delivered to the USAF, the responsibility for those patients is transferred from the Army hospital to the TAES. Upon arrival in CONUS further movement is the responsibility of the GPMRC.

6-26. All patients may not be able to be moved by air from the theater to CONUS. In that event, the MSC is used to move them by surface means. The movement authority also comes from the G/T/JPMRC or TPMC which has arranged with the Navy Service Component Command for the movement of patients by hospital ships. When the patients are moved by ships, the MEDCOM has to provide holding facilities at the port (collocating an ASMC can provide this support). Patients are delivered to these holding facilities and held there until loaded aboard the ships.

MOBILE AEROMEDICAL STAGING FACILITY

6-27. The MASF is a 13-person, mobile, tented, temporary staging facility deployed to provide supportive patient care and administration. Each MASF is capable of a throughput of 40 patients per 24 hours and should be able to sustain this tempo for 72 hours before augmentation is required and is not intended to hold patients overnight or for an extended period.

6-28. This theater system is used to evacuate patients from—

- United States Air Force operational locations within the CZ to hospital facilities outside the CZ.
- Airhead or airborne objective areas where airborne operations include USAF forward logistics support.

6-29. Mobile aeromedical staging facilities—

- Provide supportive medical care when not augmented by the CCATTs. (When augmented by CCATTs, continued stabilization of patients can be accomplished. Aeromedical evacuation crews and CCATTs fly airlift missions to provide in-flight patient care.)
- Confirm sending facility has prepared patients for evacuation.
- In coordination with the OMF, ensure patient evacuation manifests are completed. (If the OMF is an Role 2 facility [such as an FSMC augmented with an FST], the forms required to complete the manifest may not be available and the MASF will be required to complete the appropriate forms.)
- Identify patient baggage tags.

6-30. Upon deployment, the OMF provides an adequate quantity of medications for patients' transit time to the regulated destinations.

6-31. The MASF staff also establishes liaison with the OMF. The AELT is composed of one medical service corps officer for administrative assistance and a flight nurse. The team provides the initial interface between the user service and the TAES. The AELT is located at any level of the combat forces where Air Force fixed-wing requests are initiated.

LIMITATIONS OF THE UNITED STATES AIR FORCE THEATER AEROMEDICAL EVACUATION SYSTEM

6-32. There are a number of limitations that are inherent in the current system. These include the following:

- Absence of biological weapon and chemical weapon agent decontamination ability.
- The MASF does not have the capability to provide patient meals.
- The AECT ensures the initial 30-days medical resupply package arrives at the MASF. The MASF/AELT/aeromedical evacuation operations team rely on the user service for all other logistical support.
- It is the Army's responsibility to provide food and other logistical support required including moving patients back to Army facilities should USAF AE support be delayed.

ORIGINATING MEDICAL FACILITY'S RESPONSIBILITIES

6-33. Once the authorization to move the patient has been given, the OMF must complete the following administrative procedures prior to entering the patient into the TAES:

- The patient's baggage tag, patient evacuation manifest, and patient evacuation tag are the specified evacuation forms for all services and are completed as required by triservice regulation.

Note. If the OMF is a Role 2 facility (such as the BSMC), the forms may not be available; in which case, the forms will need to be completed by the MASF.

- All of the patient's medical records must be collected together and packaged. The dental records are forwarded separately in the event they are needed for identification.
- At the appropriate time, the OMF provides transportation to the MASF and assists in the offload.
- The OMF must provide the necessary medications, medical supplies, and equipment to support the patients' travel time to the regulated destination.
- Any requirements for armed guards must be met by the echelon commander.

Note. Medical units do not provide guards for prisoners or EPW in their care. When guards are required, they are provided by the echelon commander. The OMF will coordinate for this support when needed.

- A limited amount of personal baggage is authorized if each piece is properly tagged and delivered to the MASF with the patient. Patients will always be evacuated with CBRN-protective equipment, less the protective overgarment.
- Each patient must be clearly identified with a wristband or equivalent identification and properly classified as to his medical condition.
- The OMF must ensure that each patient is properly briefed and prepared for his evacuation prior to his arrival at the MASF.

MEDICAL REGULATING FOR ARMY SPECIAL OPERATIONS FORCES

6-34. As in medical evacuation, the medical regulating plan must be integrated with the ARSOF operational and logistic plan. Maximum use of opportune (operational and logistics) aircraft and command and logistics communications nets must be coordinated to expedite mission requests and ensure success.

6-35. The ARSOF medical planner must constantly coordinate with the battalion or group operations and logistics sections to obtain up-to-date information of opportune transportation assets to be used for evacuation. In an operation, or when the theater is not sufficiently developed to allow the TAES to be used effectively, the primary means of air evacuation will be those SOA or USAF SOF airframes conducting the clandestine mission. It is essential that coordination is made through the Theater Special Operations Command (TSOC) or the highest C2 element for flight medics or para rescue men to accompany the flight when backhauling the casualties. Otherwise, a medic from the SOF unit being supported may have to accompany the patient, leaving the mission without proper medical support, or the casualty may have to be transported without en route care.

6-36. For all other special operations, the supporting medical evacuation unit provides air and ground ambulances according to standard doctrinal procedures. United States Air Force MASFs or AELTs may be collocated at SOF support bases, or C2 bases, particularly during contingency operations where the build-up phase allows for prepositioning of assets.

6-37. During sustained special operations missions, the TSOC cannot afford to lose the services of low-density ARSOF skilled Soldiers who become casualties. Every effort must be made to preserve ARSOF capabilities and combat power in theater. A determination must be made as to who can be treated and returned to duty at hospitals within the EAC. As an exception to the theater evacuation policy, the combatant command may retain injured or wounded ARSOF in theater where they can be returned to limited duty without jeopardizing their recovery and health. There they can assume the support duties performed by other ARSOF Soldiers, freeing the latter for operational duties.

Appendix A
Geneva Conventions and Medical Evacuations

The conduct of armed hostilities on land is regulated by customary international law and lawmaking treaties such as The Hague and Geneva Conventions. The rights and duties set forth in the Conventions are part of the supreme Law of the Land. The US is obligated to adhere to these obligations even when an opponent does not. It is a DOD and Army policy to conduct operations in a manner consistent with these obligations. An in-depth discussion of the provisions applicable to medical units and personnel is provided in FM 4-02. This appendix discusses only those articles or actions which affect medical evacuation operations.

DISTINCTIVE MARKINGS AND CAMOUFLAGE OF MEDICAL FACILITIES AND EVACUATION PLATFORMS

This paragraph implements STANAG 2454.

A-1. All US medical facilities and units, except veterinary, display the distinctive flag of the Geneva Conventions. This flag consists of a Red Cross on a white background. It is displayed over the unit or facility and in other places as necessary to adequately identify the unit or facility as a medical facility.

AUTHORIZED EMBLEM

A-2. The Geneva Conventions authorizes the use of the following distinctive emblems on a white background: Red Cross; Red Crescent; and Red Lion and Sun. In operations conducted in countries using an emblem other than the Red Cross on a white background, US Soldiers must be made aware of the different official emblems. Although not specifically authorized as a symbol in lieu of the Red Cross, enemies of Israel in past wars have recognized the Red Star of David and have afforded it the same respect as the Red Cross. This showed compliance with the general rule that the wounded and sick must be respected and protected when they are recognized as such, even when not properly marked.

A-3. United States forces are legally entitled to only display the Red Cross. However, commanders have authorized the display of both the Red Cross and the Red Crescent to accommodate HN concerns and to ensure that confusion of emblems would not occur. Such use of the Red Crescent must be in a smaller size than the Red Cross.

This paragraph implements STANAG 2931.

Camouflage of the Distinctive Emblem

A-4. Camouflage of medical facilities (medical units, medical vehicles, and medical aircraft on the ground) is authorized when the lack of camouflage might compromise tactical operations. The marking of facilities and the use of camouflage are incompatible and should not be undertaken concurrently.

A-5. If failure to camouflage endangers or compromises tactical operations, the camouflage of medical facilities may be ordered by a NATO commander of at least brigade level or equivalent. Such an order is temporary and local in nature and is rescinded as soon as circumstances permit.

Note. There is no such thing as a *camouflaged* Red Cross. When camouflaging a medical unit or ambulance, either cover up the Red Cross or take it down. A *black cross* on an olive drab or any other background is not a symbol recognized under the Geneva Conventions.

MEDICAL AIRCRAFT

A-6. Medical aircraft exclusively employed for the removal of wounded and sick and for the transport of medical personnel and equipment shall not be attacked, but shall be respected by the belligerents, while flying at heights, times, and on routes specifically agreed upon between the belligerents concerned.

A-7. The medical aircraft shall bear, clearly marked, the distinctive emblem together with their national colors on their lower, upper, and lateral surfaces.

A-8. Unless agreed otherwise, flights over enemy or enemy-occupied territory are prohibited.

A-9. Medical aircraft shall obey every summons to land. In the event that a landing is thus imposed, the aircraft with its occupants may continue its flight after examination, if any.

A-10. In the event of involuntary landing in enemy or enemy-occupied territory, the wounded and sick, as well as the crew of the aircraft, shall be prisoners of war; medical personnel will be treated as prescribed in these Conventions.

SELF-DEFENSE AND DEFENSE OF PATIENTS

A-11. When engaging in medical evacuation operations, medical personnel are entitled to defend themselves and their patients. They are only permitted to use individual small arms.

A-12. The mounting or use of offensive weapons on dedicated medical evacuation vehicles and aircraft jeopardizes the protections afforded by the Geneva Conventions. These offensive weapons may include, but are not limited to machine guns, grenade launchers, hand grenades, and light antitank weapons.

A-13. Medical personnel are only permitted to fire in their personal defense and for the protection of the wounded and sick in their charge against marauders and other persons violating the Law of War.

ENEMY PRISONERS OF WAR/DETAINEES

A-14. Sick, injured, wounded enemy EPW, and detainees are treated and evacuated through normal medical channels but are physically segregated from US, allied, or coalition patients.

A-15. Personnel resources to guard EPW or detainee patients are provided by the echelon commander. Medical personnel do not guard EPW or detainee patients.

COMPLIANCE WITH THE GENEVA CONVENTIONS

A-16. The US is a party to the 1949 Geneva Conventions. Two of these Conventions afford protection for medical personnel, facilities, and evacuation platforms (to include aircraft on the ground). All HSS personnel should thoroughly understand the provisions of the Geneva Conventions that apply to medical activities.

Consequences of Violations

A-17. Violation of these Conventions can result in the loss of protection afforded by them. Medical personnel should inform the tactical commander of the consequences of violating the provisions of these Conventions. The consequences may include—

- Medical evacuation assets subjected to attack and destruction by the enemy.
- Health service support capability degraded.
- Captured medical personnel becoming prisoners of war rather than retained persons. They may not be permitted to treat their fellow prisoners.
- Loss of protected status for medical unit, personnel, or evacuation platforms (to include aircraft on the ground).

PERCEPTION OF IMPROPRIETY

A-18. Because even the perception of impropriety can be detrimental to the mission and US interests, HSS commanders must ensure they do not give the impression of impropriety in the conduct of medical evacuation operations. For example, if a medical commander included in the unit's TSOP rules governing the use of automatic or crew-served weapons, it would give the impression that the unit possessed and intended to use these types of weapons. Under the provisions of the Geneva Conventions, medical units are only authorized individual small arms for use in the defense of the patients under their care and for themselves. Even though the unit does not possess these types of weapons, the entry in the TSOP could be misinterpreted and a case made that the commander intended to use these weapons in violation of the Geneva Conventions.

Appendix B

Legacy Units

Currently there are two units that are still active but will be transitioning out of the Army's inventory. They are the medical evacuation battalion and the medical company, AA (UH-1V).

SECTION I — MEDICAL EVACUATION BATTALION

B-1. The HHD, medical evacuation battalion, serves as the central manager of ground and air evacuation assets within the corps and EAC.

B-2. Assignment. The medical evacuation battalion is assigned to the MEDCOM in the EAC or to the medical brigade in the corps. It is normally further assigned to a medical group for C2.

Note. Under the MRI design, the medical group was replaced by the MEDBDE.

B-3. Air and ground ambulance companies assigned to the MEDCOM or MEDBDE are attached to the medical evacuation battalion for C2.

B-4. The basis of allocation is one medical evacuation battalion per a combination of three to seven of the following units:

- Medical companies, AA.
- Medical companies, ground ambulance.

B-5. Mission and capabilities.

- The mission of the medical evacuation battalion is to provide C2 of air and ground medical evacuation units within the AO. It tactically locates in the area where it can best control subordinate air and ground ambulance companies.
- The medical evacuation battalion is designed to focus on C2, planning, patient evacuation, subordinate unit support, and vehicle management. Specific capabilities are—
  - Command and control, planning and supervision of operations and training, and administration of a combination of air and ground ambulance companies.
  - Staff and technical supervision of aviation operations, safety, standardization, and aviation unit maintenance (AVUM)-level maintenance within the attached AA companies.
  - Coordination of medical evacuation operations and communications functions.
  - Coordination of logistics and service support to attached units.
  - Aviation medicine and unit-level AHS.
- This unit is dependent upon appropriate elements of the corps or ASCC for—
  - Personnel service support.
  - Health service support, to include hospitalization.
  - Mortuary affairs support.
  - Laundry, shower, and clothing repair.
  - Communications security equipment maintenance.
  - Military police support.

B-6. Organization and functions.

- Medical Evacuation Battalion (Figure B-1). The headquarters and headquarters detachment (HHD), medical evacuation battalion, is organized into a—
  - Battalion headquarters section.
  - Adjutant (US Army) section.
  - Intelligence Officer (US Army)/S3 section.
  - Supply Officer (US Army) section.
  - Detachment headquarters.
  - Treatment team.
- Battalion Headquarters Section. This section provides C2 of the assigned and attached air and ground ambulance companies. It also assists the commander on all military intelligence matters (to include the health threat), organization, training, operations, planning, personnel service support, and logistics support. Further, it provides information on the health of the command and aviation medicine expertise, as well as providing supervision over technical and flight aspects of administration, training, and safety within subordinate aviation units.
- Adjutant (US Army) Section. This section operates according to METT-TC and is the principle coordinating staff officer responsible for the delivery of human resources support. The S1 is responsible for the execution of all of the HR core competencies to include —PRM; PASR; PIM; R5 operations; casualty operations; EPS; and postal operations. It also includes MWR operations and HR planning and staff operations. The S1 also coordinates with elements of supporting agencies for finance, legal, religious, and administrative services.
- Intelligence Officer (US Army)/S3 Section. This section assists the S2/S3 officer in the execution of his duties and is capable of sustained 24-hour operations. This section remains abreast of the tactical situation and determines future medical evacuation requirements. It plans for ground and air evacuation operations, coordinates command post operations, and maintains the status of the AA units and plans for their employment. Further, this section maintains communications systems and nets, determines intelligence requirements, coordinates with movement control elements, and prepares orders and overlays.
- Supply Officer (US Army) Section. The S4 section assists the S4 officer in the execution of his duties. This section plans, coordinates, and supervises the requisitioning, receipt, storage, issue, and accounting for all classes of supply. Further, it monitors and keeps the commander informed on all matters pertaining to maintenance on assigned aircraft, ground vehicles, and medical equipment. This section also serves as the interface with the supporting medical battalion, logistics (forward/rear) for medical nonexpendable and durable item supply transactions.
- Detachment Headquarters Section. The detachment headquarters section provides, C2, administration, and logistics support for assigned personnel. It is also responsible for company supply and armament functions, food service operations, maintenance operations, and unit administration.
- Treatment Team. The treatment team provides unit-level AHS to assigned and attached elements collocated with the detachment headquarters and to adjacent units on an area support basis. The physician is a flight surgeon and provides staff assistance to the battalion commander on all matters pertaining to aviation medicine. The flight surgeon provides care and treatment for all assigned and attached aircrew members. This physician is dual-hatted as the battalion surgeon.

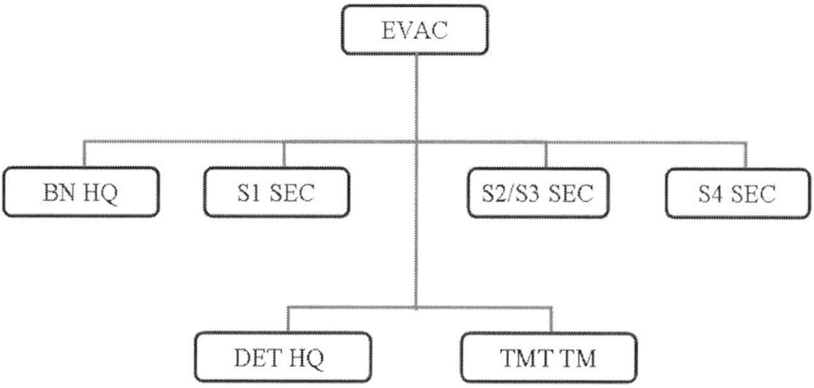

Figure B-1. Medical evacuation battalion

SECTION II — MEDICAL COMPANY, AIR AMBULANCE

B-7. The medical company, AA, provides aeromedical evacuation for all categories of patients consistent with evacuation precedence and other operational considerations. Medical evacuation is effected from as far forward as possible in the tactical AO to division- and corps-level MTFs.

B-8. Assignment.

- The medical company, AA, is normally assigned to the MEDCOM or MEDBDE and attached to the medical evacuation battalion for C2.
- The basis of allocation is one unit in support of each division or equivalent force supported. Further, one unit is in GS in the corps per two division or fraction thereof; or .333 units per separate brigades or ACRs.

B-9. Mission and capabilities.

- The mission of the medical company, AA, is to provide.
  - Aeromedical evacuation support within the theater, either DS to the divisions or GS to the corps.
  - Emergency movement of medical personnel, equipment, and supplies including whole blood, blood products, and biologicals.
- Specific capabilities of this unit are to operate on a 24-hour-a-day basis and evacuate patients based on operational capability (dependent on type of aircraft).
  - Operate 15 AAs (UH-60s). These ambulances are each capable of carrying six litter patients and one ambulatory patient, or seven ambulatory patients, or some combination thereof. Single patient lift capability is 90 litter patients, or 105 ambulatory patients, or some combination thereof. In-flight medical treatment and surveillance of patients is provided by a flight medic.
  - Operate 15 AAs (UH-1H/V). These ambulances are capable of carrying six litter, or nine ambulatory patients, or some combination thereof. Single patient lift capability is 90 litter, 135 ambulatory, or some combination thereof. In-flight medical treatment and patient surveillance are provided by a flight medic.
- Provide internal/external load capability for the movement of medical personnel and equipment.

- Perform AVUM on all organic aircraft and organizational maintenance on all organic avionics equipment. It also performs unit-level maintenance on all organic equipment less medical.
- Provide air crash rescue support, less fire suppression.
- Operate as an area support MEDEVAC section and three FSMTs to provide flexibility in supporting division, brigade, or brigade TF equivalent operations.

B-10. This unit is dependent upon:

- Support elements of corps or ASCC for—
  - Financial management, legal, and religious support.
  - Human resources support.
  - Logistics.
  - Health service support, to include medical supply and equipment.
  - Food service support.
  - Communications security equipment maintenance.
  - Mortuary affairs support.
  - Military police support.
  - Laundry, shower, and clothing repair.
  - Engineer support for heliport/landing strip construction and maintenance.
- The supporting aviation intermediate maintenance (AVIM) organization for AVIM support.

B-11. Organization and functions.

- The medical company, AA (Figure B-2), is organized into a—
  - Company headquarters.
  - Flight operations platoon consisting of a platoon headquarters, a flight operations section, and an airfield service section.
  - Aircraft maintenance platoon consisting of a platoon headquarters, a component repair section, and a maintenance section.
  - Air ambulance platoon consisting of a platoon headquarters, an area support MEDEVAC section, and three FSMTs.

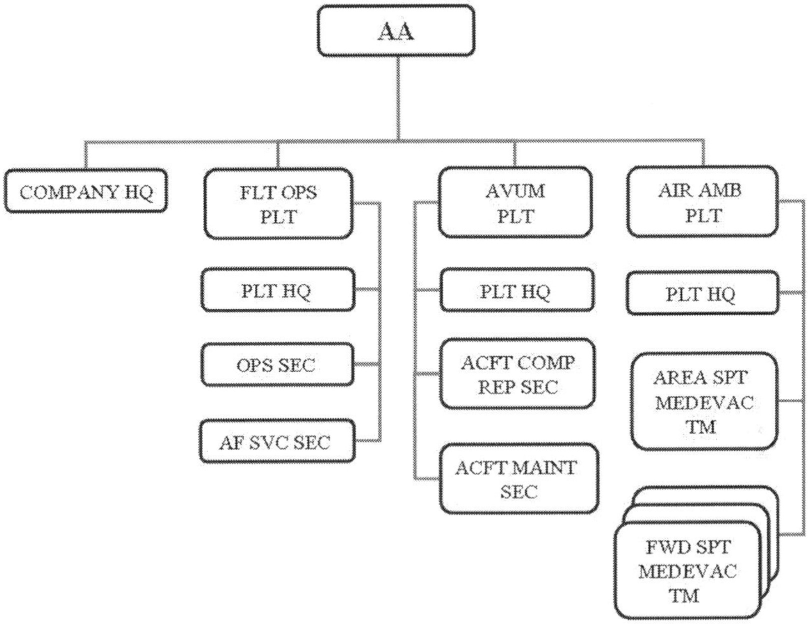

Figure B-2. Medical company, air ambulance

Appendix C

Tactical Standing Operating Procedures

This appendix provides a sample TSOP for an MMB engaged in medical evacuation missions. It should not be considered as all-inclusive. It may be supplemented with the information and procedures required for operating within a specific command or special operation.

PURPOSE

C-1. The TSOP prescribes policy, guidance, and procedures for routine support of tactical operations of a specific unit. It should cover broad areas of unit operations, but be sufficiently detailed to provide newly assigned personnel with the guidance required for them to perform their mission. A TSOP may be modified by the TSOP and OPLANs/OPORDs of higher headquarters. It applies to a specific unit and all subordinate units assigned and attached. Should a TSOP not be in conformity with the TSOP of the higher headquarters, the higher headquarters TSOP governs.' The TSOP is periodically reviewed and updated as required.

FORMAT FOR THE TACTICAL STANDING OPERATING PROCEDURE

C-2. There is not a standard format for all TSOPs; however, it is recommended that a unit TSOP follow the format used by its higher headquarters. The TSOP can be divided into sections (specific functional areas or major operational areas) and further subdivided into annexes. An annex can be further subdivided into appendixes and then into tabs. Appendixes can be used to provide detailed information on major subdivisions of the annex, and tabs can be used to provide additional information (such as report formats or area layouts) addressed in the appendix.

C-3. Regardless of the format used, the TSOP follows a logical sequence in the presentation of material. It should discuss the chain of command, major functions and staff sections of the unit, operational requirements, required reports, necessary coordination with higher and subordinate elements for mission accomplishment, programs (such as command information, preventive medicine measures [PMM], and combat stress control) and other relevant topics.

C-4. Pagination of the TSOP can be accomplished by starting with page 1 and numbering the remaining pages sequentially. If the TSOP is subdivided into sections, annexes, appendixes, and tabs, a numbering system that clearly identifies the location of the page within the document can be used. Annexes are identified by letter and are listed alphabetically. Appendixes are identified by numbers and arranged sequentially within a specific annex. Tabs are identified by a letter and are listed alphabetically within a specific appendix. After numbering the initial sections using the standard numbering system (sequentially starting with page 1 through to the end of the sections), number the annexes and their subdivisions. They are numbered as the letter of the annex, the number of the appendix, the letter of the tab, and the page number. For example, page 4 of Annex D is written as "D-4"; page 2 of Appendix 3 to Annex D is written as "D-3-2"; page 5 of Tab A to Appendix 3 of Annex D is written as "D-3-A-5." This system of numbering makes the pages readily identifiable as to their place within the document as a whole.

C-5. In addition to using a numbering system to identify specific pages within the TSOP, descriptive headings should be used on all pages to identify the subordinate elements of the TSOP.

- The first page of the TSOP should be prepared on the unit's letterhead. The remaining pages of the major sections should include the unit identification in the upper right-hand corner of the page (for example: "Multifunctional Medical Battalion").
- A sample heading for an Annex is: "ANNEX B (Command Post) to Multifunctional Medical Battalion."

- A sample heading for an Appendix to Annex B is: "APPENDIX 3 (Command Post Security) to ANNEX B (Command Post) to Multifunctional Medical Battalion."
- A sample heading for a Tab to Appendix 3 to Annex B is: "Tab A (Tactical Operations Center [TOC] Security) to APPENDIX 3 (Command Post Security) to ANNEX B (Command Post) to Multifunctional Medical Battalion."

C-6. As the TSOP is developed there may be an overlap of material from one annex to another. This is due in part to similar functions that are common to two or more staff sections. Where overlaps occur, the material presented should not be contradictory. All discrepancies will be resolved prior to the authentication and publication of the TSOP. The TSOP is authenticated by the unit commander.

Sample Tactical Standing Operating Procedures (Sections)

C-7. The information contained in this paragraph can be supplemented. It is not intended to be an all-inclusive listing. Different commands will have unique requirements that need to be included. The first section of the TSOP identifies the specific unit/headquarters that developed it.

- Scope–This document establishes and prescribes procedures to be followed by the unit identified and its assigned, attached, or OPCON units/elements.
- Purpose–This document provides policy and guidance for routine tactical operations of this headquarters and its assigned, attached, or OPCON units.
- Applicability–Except when modified by policy guidance, TSOP, or OPLANs/OPORDs of the higher headquarters, this document applies to this unit and to all units assigned, attached, or under OPCON for combat operations. These orders, however, do not replace judgment and common sense. In cases of nonconformity, the document of the higher headquarters governs. Each subordinate element will prepare a unit TSOP, conforming to the guidance herein.
- References–This paragraph can include any pertinent regulations, policy letters, higher headquarters TSOP, or any other appropriate documents.
- General information–This paragraph discusses the required state of readiness of the unit; primary, secondary, and contingency missions; procedures for operating within another command's AO; and procedures for resolution of conflicts with governing regulations, policies, and procedures.

C-8. The second section of the TSOP discusses the specific organization.

- Organization–This paragraph furnishes specific information concerning the authority for establishing the unit, such as applicable MTOE or other staffing documentation.
- Succession of command–The guidance for determining the succession of command is discussed.
- Task organization–Task organization is contingent on the mission and will be approved by the headquarters ordering deployment.
- Joint, multinational, and interagency operations–This paragraph provides guidance on any issues concerning C2 and TF organization in joint, multinational, and interagency operations. It also provides guidance on liaison officer requirements.
- Organizational charts–Contained in Annex A.

C-9. The third section of the TSOP discusses the unit functions.

- Battalion headquarters–This paragraph discusses the C2 of the assigned, attached, and OPCON units.
- Headquarters detachment–This paragraph discusses the functions of the headquarters detachment element, such as supervising movements, internal arrangements, area layout, physical security, and operation of the headquarters and staff.
- Attached, assigned, and OPCON units–This paragraph discusses the missions and functions of these units (such as providing medical evacuation of patients, emergency movement of blood and blood products, biologicals, medical logistics, and transportation of medical personnel and equipment).

- Staff responsibilities–This paragraph lists the unit's key personnel and their duties as prescribed in FM 5-0 and any command-specific duties.

C-10. The fourth section of the TSOP pertains to staff operations and is subdivided into annexes.

SAMPLE TACTICAL STANDING OPERATING PROCEDURES (ANNEXES)

C-11. Annexes are used to provide detailed information on a particular function or area of responsibility. The commander determines the level of specificity required for the TSOP. Depending upon the complexity of the material to be presented, the annex may be further subdivided into appendixes and tabs. If the annex contains broad guidance or does not provide formats for required reports, paragraphs may be used. The annex should not require further subdivision. However, as the material presented becomes more complex, prescribes formats, or contains graphic materials, the annex may require additional subdivision. This paragraph discusses the subdivision of the annex by appendixes. It does not contain examples of subdividing the information presented in the appendixes into tabs. Applicable references such as ARs, FMs, and technical manuals should be provided in each annex. The number of annexes and their subdivisions presented below are not to be considered as an all-inclusive listing. Different commands will have unique requirements; therefore, supplementation of the information presented is permitted.

C-12. Annex A. (Organizational Charts.)

C-13. Annex B. (Command Post.)

- General. The division may operate main and/or tactical command post CPs (TAC CP) depending on the mission and tactical situation. Characteristics of the CPs include—
  - Main CP–Normally located in the sustainment area. Personnel staffing is tailored to provide planning, coordinating, and C2 of assigned/attached/OPCON units. The area location for the main CP is selected by the battalion S3; the exact site is designated by the commander in coordination with the XO and battalion S3. The XO designates work areas within the main CP. The commander, HHD, serves as the headquarters commandant. His duties include coordinating for and obtaining construction, maintenance, and logistical services and support for the CP.
  - Tactical command post CP–Normally established in the objective area or at an intermediate staging base (ISB). The ISB is the forward designated staging base at which final preparation for assault operations are controlled and is generally a safe area for support operations. The TAC CP is tailored to extend the commander's span of control and to better assess the situation on the ground and coordinate the arrival and subsequent activities of AHS organizations within the AO.
- Battalion tactical operations center.
  - Definition–The TOC is the command element of the battalion containing communications and personnel required to command, control, and coordinate HSS operations. The TOC is located within a secure, controlled area whether at a main CP or TAC CP.
  - Purpose–The purpose of the TOC is to provide the commander, in a secure environment, current evaluated information and recommendations concerning HSS/FHP operations.
  - Responsibilities–The battalion commander has overall supervision and control over the TOC. The battalion XO operates the TOC and has primary staff responsibility in the absence of the commander.
  - Operations–The TOC will operate on a 24-hour basis. It is principally staffed by each primary staff section furnishing necessary manpower as required. Secure and nonsecure telephone communications connect the TOC to other staff sections within the CP area. Access to the TOC is strictly controlled by means of an access roster, and if available, security badges. Only essential personnel and authorized visitors are allowed to enter. Each staff section will maintain an SOP on the organization and operation of its element. All elements within the TOC will, when appropriate, maintain a current situational map of their specific operations. Discussion and portrayal of tactical plans outside of the security area is prohibited.

- Composition of the TOC–This is a listing of those personnel comprising the TOC. It normally includes the commander, XO, command sergeant major, principal staff members, and other specific staff members such as the S3 (air) or the battalion aviation maintenance officer.
- Tactical operations center configuration–This is a schematic representation of the physical layout of the TOC.

- Responsibilities for the main CP services–The HHD commander's main mission is to support the headquarters, medical evacuation battalion. He has the personnel and resources assigned to facilitate this mission. He plans and coordinates for providing shower, laundry, transportation, maintenance, and other required services. He is also responsible for establishing and maintaining security; supplying fuel and other POL products; establishing the support area for the main CP (orderly room, supply room, motor pool, billets, and dining facilities); providing and operating generators to meet electrical requirements.
- Camouflage–This appendix discusses what camouflage procedures are required, to include type and amount of required camouflage materials (such as nets and shrubs); display of the Geneva Conventions distinctive emblem on facilities, vehicles, and aircraft on the ground (STANAG 2931); and other pertinent policies, guidance, or procedures.
- Message center–This establishes procedures for the handling of classified messages; provides delivery and service of IMMEDIATE and FLASH messages to the appropriate staff section; establishes procedures for preparing outgoing messages; and establishes a delivery service to the servicing message center for transmission of outgoing messages.

C-14. Annex C. (Administration and Human Resources.) This annex outlines procedures relating to administrative and human resources matters and associated activities.

- Personnel accountability and strength reporting.
  - Personnel daily summary (PDS)–This provides the procedures for filling out and submitting a daily personnel status report. The instructions may include requirements for encrypting the report prior to transmission, and specific guidance on time of submission, corrections, or other administrative requirements.
- Personnel readiness management.
  - Maintain the personnel readiness of unit personnel to include projected gains and losses.
- Personnel information management.
  - Maintain 100 percent accountability of the unit's military personnel files.
  - Update records directly through the officer record brief/enlisted record brief.
  - Reception, replacement, RTD, R&R, and redeployment.
  - Track the flow and duty status of the five categories of Soldiers into and out of the unit.
  - Update deployed theater accountability software database as required.
  - Coordinate the movement of the five categories of Soldiers.
  - Replacements–Individual replacements will not be readily available during the initial phases of operations. The S1 will automatically initiate replacement requests for personnel who are reported on the PDS report as WIA, missing in action (MIA), or killed in action (KIA).
- Essential personnel services.
  - Issue identification cards and tags.
  - Personnel actions–All personnel actions will be channeled through the S1. Company XOs and 1SGs will be the company points of contact. Actions will be handled expeditiously and meet suspense dates (tactical situation permitting).
  - Efficiency reports–This paragraph provides pertinent information on the completion and submission of these reports.
  - Award recommendations–This paragraph delineates the responsibilities for and guidance concerning submitting recommendations for awards and for scheduling and conducting award ceremonies.
  - Promotions–This paragraph discusses the procedures for submitting recommendations for promotion and scheduling and conducting promotion ceremonies.

  - Correspondence–All correspondence addressed to higher headquarters will be submitted through the S1. Requirements for submission, preparation, and approval are also provided.
- Casualty operations.
  - Casualty Feeder Card–This report is submitted on DA Form 1156 (Casualty Feeder Card). Instructions on the completion of the form and submission requirements are included.
  - Tracking of casualties will require coordination with CLT and CSH.
- Postal operations.
  - This includes hours of operation and services available.
- Morale, welfare, and recreation operations.
  - Sporting activities and morale and welfare activities.
  - American Red Cross.
  - This includes hours of operation and services available.
  - Post Exchange services–This includes hours of operation and availability.
- Management, and HR planning and staff operations.
  - Preparation of the personnel estimate and personnel portion of plans or OPORD.
  - Coordinate external HR support with HR operations cell within the sustainment brigade.
- Other administrative support.
  - Financial management–This service includes disbursements and currency control, payday activities, currency conversion, check cashing, and the appointment of Class A agents.
  - Legal services–Information and specific guidance on administrative boards, courts-martial authority and jurisdiction, legal assistance, and general services should be provided.
  - Religious activities–Religious activities include chaplain support, services available from different faiths, schedule of services, and hospital visitations.
  - Distribution–Pick up and delivery schedules and any command-specific issues/procedures are provided.
- Graves registration (GRREG)–Commanders at all levels are responsible for the recovery, identification, and evacuation of US dead. This section discusses the responsibilities and procedures for unit-level GRREG activities for assigned and attached personnel.

Note. This activity is for unit members only. Medical evacuation units do not accept nor transport the remains of nonunit members. Remains will not be transported in ambulances under any circumstances.

  - Responsibilities–This paragraph discusses both unit and battalion requirements.
  - Disposition–Specific guidance on procedures, GRREG collection points, transportation requirements, and the handling of remains are provided.
  - Hasty burials–Specific requirements for conducting hasty burials, marking, and reporting of grave sites are included.
  - Personal effects–Guidance on accounting for personal effects and requirements for burial should a hasty burial be required is contained in this paragraph.
  - Disposition of civilian and EPW remains–The local civilian government is responsible for the burial of remains of its citizens. The remains of EPW should be accomplished in separate cemeteries from US and allied personnel. If this is not possible, separate sections of the same cemetery should be used.
  - Contaminated remains–This paragraph discusses handling and disposition requirements (to include protective clothing), procedures, and marking and reporting of burial site.
- Public information–This annex contains procedures for obtaining approval on the public release of information to include the hometown news release program. In overseas locations, specific guidance on the interaction of unit members with the news media should be provided.

- Maintenance of law, order, and discipline–This appendix should provide applicable regulations, policy, and command guidance on topics such as serious incident reports, notifications, and submission formats, straggler control, confinement of military prisoners, and EPW and detainees (also discussed below).
- Enemy prisoners of war–This appendix discusses the unit responsibilities of EPW and detainees surrendered to the medical unit. These procedures do not pertain to EPW and detainee patients captured by other units. Medical personnel do not guard, search, or interrogate EPW and detainees while in the AHS; guards are provided by nonmedical personnel designated by the tactical commander for these duties. Until EPW and detainee personnel can be evacuated to an EPW collection point, medical personnel should remember and enforce the basic skills: segregate, safeguard, silence, secure, speed, and tagging. (The speed portion of evacuating EPW and detainees to designated collection points is of paramount importance to medical units.)

Note. The treatment of EPW and detainees is governed by international and US law and the provisions of the Geneva Conventions. Personnel should be aware of these requirements and have ready access to the applicable regulations and policy guidance.

- Allied, coalition, and interagency personnel–This paragraph should provide guidance on the eligibility of allied, coalition, and interagency personnel for evacuation by Army ambulance and destination locations.

C-15. Annex D. (Intelligence and Security.) This annex pertains to intelligence requirements and procedures and OPSEC considerations.

- Intelligence–The S2 has the responsibility of collecting information to assist the commander in reaching logical decisions as to the best COA to pursue. Priority intelligence requirement (PIR) may include the location, type, and strength of the air defense threat; location, type, and strength of enemy air defense radars; known or suspected CBRN activity; the health threat; and issues which the commander considers to be PIR. In addition to PIR, the commander's critical information requirements (CCIR) are also considered.
- Intelligence reports–The S2 is responsible for disseminating all applicable estimates, analyses, periodic intelligence reports, and intelligence summaries generated within the battalion or received from higher headquarters. Information on submission of reports and suspense on intelligence products and reports should also be addressed in this appendix.
- Weather data–Weather conditions can disrupt ground evacuation efforts and planners must recognize their impact when coordinating AA operations.
- Reports–These include information acquired during the routine performance of duty by pilots, ambulance drivers, and medics.
- Counterintelligence.
  - Camouflage–When ordered or directed by the tactical commander, all units will initiate and continually strive to improve camouflage operations of positions, vehicles, and equipment. Noise and light discipline is emphasized at all times.
  - Communications security–These measures will be enforced at all times. The specific requirements and considerations are included.
  - Signs and countersigns–This paragraph lists the signs and countersigns to be used during hours of darkness. It also includes reporting requirements and procedures if the sign/countersign is lost or compromised.
  - Signal operating instructions.
  - Document security–This paragraph discusses the procedures for marking and safeguarding classified material, both work documents and completed documents. (Reporting requirements in the event of compromise are also included.)
- Captured or surrendered personnel, equipment, supplies, and documents–This appendix provides specific guidance on the handling of captured personnel, equipment, supplies, and

documents. The disposition of captured medical equipment and supplies is governed by the Geneva Conventions and is protected against intentional destruction.

- Security–This appendix discusses weapons security and checks, safeguarding of patient weapons, aircraft security, signal operating instructions (communications) security, TOC security, Sensitive Item Status Report, and escape and evasion.

C-16. Annex E. (Operations.) This annex establishes procedures for S3 operations within the medical evacuation battalion, and provides a basis for standardization of medical evacuation operations in a tactical environment. It is essential that these procedures be standardized to ensure common understanding, facilitate control and responsiveness, and enhance mission accomplishment. Information on readiness levels, threat levels, warning levels, camouflage, security, area damage control, operations, and leader checklists is also included.

- Operational situation report–Requirements for preparation, format, and submission of this report are discussed in this appendix.
- Operations security–This appendix provides the guidance and procedures for secure planning and conduct of combat operations.
  - Priority intelligence requirements and security classification.
  - Responsibilities–The commander is ultimately responsible for denying information to the enemy. The S3 is responsible to the commander for the overall planning and execution of operations. The S2 has the principle staff interest in assuming the required degree of OPSEC and has the primary staff responsibility for coordinating the efforts of all other staff elements in this regards. The OPSEC officer is responsible for the preparation of the PIR and providing classification guidance. Additionally, the OPSEC officer identifies the priorities for OPSEC analysis and develops OPSEC countermeasures. The S2 coordinates with the S3 in planning OPSEC.
  - Hostile intelligence threat–The different sources of intelligence (human intelligence, signal intelligence, and so forth) are discussed.
  - Operational security program–This includes physical security, information security, and signal security.
  - Document downgrading/declassification and classification authority.
- Terrorist threat and countermeasures–This appendix provides guidance on security measures to counter the terrorist threat, both on an individual and unit level.
- Minefield threat–This appendix provides guidance on the potential of minefields being established in the AO and delineates reporting requirements. (Specific minefield extraction techniques are contained in Annex L of this TSOP.)
- Operations security and countermeasures–This appendix discusses camouflage, light discipline, noise discipline, physical security, information security, and signal security.
- Unit location update–his appendix provides timely information on the location of main and forward CPs, location of subordinate unit CPs, location of helipads, and location of POL points.
- Flight operations–This appendix provides information concerning Army aviation LZs throughout the corps area and division areas; required reports; airdrop information; and mission debriefing.
- Communications-electronics–This appendix establishes communications policies, procedures, and responsibilities for the installation, operation, and maintenance of CE equipment.
  - Responsibilities of the battalion CE noncommissioned officer in charge.
  - Concept of operations.
  - Command and control.
  - Radio communications.
  - Radio teletypewriter communications.
  - Message/communications center service.
  - Message handling procedures.
  - Wire communications.

- Switchboard operations.
- Communications security and operations.
- Intelligence security.
- Meaconing, intrusion, jamming, and interference (MIJI) reporting and electronic communications countermeasures.
- Security violations–This prescribes procedures for reporting any event of action which may have jeopardized the security of communications.
- Destruction of material.
- Daily shift inventory.
- Physical security.
- Transmission security.
- Security areas–This discusses access procedures and rosters, access approval requirements, and prohibited items.
- Inventory of classified materials.
- Communications security officers and custodians–The appointment procedures, orders requirements, and duties of personnel are described.
- Safety–This discusses requirements for grounding, handling, and storing COMSEC equipment.
- Power units.

- Rear battle responsibilities–This appendix discusses rear battle responsibilities, task organization, and support for reaction forces.

C-17. Annex F. (Chemical, Biological, Radiological, and Nuclear Operations.) This annex prescribes the policy, guidance, and procedures for CBRN defensive operations.

- Responsibilities.
- Chemical, biological, radiological, and nuclear reporting requirements and procedures.
  - Contamination avoidance.
  - Protection–Protection pertains to those measures each Soldier must take before, during, and after a CBRN attack to survive and continue the mission.
  - Decontamination–This discusses equipment requirements, procedures, and types of decontamination (such as hasty). This paragraph also provides guidance on deploying clean ambulances into a contaminated area.
  - Mission-oriented protective posture–This appendix provides guidance on the garments required for the different MOPP levels and identification procedures for personnel in MOPP.
  - Radiation exposure guidance–Establishes OEG. Discussion in this appendix includes determining what constitutes a radiological hazard, prescribing acceptable limits of potential casualty producing doses of radiation, minimizing exposure, and protecting against electromagnetic pulses.
  - Masking and unmasking procedures.
  - Radiological monitoring and survey operations.

C-18. Annex G. (Logistics.) This annex establishes logistics procedures for subordinate units when operating in a field environment.

- Nonmedical supply and services–A discussion of applicability, responsibilities, policy, classes of supply, requisition and delivery procedures, hours of operation, and other supply-relevant topics and available services (such as laundry and bath) can be addressed in this section.
- Medical logistics–This paragraph discusses supply and resupply procedures for Class VIII; policies and procedures for the backhaul of emergency Class VIII on medical vehicles; and hours of operation for the supporting medical logistics facility.
- Food service–This appendix discusses responsibilities, hours of operation, Class I supplies, sanitation requirements, layout of field kitchen, fuel storage, maintenance, safety precautions,

and administration, such as headcounts, required reports, shift schedules, meals, ready-to-eat (MRE), and inspections/visits of subordinate unit kitchens.

- Transportation/movement requirements–This appendix may cover the following areas: deployability data, applicability; responsibilities; policies on speed, vehicle markings, transporting flammable materials, transporting ammunition and weapons, and so on; convoy procedures; safety; and accident reporting.
- Fire prevention and protection–Guidance on the use of the tent stoves, flammable materials, use of cigarettes, matches, and lighters, electrical wiring and appliances, safety of tents and occupants, spacing of tents, stoves and ranges, installation of British thermal unit (BTU) heaters, and fire fighting equipment are presented in this appendix.
- Field hygiene and sanitation–This appendix provides uniform guidance and procedures for the performance of functions related to field hygiene and sanitation. It includes policies, communicable disease control, field water supply, water containers and cans, water purification bags, food sanitation, latrines, liquid waste disposal, and garbage and rubbish disposal.
- Conventional ammunition down/upload procedures–This appendix delineates responsibilities and provides guidance and procedures for the requisition, storage, and distribution of ammunition and weapons, reporting requirements, and safety.
- Petroleum, oils, and lubricants accounting.
- Medical logistics support. The medical logistics concept of operations, requisition and distribution procedures, accountability, and reports are provided in this appendix.
- Maintenance–This appendix includes information on the maintenance requirements of the battalion and the location and hours of operation of maintenance units and collection points. Maintenance for medical equipment, vehicles, aircraft, and communications and other categories of equipment are discussed.

C-19. Annex H. (Safety.) This annex establishes minimum essential safety guidance for commanders and units. It includes risk assessment, accident reporting, safety measures, emergency procedures, vehicle safety, ground guide procedures, fire prevention and protection, antennas, climate, survival training, animal and arthropods hazards, personal protective measures, hearing conservation, carbon monoxide poison, helicopter safety, and refueling operations.

C-20. Annex I. (Civil-Military Operations.) This annex discusses the participation in CMO. Medical elements are often involved in CMO, humanitarian and civic assistance, and disaster relief operations. The activities which may be covered include providing—

- Direct support medical evacuation for the operation.
- Guidance on developing a medical evacuation system in a HN.
- Training to a HN's medical infrastructure.

C-21. Annex J. (Mass Casualty Situations.) This annex discusses the procedures for providing medical evacuation support to MASCAL situations, to include coordination for nonmedical transportation assets and the augmentation of these assets with medical personnel to provide en route patient care. It discusses evacuation capabilities of US forces.

C-22. Annex K. (Minefield Operations.) This annex discusses notification procedures, training requirements, and techniques for clearing pathways to casualties, marking mines, and extracting personnel and equipment.

C-23. Annex L. (Religious Support.) This annex discusses procedures for processing unit requests for religious support and coordination of religious support to casualties, declared MASCALs, mortuary affairs support, unit memorial ceremonies/services, EPW, detainees, religious support logistics and supplies, and reports and administration.

Appendix D

Example of the Medical Evacuation Plan and of the Operations Order

SECTION I — EXAMPLE FORMAT OF THE MEDICAL EVACUATION PLAN

D-1. Figure D-1 provides an example of the medical evacuation plan.

(Classification)

Headquarters
Place
Date, time, and zone

Time Zone Used Throughout the OPLAN: *The time zone used throughout the OPLAN (including attachments) is the time zone applicable to the operation. Operations across several time zones use ZULU time. Place the classification and short title of the OPLAN at the top of the second and any subsequent pages of the base OPLAN.*

Task Organization: *List the number and coordinates of medical evacuation units here or in a trace or overlay. If you do not list units here, omit this heading.*

1. SITUATION. *(State the general medical evacuation factors affecting support of the operation. Include any information essential to understanding the current situation as it influences medical evacuation.)*

a. Enemy forces. *(Refer to the intelligence annex if it has been published or is to be published. List information about the composition, disposition, location, movements, estimated strengths, and identification of enemy forces. List enemy capabilities that could influence the medical evacuation mission.)*

(1) Strength and disposition. *(The size of the opposition force and its placement on the battlefield is important during the planning process for medical evacuation operations. When evacuation routes are selected, caution must be exercised to ensure medical evacuation assets are not compromised by going through enemy-held territory or by being ambushed by isolated pockets of resistance. Further, this information is vital in determining if medical evacuation assets [both ground and air] will require a security escort provided by movement and maneuvers forces before entering areas of the battlefield.)*

(2) Combat efficiency. *(Information on actual combat units or other forces, their training status, and their level of expertise and experience can be identified here. The level of AHS training and the development of a health care delivery system can also be included.)*

(3) Capabilities. *(This paragraph should discuss the potential capability to interfere with or disrupt medical evacuation operations.)*

(4) Logistics situation. *(This can include information on how well supplied the enemy/ opposition force is with food, clothing, or other vital logistics factors. It may also include the financial backing and availability of future support from outside individuals/groups/nations.)*

(5) Weapons. *(This should include a discussion of the enemy's weapons that present the greatest threat to air and ground evacuation personnel and vehicles. Ground assets are more likely to face a small arms threat, rocket propelled grenades (RPGs), and improvised explosive devices (IEDs) while performing their mission; air assets are vulnerable to surface-to-air weapons and when on the ground to the same threats as ground elements.)*

b. Friendly forces. *(List pertinent information concerning friendly forces that might influence the medical evacuation mission.)*

(1) Strength and disposition. *(This should include all forces [US, allied, coalition, and HN] and should be maintained on overlays. It may also include liaison officers, interpreter support, and coordination requirements.)*

(2) Combat efficiency. *(The health of the command has a significant impact on this factor. Additionally, when there are significant numbers of DNBI casualties, the medical evacuation workload will increase.)*

(3) Present and projected operations. *(Medical evacuation planners must be familiar with plans for current and projected operations. A risk management assessment of current and projected operations is conducted. Casualty collecting points and AXPs must be designated during the planning process. Medical evacuation planners must also be able to anticipate changing requirements to ensure continuous medical evacuation is available to the supported force.)*

(4) **Logistics situation.** *(Since medical evacuation vehicles conduct emergency resupply of forward deployed medical units, the medical evacuation planner must maintain visibility of the current medical logistics situation.)*

(5) **Sustainment area protection plan.** *(Medical units play an important role in the sustainment area protection plan and should be included in the planning process.)*

(6) **Weapons.** *(Medical units only have small arms for self-defense and defense of their patients. Medical vehicles and aircraft should not carry any automatic or crew-served weapons; to do so would result in the loss of the protections afforded under the Geneva Conventions for the unit, medical personnel, and patients under its care.)*

(7) **Health of the command.**

(a) **Acclimation of troops.** *(Medical evacuation crews require acclimation when introduced into an AO where the temperature range and elevation are different from their home station. Medical evacuation operations involve heavy lifting and may require extended periods of time to complete. To reduce the risk of injury to litters bearers and to facilitate the evacuation effort, medical and nonmedical personnel engaged in these operations should be acclimated to the AO and a work/rest schedule should be developed and implemented. Air ambulance crew rest requirements impact aircraft availability and must also be considered.)*

(b) **Presence of disease.** *(The presence of disease impacts medical evacuation operations in two areas. The presence of disease in the AO [usually at subclinical levels in the native population] contributes to the incidence of disease manifesting itself within the supported force. As DNBI rates increase, so do the requirements for medical evacuation. Medical evacuation personnel are also susceptible to the endemic and epidemic diseases within the AO and/or multinational force. High rates of DNBI for medical personnel will adversely impact the medical evacuation capability.)*

(c) **Status of immunizations and/or chemoprophylaxis.** *(Commanders must ensure appropriate measures are taken to protect their Soldiers from DNBI. The records of replacement personnel need to be screened to ensure all required immunizations have been received and appropriate chemoprophylaxis/barrier creams for the AO initiated/provided.)*

(d) **Status of nutrition.** *(The nutrition status of the troops involved impacts on the susceptibility to disease and environmental injuries, morale, and fatigue. The medical evacuation commander must also ensure that his evacuation crews carry sufficient supplies of MREs to sustain themselves and their patients should delays in evacuation be experienced [such as delays resulting from shifts in the battle or inclement weather].)*

(e) **Clothing and equipment.** *(Special requirements for clothing and equipment to support a particular operation should be obtained prior to the beginning of the operation [for example, cold weather clothing for extreme cold weather operations; additional blankets for patients being transported (either to add padding or warmth); or block and tackle equipment for medical evacuation operations conducted in mountainous terrain or in urban areas].)*

(f) **Fatigue.** *(Mandatory work/rest schedules and sleep plans for air crews and ground evacuation personnel must be developed and implemented.)*

(g) **Morale.**

(h) **Status of training.** *(This can include any specialized training required for the conduct of a specific operation [for example: helicopter crews require deck-landing qualifications to perform shore-to-ship evacuation with US Navy ships or orientation to the social, political, economic, religious, and ethnic issues of a HN or supported population, or training nonmedical personnel in the proper techniques for carrying litters].)*

c. **Environment.**

(1) **Terrain.** *(This paragraph should discuss any aspects of the terrain that will either hinder or enhance the execution of the evacuation mission. It should discuss both natural and man-made terrain, as medical evacuation urban areas can pose significant challenges not found on a natural battlefield. Medical personnel must be able to recover injured and wounded Soldiers from below ground and from upper levels of man-made structures. If UO are planned, armor ambulances provide added protection for medical personnel and their patients; augmentation should be requested if armor ambulances are not organic to the unit. If additional resources will be required to accomplish evacuation due to impassable or difficult terrain [such as in mountain operations where additional litter bearers and medical personnel are required], it should also be addressed here. Further, the type of terrain to be traversed, such as rugged mountain or jungle swamps may require a patient [who would otherwise be ambulatory] to be transported on a litter until easier terrain is encountered.)*

(2) **Weather.** *(This should include a discussion of current weather conditions and seasonal variants. Weather conditions impact both ground and air evacuation operations; however, the most significant impact may be on AAs as severely inclement weather can ground all aircraft. It should also discuss the impact that the weather has on the terrain [such as rivers being frozen in winter or tundra becoming impassable in spring]. The climate may pose problems with acclimation as well as place additional requirements to sustain personnel during evacuation on evacuation assets [litter evacuation in the mountains in extreme cold weather operations may require warming tents and the capability to sustain the patient during nighttime when evacuation is difficult].)*

d. **Civil considerations.** Dislocated civilian population, EPW, and detainees. *(Dislocated civilians fleeing an area of hostilities can clog roadways leading away from the conflict area. This congestion on road networks may make evacuation along these routes almost impossible. The establishment of camps to sustain these categories of personnel may interrupt the road network requiring detours and lengthened evacuation times. Injured, ill, or wounded EPW or detainees are evacuated using the same evacuation means but are segregated from US, allied, or coalition patients. Coordination for nonmedical guards for EPW/detainee* *patients being evacuated through medical channels must be accomplished with the echelon commander. A determination of eligible beneficiaries for evacuation should also be developed and disseminated.)*

2. **MISSION.** *(Statement of the medical evacuation mission.)*

3. **EXECUTION.**

a. **Concept of operations.** *(The concept of operations describes how the commander sees the actions of subordinate units fitting together to accomplish the mission. Commanders ensure that their concept of operations is consistent with their intent and that of the next two higher headquarters. The concept of operations describes any other details the commander considers appropriate to clarify the concept of operations and ensure unity of effort. If the integration and coordination are too lengthy for this paragraph, they are addressed in the appropriate annexes. When an operation involves two or more clearly distinct and separate phases, the concept of operations may be prepared in subparagraphs describing each phase.)*

b. **Coordinating instructions.** *(List only instructions applicable to two or more units and not covered in unit SOPs.)*

c. **Patient estimates.** *(Indicate rates and numbers by type unit/division/corps/EAC.)*

(1) **Number of patients anticipated.** *(The anticipated number of patients affects the number and type of medical evacuation resources required.)*

(2) **Distribution within the area of operations.** *(Distribution within the AO is an important consideration because varying types of terrain will generate different requirements for support. Operations conducted in mountainous terrain rely heavily on manual and litter evacuation techniques, are labor intensive, and require more time to complete. Differences in road composition, traffic density, and abundance/absence of paved roads will impact differently on the type of support required. Further, if the distance between Role 2 and Role 3 MTFs exceeds the capability of US Army AAs, the use of the USAF aircraft should be anticipated and preplanned. Coordination with the USAF is required.)*

(3) **Distribution in time during the operation (evacuation time).** *(As the battle progresses over time and space, medical evacuation assets may be reallocated to support the units in contact. The medical evacuation plan must be sufficiently flexible to enable the commander to shift resources as required.)*

(4) **Areas of patient density.** *(Areas with a high patient density may require augmentation of the medical evacuation assets supporting that location.)*

(5) **Possible mass casualties.** *(MASCAL situations should be anticipated when possible. Evaluation of the general threat and health threat in the AO can facilitate the medical evacuation planner in forecasting evacuation requirements. When dealing with MASCAL situations, the use of nonmedical transportation platforms should be included. Augmentation of these vehicles by medical personnel to provide en route care or CLS to provide first aid will assist in reducing the deterioration of the patient's medical condition until arrival at an MTF.)*

(6) **Lines of patient drift.** *(This indicates what routes injured Soldiers are most likely to take from the battlefield. It is usually the most direct route over the least demanding terrain, such as at the base of the hill rather than over the hill where climbing would be required.)*

(7) **Evacuation routes/corridors.** *(Evacuation routes should be preplanned, indicated on the medical evacuation overlay, and reconnaissance accomplished. Routes that provide lucrative targets, have significant obstacles to circumvent, or will be unduly congested due to fleeing refugees or other displaced persons should only be used if no other routes are available. Air corridors are provided by the supporting A2C2 element. Due the placement of hospitalization resources within the JOA, AA may be required to operate out of fixed sectors and coordination for air escorts must be coordinated.)*

(8) **Support requirements.** *(This paragraph discusses the type of direct support, general support, or operational control required for the operation and its command relationship. Specific guidance on trigger points and release points from a particular category of support should be included.)*

(9) **Medical evacuation procedures.**

(a) **Evacuation overlays.** *(Evacuation overlays must be developed to facilitate the evacuation effort. Both supporting and supported units must maintain and update [as required] overlays throughout the operation.)*

(b) **Communications.** *(Medical evacuation frequencies must be designated at the outset of the operation. Requirements for liaison officers, A2C2, and interpreters must be anticipated.)*

(c) **Casualty collecting points.** *(Location, staffing, and activation trigger [such as crossing a phase line] must be known to supported and supporting units.)*

(d) **Ambulance exchange points.** *(Location, staffing, and activation trigger [such as crossing a phase line] must be known to supported and supporting units).*

Note: If an AXP is staffed with a treatment element, it is no longer considered to be an AXP but rather it becomes the forward element of a BAS or Role 2 MTF. The AXP is only a location on the ground where the patient is transferred from one evacuation platform to another. If it is staffed, it is usually staffed by one member of the evacuation platoon to direct returning ambulances to the area where the last ambulance departed from. The element designating the AXP is responsible for staffing it. Ambulance exchange points may be rendezvous points to be used only once during the operation.

Note: Ambulance exchange points should not be used as resupply points for Class VIII. (Planning on pushing Class VIII to these points may result in the loss of the Class VIII materiel, as there may not be anyone at the point to receive the supplies.)

(e) Ambulance shuttle system. *(The ambulance [or litter] shuttle system is a management tool to facilitate the medical evacuation of forward areas.)*

(10) Casualty evacuation (transportation).

(a) Medical augmentation. *(Units and/or facilities capable of providing medical augmentation support should be identified and tasked as appropriate.)*

(b) Manual evacuation.

(c) Litter evacuation. *(A trauma specialist or CLS should be included as a member of the litter team when there is a significant distance to be covered.)*

(d) Pack animals. *(In some remote locations, the use of pack animals may be the most feasible form of transportation from the POI to a place where vehicles, aircraft, or watercraft become accessible. Whenever possible, medical personnel should accompany patients evacuated in this manner.)*

(e) Nonmedical vehicle. *(Nonmedical vehicles which could be preplanned for use for casualty transport should be identified. Further, guidance on the use of vehicles of opportunity should also be discussed. When possible nonmedical vehicles used for these purposes should be augmented with medical personnel or CLS.)*

(f) Nonmedical aircraft. *(Nonmedical aircraft which could be preplanned for use for casualty transport should be identified. Guidance on the use of aircraft of opportunity should also be discussed. When possible nonmedical vehicles used for these purposes should be augmented with medical personnel or CLS.)*

(g) Nonmedical watercraft. *(Nonmedical boats, ships, or other watercraft which could be preplanned for use for casualty transport should be identified. Further, guidance on the use of these platforms of opportunity should also be discussed. When possible nonmedical vehicles used for these purposes should be augmented with medical personnel or CLS.)*

(11) Resources available.

(a) Mode of transportation.

(b) Personnel. *(This includes a discussion on personnel strengths and the impact shortages have on the medical evacuation operations.)*

(c) Supplies and equipment. *(This includes both medical and nonmedical supplies and equipment. Litter and blanket exchanges should be covered here. If PMI are an issue, they should be discussed here.)*

(12) Use of smoke and obscurants. *(In most cases, medical units will not have a high priority for the use of smoke and obscurants; however, if the supported units have planned for its use, medical evacuation units can take advantage of the situation to clear the battlefield of casualties. Guidance on the use of colored smoke to identify the pickup location for patients can also be included.)*

d. Warfighting functions.

(1) Medical planners must take the following warfighting functions into consideration.

(a) Intelligence. (Known or suspected enemy situation. Terrain and weather. What kind of road network is in the AO? How will the predicted weather impact the traffic capability on and off the roads? Are there LZs along the axis of advance and in or around the urban area? How will the weather impact on the availability of aircraft?)

(b) Movement and maneuver. *(What is the combat scheme of maneuver and are there medical assets readily available to care for potential casualties? Particular attention should be paid to the employment of the armored ambulances.)*

(c) Fire support. *(Are there any buildings specifically targeted? If so, medical planners should ensure CCPs, evacuation routes, and treatment elements remain at a safe distance away from these buildings to prevent potential fratricide incidents.)*

(d) Protection. *(Preserving the force includes protecting personnel (combatant and noncombatant), physical assets, and information of the US and multinational partners.)*

(e) Sustainment. *(Typically, the BCT moves a forward logistics element toward the battle area to provide support. If this is the case, then how are medical assets integrated into the plan, and, more importantly, where are they positioned in the convoy and how are they secured?)*

(f) Command and control. *(The senior military on-site or the senior medical person at the scene, if available, assigns the evacuation precedence and determines if the patient should be evacuated by ground or air resources. What are the primary and secondary frequencies? This element is the central point for prioritizing assets and medical evacuation requests. It is imperative that this element is capable of continuous situational awareness through accurate battle tracking.) From information provided by the S2, medical planners must understand the tactics of the enemy and be prepared for casualties at all possible areas of contact. The medical planners should be able to gain insight into this matter during the war gaming process. The medical planners should provide input into the counterreaction portion of the action, reaction, counterreaction drill.)*

(2) Additional planning information.

(a) *Availability and numbers of ground and AA assets, evacuation routes and flight windows.*

(b) *Nonstandard ground and aircraft availability (for MASCAL situations) and how to trigger their involvement.*

(c) *Additional medical assets from higher levels such as treatment teams, ambulances, and combat configured loads of Class VIII.*

(d) *Existing and proposed locations of all available MTFs.*

(e) *Class VIII resupply plan.*

(f) *The need for litter bearer teams and special equipment for extracting casualties from buildings and rubble.*

e. Casualty evacuation synchronization.

(1) *Medical evacuation and CASEVAC assets must be incorporated into the evacuation plan. Ideally there should be a single dedicated controlling authority. This prevents duplication of effort and confusion and enables the battlefield to be cleared faster and more efficiently.*

(2) *In MASCAL situations CASEVAC assets once put into operation should be dedicated versus designated (lifts of opportunity). Potential problems that arise arising from relying on designated versus dedicated CASEVAC include excessive time spent acquiring the information needed to transport casualties. Medical regulating to higher roles of care facilities may have to be accomplished by blocks of beds rather than on a by name basis. If a facility is designated for certain specialty care [such as head trauma] the evacuator must know which facilities have been identified and where they are located.).*

(3) *Another consideration is aircraft configuration. Casualty evacuation may be a seats-out operation. By dedicating aircraft and crews, they have time to prepare for the mission. Casualty evacuation crews must be briefed on the details of their mission including LZs, PZs, air corridors, communications architecture, and supported units.*

(4) *Any CASEVAC plan should clearly articulate the following information.*

(a) *Who is responsible (such as the company 1SG) for coordinating the company CASEVAC mission?*

(b) *Combat lifesavers role.*

(c) *Class VIII and blood resupply procedures.*

(d) *Personnel to act as litter bearers.*

(e) *Locations of CCPs, forward treatment teams, BAS, and AXPs.*

(f) *Location and designation of MTFs.*

(g) *Medical regulating officer coordination.*

(h) *Medical treatment facility and MRO contact frequencies.*

4. SERVICE SUPPORT.

a. Supply. *(Refer to TSOP or another annex whenever practical.)*

(1) General supply. *(Provide special instructions applicable to medical evacuation units.)*

(2) *Consider supply levels for all classes of supply, in the event of mission requirements in an austere environment and at extended distances from the full compliment of logistics and sustainment resources.*

b. Medical logistics. *(Provide special procedures applicable to the operations.)*

(1) Requirements. *(For sustaining the ground and AA operations and personnel.)*

(2) Distribution. *(This should include the method of distribution and any limitations or restrictions that are applicable. Additionally, if special transportation requirements exist, they should also be noted.)*

(3) Medical logistic activities. *(This includes the location of the medical supply activity supporting the AO and the means of communicating requests for resupply.)*

c. Supplies required for ground and AA operations and personnel. *(Includes estimates of the number of patients anticipated to be evacuated.)*

d. Health service support. *(Include information and instructions for supported units. Prescribe the plan for ground and air medical evacuation.)*

(1) Evacuation. *(List CCPs and AXPs. Establish ambulances shuttles, routes, means, and schedules of evacuation and en route treatment policies for the use of nonmedical transportation assets; policies for evacuation by ground and air; and evacuation of CBRN-contaminated patients. Include information about medical evacuation request procedures and channels, and holding policies.)*

(2) Treatment.

(a) Treatment of supported forces.

(b) Policies. *(Should address treatment and hospitalization policies.)*

(c) Units. *(This includes information concerning the location, capabilities, and communications means of units providing support.)*

(d) Other. *(This can be information on other medical assets that may be used in emergencies.)*

e. Personnel.

(1) Current subordinate unit manning levels/critical MOS shortages (consider all manning influences—Task Organization, R&R flow, Boots On Ground data, and so forth).

(2) New personnel requirements resulting from the operation (language skills, additional skill identifier [ASI], and so forth).

(3) Casualty estimates (as developed).

(4) Forecasted replacement availability.

(5) Evacuation policy for the operation.

(6) Supporting HR organizations (location, command/support relationship, controlling element).

(7) Supporting medical elements

(8) Replacement fill priority (Coordinate w/S3/G3).

(9) Crew/key leader replacement.

(10) Projected postal flow/limitations.

(11) Change to established PASR reporting flow/times.

(12) Change to casualty reporting flow (CLT change, reporting changes).
(13) RSOI Reception HR impacts.
(14) R&R schedule/flow operational impact (during operations conducted during sustained operations).
(15) Civilian and JIM manning impacts on the operation focusing on strengths and personnel service requirements.
(16) Specific Army G1 personnel planning guidance impacts on the operation.
(17) Status of other personnel service or personnel support factors (postal, MWR, legal support, military pay support).

f. Coordinating instructions. *(List instructions applicable to two or more units and not routinely covered in unit SOPs. This is always the last subparagraph in paragraph 4. Place complex instruction in an annex.)*

(1) Time or condition when a plan becomes effective. *(Include the time or the conditions under which the plan is to be placed in effect.)*

(2) Commander's critical information requirements. *(List only in coordinating instructions and not in annexes.)*

(3) Risk mitigation control measures. *(These are measures unique to this operation and not included in unit SOPs. They must include MOPP, OEG, troop safety, vehicle recognition signals, and fratricide prevention measures.)*

(4) Rules of engagement. *(If required.)*

(5) Boundaries. *(Address measures established to protect units. Specify which tactical units are to provide protection.)*

g. Special reports. *(List reports requiring special emphasis that is not addressed elsewhere.)*

5. COMMAND AND SIGNAL.

a. Command. *(State the map coordinates for the CP locations and at least one future location for each CP. Identify the chain of command if not addressed in unit SOPs.)*

b. Signal. *(Include the headquarters location and movements, liaison arrangements, recognition and identification instructions, and general rules concerning the use of communications and other equipment, if necessary. Use an annex when appropriate.)*

(Classification)

Figure D-1. Example of a medical evacuation plan

SECTION II — EXAMPLE FORMAT FOR AN OPERATIONS ORDER

D-2. Figure D-2 provides an example of the medical operations order.

(Classification)

Place the classification at the top and bottom of every page of the OPLAN.

(Change from verbal orders, if any)

The phrases "No change from verbal orders" or "No change from verbal orders except paragraph #" are required. This statement is only applicable if the commander issues a verbal order.

Copy ## of ## copies
Issuing headquarters
Place of issue

Show the name of the town or place in capital letters, coordinates in parentheses, and the country in capital letters. You may encode both.

Date-time group of signature

(The effective time for implementing the plan or order is the same as the date-time group of signature unless the coordinating instructions state otherwise. Use time zone Z unless the order states otherwise. When orders apply to units in different time zones, use Z time. When an OPORD does not specify the actual time to begin an operation, state that time in terms of an event.)

OPERATION ORDER 00-001, Operation Liberation
References: OPORD 00-000
Time zone: Zulu
Task Organization

Ground Ambulance Units participating in Operation Liberation
123 MED CO (GA) assigned to 1st MMB and provides emergency support to the 1st Marine Expeditionary Force (MEF)
234 MED CO (GA) assigned to 1st MMB
Ambulance Platoons assigned to:
C CO BSMC
C CO BSMC
C CO BSMC
C CO BSMC

Air Ambulance Units participating in Operation Liberation
C 2-1 MED CO (AA) CAB assigned to 1st Infantry Division
C 2-2 MED CO (AA) CAB assigned to 2nd Infantry Division
C 2-3 MED CO (AA) DS to the 1st MEF
C 2-4 MED CO (AA) CAB assigned to 1st Corps support

Air Force aeromedical evacuation assets
2d AF Combat Search and Rescue (CSAR), 1st AF Squad

1. **SITUATION.**

a. **Enemy forces.** *As addressed in OPORD 00-000, the ground-to-air threat is significant in the all enemy sectors. The air–to-air threat is minimal, in all sectors. All ground planned evacuation routes will be targets for ambush by enemy forces and sympathetic host nationals.*

b. **Friendly forces.** *A six (6) aircraft (HH-60) Air Force CSAR unit will be staged from the APOD. They will respond to downed aircrew evacuation recovery requests from the Air Force Combat Control Center.*

c. **Attachments and detachments.** *None.*

d. **Assumptions.** *None.*

2. **MISSION.** *On order provide tactical and operational ground and air intratheater medical evacuation or patient movement in support of Operation Liberation and continue coverage of all combined joint task force (CJTF) assets until relieved.*

3. **EXECUTION. Intent: State the commander's intent.**

a. **Concept of support operations.** *This is a three phase operation.*

Phase 1—Will be prior to initiation of the attack. *All US forces will be in the staging areas and in preattack positions. There will be no or very limited direct contact with enemy forces. C 2-1, C 2-2, and C 2-3 will not be required to* *provide evacuation support until D-3 days. Units in the staging area will be directed to call C 2-4 for evacuation at all times prior to D-3 days. All other evacuation units will concentrate on training and preparation for hostilities as directed by their commanders.*

Phase 2—Will be the attack. *C 2-1, C 2-2, C 2-3 will provide DS to their assigned, GS to the corps sustainment area and the backhaul mission.*

Phase 3—Will be from the conclusion of the attack until stability operations commence. *Evacuation assets will begin transferring to an area support mission during this phase.*

Note. The concept of support operations takes into account both the tactical OPORD and the medical Annex to the OPORD. Information in both documents will shape the evacuation plan. Simultaneous planning should therefore occur.

---

b. Tasks to maneuver units: *Not applicable.*

c. Tasks to movement and maneuvers units:

(1) Evacuation responsibilities.

(a) Phase 1.

C 2-1, C 2-2, C 2-3 will provide direct support to their division assets from D-3 on. C 2-1, C 2-2, C 2-3 will respond to any request for evacuation that is geographically close to their positions even if not in their division area.

C 2-4 will provide support to the entire staging area (AO Buffalo). C 2-4 will perform the ship-to - shore mission as directed by the 11th MEDCOM TPMC

The 123 MED CO (GA) will operate in the 1st MEF sector in a direct support role. 1st MMB has will cover the 1st and 2nd Division footprint.

The 234th MED CO (GA) will provide support to Camps Jones, Smith, and Franks and perform patient transfers from camps to the APOD.

(b) Phase 2.

C 2-1, C 2-2, C 2-3 will provide direct support to their division assets. C 2-1, C 2-2, C 2-3 will be responsible for moving patients from within the division area back to the staging AO Buffalo until FLOT is beyond phase line alpha.

After FLOT is beyond phase line alpha (at which time one way flight time for the H-60 back to AO Buffalo will exceed 2 hours), C 2-1 will designate 1 FSMT to stage at XYZ Airfield to perform the backhaul and the area support mission until relieved. United States AF aeromedical transports will be coordinated for by 11th MEDCOM flying from XYZ airfield.

Beyond phase line Alpha C 2-2 will provide one FSMT to support the greater casualties expected in C 2-2 sector. Coordinate internally for link-up.

C 2-4 will provide support to the entire staging area (AO Buffalo). C 2-4 will perform the ship-to-shore mission as directed by the 11th MEDCOM MRO. C 2-4 will designate 4 aircraft to go forward as required to backhaul patients to AO Buffalo.

(c) Phase 3.

C 2-1, C 2-2, C 2-3 will provide direct support to their Divisions' assets.

The 1st MEF will identify areas of high casualty density conducive to ground evacuation (urban areas) and divide the resources of the 123rd MED CO (GA) to cover these areas efficiently, if required.

The 234th MED CO (GA) will conduct convoy support and provide coverage for AO Buffalo.

d. Evacuation mission guidelines.

(1) Phase 1, 2, and 3. *The senior military person advised by medical personnel on-scene will assign the evacuation precedence and determine the mode of evacuation (ground or air). Launch authority, a separate issue, based on risk will be according to the aviation SOP.*

(a) *Weather below acceptable minimums.*

(b) *No acceptable LZ in the immediate vicinity, except for hoist missions. Ground evacuation may be used for part of the evacuation.*

(c) *Chemical, biological, radiological and nuclear conditions at LZ would contaminate the aircraft. Ground evacuation may be used for part of the evacuation, followed by an air pickup after decontamination is complete.*

(d) *Patients are located at a LZ in direct enemy contact. Ground evacuation may be used for part of the evacuation, followed by an air pickup once clear of the enemy.*

(e) *Air evacuation assets are unavailable.*

(f) *Mission flight route requires an armed escort and one is not available.*

(2) The following are additional guidelines for evacuation:

(a) *Ground evacuation is the preferred method to clear casualties out of hostile urban areas. Air ambulance pick-up at a secured LZ outside of the urban area is preferred. For large urban areas, this may not be possible.*

(b) *If a mission requires the use of a hoist (such as thick forest or jungle) and the pick-up area has suspected enemy contact, ground evacuation is advised to a secure pick-up site.*

(c) *Prescheduled movement of ROUTINE or CONVENIENCE patients along secure routes of evacuation should be conducted using ground evacuation if the distance of evacuation is short.*

(3) Coordinating instructions:

(a) Medical regulating. *The 11th MEDCOM will oversee medical regulating during the operation.*

(b) Tentative locations of MTFs.

Phase 1

- *2nd CSH AB 1234 5678*

- *1st CSH AB 2345 6789*

Phase 2
- *2nd CSH AB 1234 5678*
- *1st CSH AB 2345 6789*

Phase 3
- *2nd CSH AB 1234 5678*
- *1st CSH AB 2345 6789*

(c) **Reporting.** *Evacuation units, through division G3 if assigned to a division, will submit a daily report of evacuation missions by 1200 the following day.*

(d) *If a MASCAL develops which overwhelms the resources of a single AA CO, contact GSAB for assistance.*

(e) **Casualty collection.** *Military treatment facilities can call 9 line missions to any medical evacuation asset for URGENT and PRIORITY missions however MTFs that need to clear beds will coordinate mass patient movements through CJTF G3 air for possible USAF support.*

- *1st CSH will call C 2-4 for evacuation requests during Phase 1.*
- *2nd CSH will call C 2-4 for evacuation requests during Phase 1.*
- *1st CSH will call C 2-1 for evacuation requests during Phase 2/3.*
- *2nd CSH will call C 2-2 for evacuation requests during Phase 2/3.*

(f) **Ground evacuation.** *Seasonal monsoons are expected to washout many of the main planned MSRs and cause flooding.*

(g) **Air evacuation.** *If air evacuation routes become so long as to seriously degrade the capabilities of an AA company, contact CJTF G3 Air for additional support and possible reallocation of evacuation assets. An AA CO is considered to be operating beyond its capability or inefficiently if:*

- *More than 50 per cent of missions require a one way trip time exceeding 2 hours.*
- *On more than 50 per cent of missions flown, routes of evacuation are flown through another AA company's area of direct support coverage.*

(h) **Medical considerations for evacuation.**

- US Army evacuation assets will be used to evacuate all injured personnel on the battlefield without prejudice.
- Human immunodeficiency virus is at epidemic proportions in Simula population. Take appropriate precautions during treatment and when dealing with biologicals.
- A significant number EPW and displaced civilians and detainees are predicted to be evacuated. Estimated to be a much as 75 per cent of total patient load.

(i) **United States Air Force aeromedical evacuation.**

- Combined joint task force G3 Air will coordinate for USAF intratheater aeromedical missions.

4. SERVICE SUPPORT.

a. Casualty estimates. Reference the Medical Annex for specific casualty estimates.

b. Aviation maintenance support. C 2-3 will receive AVUM/AVIM support from the 2-4 CAB.

c. Divisions G3 will report the following CCIR to CJTF G3 Air. Use the CJTF CCIR format.

- Any MASCAL.
- Any evacuation company with degraded capability for—
- Evacuation distance(defined above)
- Maintenance: below 70 percent operational readiness rate
- Personnel: number of crews less than available aircraft
- Any aircraft Class A accidents/events or shoot downs.

5. COMMAND AND SIGNAL.

a. Fragmentary orders (FRAGOs) to this order will be published to adjust the corps evacuation plan as necessary.

b. 11th MEDCOM MIRC Chat net name, Phone xxx-xxxx

- 1st Division Surgeon (G3), MIRC Chat net name
- 2-1 CAB MEDEVAC Cell, MIRC Chat net name
- 2nd Division Surgeon (G3), MIRC Chat net name
- 2-2 CAB MEDEVAC Cell, MIRC Chat net name
- 1st MEF, MEDEVAC Cell, MIRC Chat net name
- 1st Corps CAB MEDEVAC cell, MIRC Chat net name, phone xxx-xxxx

ACKNOWLEDGE:
PATRICK, LTG

Figure D-2. Example of a medical operations order

Appendix E

Use of Smoke and Obscurants in Medical Evacuation Operations

This appendix provides guidance for AMEDD personnel in the use of smoke and obscurants in medical evacuation operations.

TACTICAL COMMANDER

E-1. Throughout the battlefield, forces acquire and engage targets based on visual, laser, and microwave technologies. Friendly and enemy units use smoke and obscurants across the battlefield as a combat multiplier. The use of smoke and obscurants to mask combat operations is dictated by the tactical commander. He normally provides the operational guidance for units or elements operating in an area requiring obscuration. Permission to employ smoke and obscurants solely to mask medical evacuation operations may not be approved. However, if the tactical commander's plan indicates that smoke operations are to be employed in a given AO, the medical planner should consider both the advantages and disadvantages posed by their employment.

FACTORS TO CONSIDER

E-2. The medical planner should consider the factors which might impact the use of smoke and obscurants in medical evacuation operations. Factors to consider are the—

- Phase of the tactical operation in which smoke and obscurants will be employed.
- Effect on ground and air evacuation routes when operating in an obscured environment (such as limited hours of use, checkpoint or convoy requirements, or the elimination of NOE approaches).
- Potential for exploiting the use of the cover and concealment provided for clearing the battlefield of casualties, especially in defensive operations.
- Potential requirements for smoke generation to perform the medical evacuation mission which would not detract from the tactical capability and requirements.

E-3. Smoke can also be used to identify unit areas or LZs for which a medical evacuation request has been received. Further, smoke can indicate wind direction at a landing site for AA operations.

EMPLOYMENT OF SMOKE AND OBSCURANTS

E-4. Smoke and obscurants are employed to protect friendly forces from attack during the offense or defense. Smoke and obscurants disrupt enemy combat operations throughout the depth of the battlefield and across the operational continuum. They—

- Disrupt the ability to communicate.
- Conceal friendly forces.
- Deceive the enemy.
- Identify and signal.
- Degrade the effects of directed-energy weapons.

E-5. The benefit to AHS forces is derived through the tactical commander's use of smoke to obscure friendly tactical maneuvers. This obscuration—

- Prohibits the enemy from knowing how many casualties have been inflicted.
- Aids the movement of medical units and equipment.

- Enhances the ability to resupply forward deployed AHS elements.
- Aids in the tactical deception plan.

GENEVA CONVENTIONS AND THE USE OF SMOKE AND OBSCURANTS IN MEDICAL EVACUATION OPERATIONS

E-6. As discussed above, the 1949 Geneva Conventions for the Amelioration of the Condition of the Wounded and Sick in Armed Forces in the Field (GWS) provides protection of medical personnel and units from intentional attack so long as they carry out no duties harmful to the enemy (Article 21, GWS). In order to facilitate their identification so as to prevent their intentional attack, medical units, equipment, and personnel are authorized to display the distinctive emblem of the Red Cross (Article 41 and Article 42, GWS). Under tactical conditions, when requirements for concealment outweigh those for recognition, all distinctive emblems may be obscured or removed from medical equipment if ordered by competent military authority and authorized by Army regulations. Display of the distinctive emblem is not required to afford the right against intentional attack; attack of medical units, equipment, and personnel not displaying the distinctive emblem is prohibited if opposing forces realize that the forces about to be attacked are medical units performing humanitarian duties.

E-7. The use of smoke or obscurants in medical evacuation operations does not differ from the use of camouflage techniques and is not prohibited by the GWS. Its only effect will be to obscure the identity of units as they perform their humanitarian mission. Given the lethality of the modern battlefield, however, it would be difficult, if not impossible, to say that such obscuration of these units, equipment, and personnel would necessarily increase their risk from unintentional attack.

E-8. It is recognized that, with the advent of precision-guided munitions and electro-optical or laser target acquisition devices, there will be a substantial use of smoke and other obscurants on the modern battlefield as a result of normal combat operations. The legitimate use of obscurants by combatants to thwart the accuracy of precision guided munitions may increase the risk to units and equipment not employing obscurants. This may possibly place medical units and equipment at greater risk if they fail to employ them. Further, medical evacuation operations will have to be carried out on the battlefield as medical personnel find it, which will include obscurants employed for normal combat operations.

USE OF SMOKE IN AEROMEDICAL EVACUATION AND HOIST RESCUE OPERATIONS

EFFECTIVE USE

E-9. Smoke can be used effectively in aeromedical evacuation and overland hoist rescue operations to—

- Identify the landing site. Colored smoke is an excellent daytime marking method. The smoke generated from a smoke grenade is difficult to detect more than 2 to 3 miles away, but an aircraft in the area should have little difficulty in noting its location. As more than one unit may be operating in a given area, it is important that the unit requesting an aeromedical evacuation mission be able to signal the aircraft as to the correct landing site to use. Radio communications produce an electronic signature. The electronic signature created from a prolonged transmission to guide an AA to the landing site may not be an acceptable tactical risk.
- Ensure the LZ is controlled by friendly forces. When a unit employs colored smoke to mark a landing site, the aircrew should identify the color and confirm it with the ground personnel. The transmission time required for this procedure is limited, thereby reducing the electronic signature.
- Determine surface wind direction. The employment of smoke at the landing site also enables the aircrew to determine the wind direction.
- Provide cover and concealment. In some environmental conditions (such as desert operations), the phenomenon of inversion occurs. When this occurs, the smoke and obscurants used in

normal combat operations may provide an upper layer of smoke under which the AA can operate.

DISADVANTAGE

E-10. The use of smoke on aeromedical evacuation operations can be a disadvantage if incorrectly employed or if the smoke generated in the tactical operation interferes with the medical evacuation mission. Smoke can obscure the landing site and make NOE approaches unusable. Further, smoke on the battlefield can force aircraft to fly at higher than planned heights. This increases the risk of being acquired by the enemy.

OVERWATER HOIST OPERATIONS

E-11. In overwater hoist rescue operations, the employment of smoke for marking the patient pickup area, for determining surface wind conditions, and for spatial orientation is essential. The smoke employed by the aircrew must not interfere with the conduct of the operation or mask the location of the individual to be rescued.

EMPLOYMENT OF SMOKE IN GROUND MEDICAL EVACUATION OPERATIONS

E-12. The employment of smoke during ground evacuation operations must be in consonance with the tactical commander's plan. Smoke can mask medical evacuation operations on the battlefield, but must not interfere with the tactical mission. In all combat operations, but especially in UO, smoke can be employed to cover and conceal—

- Movement across open areas.
- Extraction of casualties from vehicles and/or buildings.
- Entry and exit into/out of structures.

Appendix F

Evacuation Capabilities of United States Forces

This appendix provides the evacuation capabilities of US forces and aircraft.

EVACUATION CAPABILITIES OF UNITED STATES AIR FORCE AIRCRAFT

F-1. Evacuation capabilities of USAF aircraft are provided in Table F-1. As the majority of USAF aircraft are not dedicated medical evacuation platforms, when used for medical evacuation the crew must be augmented with medical personnel to provide in-flight care.

F-2. The evacuation capability of Civil Reserve Air Fleet aircraft is provided in Table F-2.

Table F-1. Evacuation capabilities of United States Air Force aircraft

Transport Aircraft	*Litter*	*Ambulatory*	*Combination*
C-130 Hercules	74	92	Variety
C-141 Starlifter	48	170	Variety
C-5 Galaxy	70		
C-17A	*36	*54	
KC-135 and KC-10	8	24	
*Normal configuration			

Table F-2. Civil Reserve Air Fleet capabilities

Transport Aircraft	*Litter*	*Ambulatory*	*Combination*
Boeing B-767	*111	22	87 Litter 22 Ambulatory
* When activated, these aircraft are preconfigured for medical evacuation. The crew must be augmented with medical personnel			

EVACUATION CAPABILITIES OF UNITED STATES ARMY VEHICLES AND AIRCRAFT

F-3. The evacuation capabilities of US Army vehicles and aircraft are provided in Table F-3.

Table F-3. Evacuation capabilities of United States Army vehicles and aircraft

Vehicle/Aircraft	*Litter*	*Ambulatory*	*Combination*
Ground Vehicles			
M996 Truck, Ambulance	2	6	1 Litter 3 Ambulatory
M997 Truck, Ambulance	4	8	2 Litter 4 Ambulatory

M1010 Truck, Ambulance	4	8	
M792 Truck, Ambulance	3	6	2 Litter 3 Ambulatory
M1133, Stryker MEV	4	6	
Bus, Motor, 44 Passenger	18	37	
M113 Carrier, Personnel	4**	10	
M998 Truck, Cargo/Troop Carrier (Four Man)	3	4	
M998 Truck, Cargo/Troop Carrier (Two Man)	5		
Truck Cargo, 2 1/2 Ton, 5 Ton	12	16	
M997 Heavy Expanded Mobility Tactical Truck	9		
M871 Semitrailer, Cargo	16		
M1085 Truck, Cargo, Medium Tactical Vehicle, Long Wheel Base, 5 Ton	12	22	
M1093 Truck, Cargo, Medium Tactical Vehicle Air Drop/Air Delivery, 5 Ton	8	14	
M1081 Truck, Cargo, Medium Tactical Vehicle Light Air Drop/Air Delivery, 2 1/2 Ton	7	12	
Rotary-Wing Aircraft			
***UH-60 Blackhawk	6	7	*4 Litter 1 Ambulatory
****UH-60A Blackhawk	3	4	4 Litter 1 Ambulatory
UH-1H/V Iroquois	6	9	*3 Litter 4 Ambulatory
CH-47 Chinook	24	31	#
Fixed-Wing Aircraft			
C-12 Huron		8	

LEGEND
* Normal configuration
** Spall liner must be removed
*** without internal hoist installed, requires 6 man litter configuration kit
**** With internal hoist installed
# See Table F-5

Railway Car Capabilities

F-4. Although railway cars are not available within the US Army inventory, it is important to know the approximate capacities of rail transport in the event they become available and/or required in civil support operations or through wartime HN support agreements. The approximate capabilities of railway cars are provided in Table F-4.

Table F-4. Capabilities of railway cars

Rail Transport	*Litter*	*Ambulatory*	*Combination*
Pullman Car (US)	32	48	
Sleeping Car (NATO/HN)	32	32	
Ambulance, Railway Car (NATO/HN)	24	30	
Ambulance, Railway Car, Personnel	21	21	
Railbus			40 Litter 16 Ambulatory

EVACUATION CAPABILITIES OF UNITED STATES NAVY SHIPS, WATERCRAFT, AND ROTARY-WING AIRCRAFT

F-5. The evacuation capabilities of USN ships, watercraft, and aircraft are provided in Table F-5. The entries for ambulatory and litter patients on ships are the same because all patients require a bunk.

Table F-5. Evacuation capabilities of United States Navy ships, watercraft, and aircraft

Ships/Watercraft/Aircraft	*Litter*	*Ambulatory*
Hospital Ships		
T-AH 19, US Naval Ship Mercy	1000	1000
T-AH 20, US Naval Ship Comfort	1000	1000
Casualty Receiving And Treatment Ships		
Amphibious Assault Ship (LHD)(Multipurpose)	604	604
Amphibious Assault Ship (LHA)(General Purpose)	367	367
Amphibious Assault Ship (LPH)	222	222
Amphibious Transport Dock (LPD)	14	14
Dock Landing Ship (LSD)	108	108
Limited Medical Capability Watercraft		
Amphibious Cargo Ship (LKA)	12	12
Amphibious Command Ship (LCC)	24	24
Rotary-Wing Aircraft		
CH-46 Sea Knight	15	25
CH-53 Sea Stallion	24	*55
V22 Osprey	12	24

* With centerline seating installed

Appendix G

Selection of Patients for Aeromedical Evacuation and Patient Classification Codes and Precedence

Patient classification codes provide information to evacuators and treatment personnel in an abbreviated form. They can be used to complete administrative reporting requirements pertaining to the evacuation of patients.

Paragraphs G-2 through G-5 implement STANAG 3204.

BRIEFING OF PATIENTS PRIOR TO AEROMEDICAL EVACUATION

G-1. When patients are evacuated by aircraft (routinely in Roles 2, 3, 4, and 5), they should be briefed on the following points—

- A number of ambulatory patients will be detailed to assist with the evacuation of litter patients in any emergency.
- Safety belts and litter straps are to be properly fastened according to orders given by the pilot.
- Patients are instructed on the proper position to assume in preparing for an emergency. Flight crews and CCATT ensure that seat safety harnesses have been tightened.
- Ambulatory patients, with the exception of those designated to assist litter patients, are the first to leave a downed aircraft.
- Immobilized litter patients are freed from litters and assisted in leaving the aircraft. Litters will not normally be removed from their fastenings in view of the limited time available to evacuate the aircraft.
- Mentally disturbed patients should be quieted so that the orderly removal of other patients will not be jeopardized.

INTERNATIONAL STANDARDIZATION EVACUATION AGREEMENT

G-2. Table G-1 provides the patient classification codes defined in international standardization agreements.

Table G-1. Patient classification codes

Code	*Classification*
Class 1. Psychiatric Category	
1A	Stretcher patients who are frankly disturbed and inaccessible, requiring close supervision and sedation, or the use of restraint equipment.
1B	Patients who are not at the time of request grossly disturbed, but who may react badly to air travel, or who may commit acts likely to endanger themselves or the safety of the aircraft and its occupants.
1C	Sitting patients who are cooperative and have proved themselves to be reliable under preflight observation.
Class 2. Stretcher Patients Other Than Psychiatric	
2A	Patients, who in an emergency, would be unable to leave the aircraft without assistance.
2B	Patients, who in an emergency, would be able to leave the aircraft unaided.
2C	Patients who can walk onto an aircraft but on a long flight would benefit from a litter.
Class 3. Sitting Patients Other Than Psychiatric	
3A	Sitting patients, including handicapped persons, who may need medical or nursing attention en route and who, in an emergency, would require assistance to escape.
3B	Sitting patients who may need medical or nursing attention en route and who would be able to escape unassisted in an emergency.
Class 4. Sitting Patients Other Then Psychiatric	
4	Patients who will not need medical or nursing attention en route and are capable of traveling unescorted. Class 4 patients are normally required to make their own way from the deplaning airfield to their destination.

INTERNATIONAL STANDARDIZATION EVACUATION PRECEDENCE

G-3. Patients for AE will be given appropriate degrees of priority so that, if aircraft space is limited, the more urgent patients may be evacuated before those whose conditions are less serious. The degrees of priority are depicted in Table G-2. (The evacuation precedence used by the USAF is essentially the same as this listing. It contains a few word changes and introduces specific time limits. It does not contain Priority 4.)

Table G-2. Patient priorities as designated in STANAG 3204

PRIORITY 1/URGENT	Emergency patients for whom speedy evacuation is necessary to save life or limb, to prevent complication of serious illness, or to avoid serious permanent disability.
PRIORITY 2/PRIORITY	Patients who require specialized treatment not available locally and who are liable to suffer unnecessary pain or disability unless evacuated with the least possible delay.
PRIORITY 3/ROUTINE	Patients whose immediate treatment requirements are available locally but whose prognosis would definitely benefit by air evacuation on routine scheduled flights.
PRIORITY 4	Patients for whom air evacuation is a matter of medical convenience rather than necessity.

PATIENT CLASSIFICATION

G-4. Table G-3 provides the patient classification codes used aboard USAF aircraft and that can be used in completing DD Form 601 (Patient Evacuation Manifest). These codes are expanded to include categories of patients and other personnel which may or may not apply on the battlefield (such as infants, relatives, or friends).

UNITED STATES AIR FORCE EVACUATION PRECEDENCE

G-5. The evacuation precedence used by the USAF is dramatically different than that employed by the US Army medical evacuation system. These precedence should not be confused. Table G-4 (page G-5) provides the evacuation precedence and time frames used by the USAF.

Table G-3. Patient classification codes

Code	*Classification*
Class 1.	***Neuropsychiatric Patients***
1A	Severe psychiatric litter patients requiring the use of restraining apparatus, sedation, and close supervision at all times.
1B	Psychiatric litter patients of intermediate severity requiring tranquilizing medication or sedation, not normally requiring the use of restraining apparatus, but who react badly to air travel or who may commit acts likely to endanger themselves or the safety of the aircraft. Restraining apparatus should be available for use.
1C	Psychiatric walking patients of moderate severity who are cooperative and who have proved reliable under observation.
Class 2.	***Litter Patients (Other Than Psychiatric)***
2A	Immobile litter patients unable to move about of their own volition under any circumstances.
2B	Mobile litter patients able to move about of their own volition in an emergency.
Class 3.	***Walking Patients (Other Than Psychiatric)***
3A	Nonpsychiatric and nonsubstance abuse patients who require medical treatment, assistance, or observation en route.
3B	Recovered patients who are returning to their units and require no medical attention en route.
3C	Ambulatory drug or alcohol substance abuse patients.
Class 4.	***Infant Category***
4A	Infants under three years, occupying a seat or in a bassinet or a car seat secured in an ambulatory seat.
4B	Recovered infants under three years, occupying a seat or in a bassinet or car seat secured in an ambulatory seat.
4C	Infants in an incubator.
4D	Infants under three years on a litter.
4E	Outpatients under three years on a litter for comfort.
Class 5.	***Outpatient Category***
5A	Ambulatory, nonpsychiatric, and nonsubstance abuse outpatients who are traveling for an outpatient visit and do not require a litter or medical assistance in flight.
5B	Ambulatory drug or substance abuse outpatients going for treatment.
5C	Psychiatric outpatients going for treatment.
5D	Outpatients on a litter for comfort or safety.
5E	Returning outpatients on a litter for comfort or safety.
5F	Other returning outpatients.
Class 6.	***Attendant Category***
6A	Medical attendants, either physician, nurse, or technician, who are assigned to give specialized medical treatment or nursing care to a particular patient.
6B	Nonmedical attendants, either relatives or friends, who may assist with the patient's care and who may also require support.

Table G-4. Evacuation precedence used by the United States Air Force

Evacuation Precedence	*Definition And Time Frames*
URGENT PRECEDENCE	Applies only to the need for immediate life, limb or eyesight saving. These patients should be picked up and delivered to the designation facility as soon as possible.
PRIORITY PRECEDENCE	Applies to the need for prompt medical care not available locally. These patients should be picked up within 24 hours and delivered to the destination facility with the least possible delay.
ROUTINE PRECEDENCE	Applies to all other patients. Routine patients will be picked up and delivered on regularly scheduled flights.

Appendix H

Troop Leading Procedures, Precombat Checks/Precombat Inspections, and Leader Checklists

Troop leading procedures (TLPs) are universal in their application. They are used by leaders at all levels regardless of branch or MOS. For purposes of this appendix the leader titles of ambulance platoon leader and ambulance platoon sergeant are used for familiarity.

SECTION I — TROOP LEADING PROCEDURES

H-1. Leaders must conduct detailed planning and preparation before each mission in order to be successful. That is why the Army developed an eight-step process of TLP for leaders to follow.

H-2. Each step of the TLP process is designed to give leaders a framework on which to thoroughly develop their plans and orders. If these steps are not followed the resulting plans and orders may be missing information which may be critical to successfully execute the mission.

THE EIGHT STEPS OF TROOP LEADING PROCEDURES

STEP ONE: RECEIVE THE MISSION

H-3. Once staff planning is initiated the ambulance platoon leader gathers information concerning the operation from the company commander and disseminates to his subordinate leaders. Examples of this information include but are not limited to intelligence preparation of the battlefield (IPB) and METT-TC. If the company commander has not completed a full plan, he may still issue a warning order (WARNO) to the ambulance platoon leader.

- In keeping with the rules of time management, the ambulance platoon leader uses one-third of his overall time for planning the operation. The reverse planning sequence is recommended.
- The ambulance platoon leader's OPORD must also be understandable to subordinate personnel.
- Throughout the planning and development process the ambulance platoon leader may find it necessary to change or refine the plan as the tactical situation changes, as long as it follows the commander's intent.

STEP TWO: ISSUE A WARNING ORDER

H-4. The ambulance platoon leader issues a WARNO based on information received from the company commander's WARNO or OPORD.

- The ambulance platoon leader can present the order in any format.
- As a minimum, the platoon should use the company or battalion tactical standing operating procedures (TACSOP). The best method is to use already developed and implemented platoon level TACSOP that cover preparation for offensive or defensive missions. These TACSOPs should be thoroughly trained and rehearsed.
- Whatever the mission, the platoon can begin work on its preparation by assessing equipment needs through the various military supply classes (such as Class I, III, IV, V, or VIII).
- Every Soldier in the platoon should have a specific task and purpose (identified in the platoon SOP) once the ambulance platoon leader has issued a WARNO.

Step Three: Make a Tentative Plan

H-5. The ambulance platoon leader considers the following while preparing the tentative plan:

- Mission analysis, IPB, METT-TC considerations, higher mission and intent (two levels up).
- The ambulance platoon leader must know what systems are available for his use and how to employ them. He must fully understand his own equipment and capabilities as well as those of his enemy.
- Consider **Observation** and Fields of Fire, **Avenues** of Approach, **Key** Terrain, **Obstacles** and **Cover** and Concealment, (man-made and natural), (OAKOC), weather, morale, and flexibility to react to last-minute changes. As the commander updates the ambulance platoon leader, the ambulance platoon leader, in turn, continually updates the platoon with FRAGOs.
- Experience is a valuable asset for mental preparation. Knowing how Soldiers are going to react to situations, based on past experiences, can help any officer or NCO develop his plan.
- During the planning process, the leader should have a *vision* of the enemy and understand the threat using available intelligence. The better the ambulance platoon leader understands the enemy, the greater his opportunity for success.

Step Four: Start Necessary Movement

H-6. Even though listed as Step Four, necessary movement can begin as soon as the mission is received.

- If movement is required, the ambulance platoon sergeant, with the assistance of squad leaders, can move the platoon to a new tactical assembly area (TAA). This allows the ambulance platoon leader more time to prepare his order and graphics.
- At the new TAA, preparations continue with precombat checks (PCCs). This is a good opportunity to conduct PCCs on squad equipment by using the platoon SOP checklists and to certify squad rehearsals. If the ambulance platoon sergeant is unable to conduct squad inspections, the platoon SOP should designate someone else, such as the senior squad leader, for the task.
- If platoon linkups are time driven, movements and time management become critical. Failure to plan or allow for ample movement time creates late linkups and results in late rehearsals, late PCCs, and various other events that will not be accomplished on time or to standard.
- If movements were conducted late or linkups were late, the mission timeline will suffer.

Note. Time management is mission essential!

Step Five: Reconnoiter (Conduct Reconnaissance)

H-7. Although listed fifth, this step may be implemented at any time and as often as necessary during TLP. When the risk of enemy contact is too great, or establishing additional orders for reconnaissance parties is time consuming leaders may have to rely on map reconnaissance and or photographic imagery to reconnoiter an AO. Ambulance platoon leaders must ensure that ground evacuation routes are reconnoitered and strip maps developed for security escort vehicle drivers and ambulance drivers.

- A leader's reconnaissance must be performed! Leaders who see more of the AO are better able to develop more detailed and useful plans.
- Information gained by the leader's reconnaissance allows more accurate terrain models to be built and greater detail can be provided when the ambulance platoon leader is briefing the OPORD.

Step Six: Complete the Plan

H-8. Complete the plan—

- With reconnaissance completed, movement accomplished, and the initial plan established, the ambulance platoon leader should have most of the information that is needed to finish writing the OPORD.
- Any additional resources needed should be en route or already on-hand so the ambulance platoon leader, when briefing the OPORD, has positive control of additional equipment or manpower.
- Graphics must be finished before the OPORD is briefed to the platoon and quickly handed down to squad leader level. Logistics and sustainment graphics must include CBRN decontamination points, MTFs locations, and supported unit CCP and AXP.

STEP SEVEN: ISSUE THE ORDER

H-9. Avoid rushing through the OPORD briefing as this may result in pertinent information being left out of the brief.

- When completed, distribute the OPORD with graphics to each squad and team leader. Also include a terrain model or sand table to enhance visualization of the terrain.
- Minimize distractions. Having subordinate leaders in the right mindset to receive the order is critical to allow them to collect and retain all of the information presented at the OPORD brief.
- If possible, present the OPORD with a representation of the terrain in the background to enhance visualization of the mission.
- After the OPORD has been presented, the ambulance platoon leader ensures that the information has been disseminated to every Soldier. Backbriefs from Soldiers and NCOs to the ambulance platoon leader should reveal any portion or portions of the OPORD that were missed or not fully understood. Everyone leaving the OPORD brief should have a clear understanding of the mission.
- Contingencies need special consideration. If, during the OPORD brief, a potential activity is discussed that has no clear plan, subordinates can help develop a contingency plan, unless the activity is already specified in the platoon SOP. Defer making a decision on topics discussed during the OPORD brief which require more thought and possibly guidance from higher headquarters. If at all possible, allow a follow-up period to develop clarity on the situation and the techniques needed to accomplish the mission.

STEP EIGHT: REHEARSALS

H-10. Supervise rehearsals and plan to run as many rehearsals as time permits. After each rehearsal, conduct an after-action review (AAR). When conducting platoon rehearsals with attachments, ensure that the ambulance platoon leader/sergeant supervises the AAR with attachments. When conducting the AAR, ensure Soldier involvement. By this time all Soldiers understand the concept of the operation.

- When rehearsing the mission, ensure that the rehearsal lane coincides with what was briefed during the OPORD brief. Based on the latest intelligence, the rehearsal lane should match (as much as possible) what will be encountered during the actual mission. Do not allow for complacency. If there is information about the terrain and the obstacle that can be placed in the rehearsal lane to add realism, ensure that obstacle information is built into the lane. A company-level rehearsal is of limited value if the enemy obstacle does not resemble what was reconnoitered or templated.
- If time does not allow for a full rehearsal, backbriefs must be conducted. A successful understanding of the mission can be accomplished with key personnel by walking troops through the mission on the terrain model and asking questions. The backbrief is never a substitute for a rehearsal, but is an acceptable recourse when time is limited.
- Conduct final precombat inspections (PCI).
- If there are any last-minute concerns, they should be satisfied during this time.
- Conclude the brief.

H-11. Some steps of the TLPs require more development than others. Regardless of the mission, the ambulance platoon leader must not, for any reason, omit any step from the TLP.

SECTION II — PRECOMBAT CHECKS AND PRECOMBAT INSPECTIONS

H-12. Precombat checks and PCIs are tools which leaders have to ensure that their Soldiers—

- Fully understand the mission and their role.
- Have the equipment that is necessary to accomplish the mission on hand and that it is serviceable.

H-13. Leaders must conduct PCCs and PCIs for every mission. Failure to perform these vital checks could result in unnecessary casualties, loss of equipment, and failure to accomplish the mission.

H-14. When conducting PCCs/PCIs, as deficiencies are identified corrections must be made immediately.

H-15. Who performs PCCs and PCIs are both intended to accomplish the same thing, which is to have a leader ensure that personnel and equipment are ready. The only difference is who is doing the inspecting.

- Commanders conduct PCIs. Ambulance platoon leaders, ambulance platoon sergeants, and squad leaders conduct PCCs.
- Why do we identify a commander's inspection separately? Equipment readiness is essential for mission success. Squad leaders and ambulance platoon leaders are certainly capable of ensuring equipment readiness, but the commander's first-hand knowledge of mission requirements cannot be substituted.

H-16. Units need to establish standard PCC/PCI checklists based on their units mission profile. They should be included in the TACSOP and issued to all leaders. They should include routine daily checks and checks that are mission specific. They do not need to be extremely detailed; a list of items to check and the leader's knowledge of the equipment and the appropriate manuals should suffice. The PCCs checklists provided below are examples that units may use to develop their own or they can be modified as appropriate to meet their own requirements.

SECTION III — LEADER CHECKLISTS

H-17. This section provides sample checklists which may be used as is, or modified, to meet unit specific requirements. As with any checklist they are only effective if they are incorporated into a unit's TACSOPs and then used routinely.

GENERAL CONSIDERATIONS

H-18. Leader receives the mission from the next higher headquarters.

___ Clarifies any questions.

___ Coordinates with next higher headquarters as required.

___ Conducts mission analysis.

___ Produces the estimate.

___ Revises mission statement.

___ Produces tentative time schedule.

___ Issues a WARNO.

H-19. Unit members perform readiness, maintenance, and functional checks under the supervision of unit leaders which includes:

___ Medical equipment sets are present and all perishable and dated items are current and on-hand.

___ Vehicles/aircraft/generators.

___ Night vision devices.

___ Communications equipment.

___ Weapons and ammunition.

___ Field sanitation equipment and supplies.

___ Special equipment (such as hoist and forest penetrator).

___ Common table of allowances equipment.

H-20. Leader makes a tentative plan.

___ Uses estimate of the situation to depict plan of support.

___ Develops courses of action (COA).

___ War games COA.

___ Determines best COA.

___ Completes his plan.

H-21. Leader issues OPORD.

H-22. Leader uses sand table or sketches to depict plan of support.

H-23. Leader affects coordination for the mission.

___ Support requirements.

___ Current intelligence (to include health threat) update.

___ Control measures.

___ Communications and signal information.

___ Time schedule.

H-24. Leader receives attachments/augmentation (personnel), if appropriate.

___ Attachments are oriented to unit.

___ Attachments are briefed on the mission.

H-25. Leader supervises AHS mission preparation.

___ Key leaders brief back unit leader.

___ Key personnel rehearsals are conducted.

H-26. Unit leaders:

___ Supervise.

___ Inspect.

___ Ensure adequate security.

___ Conduct brief backs.

___ Rehearse.

___ Continue coordination.

H-27. Unit plans for support of combat operations.

___ Analyze mission.

___ Patient acquisition and medical evacuation requirements.

___ Area support requirements.

___ Requirements for water, Class VIII, and other supply classes.

___ Requirements for CBRN defense.

___ Transportation requirements.

H-28. Unit leaders:

___ Execute a work/rest plan based on work priorities and statutory (crew rest) requirements.

___ Monitor current situation.

___ Issues appropriate FRAGOs based on intelligence or operational updates.

___ React to messages or orders from higher headquarters.

___ Execute any actions and coordination resulting from change.

H-29. Unit headquarters remains current on positions and missions of higher, adjacent, and subordinate units.

Ground Ambulances Inspected For

___ Authorized MES are on hand.

___ Medical equipment is complete and serviceable.

___ Oxygen is on hand and serviceable.

___ Authorized medications are on hand and current.

___ Packing list is available.

___ Strip maps and/or road maps are available (with overlays).

___ On-vehicle equipment is on hand.

___ Log book is present and current.

___ All drivers are licensed.

___ Situational awareness equipment (position locator) is on hand and serviceable.

___ Communications equipment is on hand and serviceable and set to correct frequency.

___ Medical unit identification markers (according to the Geneva Conventions) are displayed.

Note. Markers are red on a white background only; camouflaged or subdued markers are not authorized.

Chemical, Biological, Radiological, And Nuclear Equipment

___ Individual protective equipment is on hand and serviceable. (One set is issued; the other is maintained in support.)

___ Protective masks are issued and serviceable.

___ Nerve Agent Antidote Kit is available and distributed, if required.

___ Convulsant antidote for nerve agent is available and distributed, if required.

___ Decontamination apparatus is complete and serviceable.

___ Basic load of decontamination supplies is on hand.

___ Chemical agent alarms are on hand and serviceable.

___ M256A1 detector kits are issued.

___ Chemical, biological, radiological, and nuclear contamination marking kits are distributed.

___ Chemical agent monitors are on hand, if authorized.

___ Replacement filters for protective masks are on hand.

___ Biological and chemical warfare agents prophylaxis, immunizations, and barrier creams have been accomplished, if appropriate.

___ Nerve agent pretreatment packets are available.

___ Radiac sets are on hand.

MISCELLANEOUS EQUIPMENT

___ Inspect binoculars.

___ Inspect camouflage nets and support systems, if appropriate.

___ Inspect night vision devices.

___ Ensure batteries are on hand and serviceable.

___ Inspect tentage.

___ Inspect global positioning systems, if available.

PERSONNEL

___ Ensure Soldiers are in the correct uniform.

___ Ask questions to ensure that Soldiers have been briefed on mission and situation.

___ Implement appropriate MOPP level.

___ Check for drivers licenses.

___ Brief Soldiers on operations safety and environmental injuries.

___ Individual equipment is on hand and stowed properly.

___ Soldier has eaten and is briefed on future field feeding.

___ Identification cards and tags are on hand and serviceable.

___ Camouflage self and equipment, when directed. (If medical vehicles are camouflaged, the Geneva Convention emblem must be removed.) (Refer to STANAG 2931.)

___ Work/rest plan (to include crew rest) is implemented.

___ Water discipline plan is implemented, if appropriate.

COMMUNICATIONS EQUIPMENT

___ Radios are operational (communications check completed).

___ Telemedicine equipment is available and operational, if available.

___ Speech security equipment functions, if available.

___ Frequencies are set.

___ Matching units are operational.

___ Antennas are tied down properly.

___ Connectors are clean and serviceable.

___ Field telephones (TA-312) are on hand and serviceable, if appropriate.

___ Batteries are on hand and charged.

___ Man-pack radio sets are complete.

___ Switchboard is on hand and serviceable.

___ Communications wire (WD-1) is on hand and serviceable.

___ Antennas and remotes are present and serviceable.

___ Signal operating instructions are available and secured.

___ Call signs, frequencies, and challenge passwords have been disseminated.

___ Perform communications check again.

VEHICLES

___ Loads are according to load plan; load plan is posted in the vehicle.

___ Hazardous cargo is properly identified and stored toward the rear of the vehicle for easy access and inspection.

___ Ammunition is issued and properly stored.

___ Vehicle fuel tank is topped off.

___ Package POL products and small arms lubricant are present.

___ Water cans are full.

___ Meals, read-to-eat are issued and stowed.

___ First-aid kits are present and complete.

___ Operator's manuals and lubrication orders are present for the vehicle, radios, and associated equipment.

___ Critical on-vehicle equipment and basic issue items are present.

___ Vehicle dispatch is complete.

___ Department of the Army (DA) Form 2404 (Equipment Inspection and Maintenance Worksheet) is complete.

___ No deadline deficiencies exist.

___ Before operation PMCS has been completed.

INDIVIDUAL WEAPONS

___ Clean and functional.

___ Cleaning tools/kits, bolts, and ruptured cartridge extractors are present.

___ Range cards are on hand.

___ Ammunition is issued, accounted for, and secured.

___ Magazines are issued.

___ Blank adapter installed, if appropriate.

___ Function check performed.

Medical Evacuation Support Plan

___ Mission statement/commander's intent/task organization.

___ Intelligence preparation of the battlefield considerations.

___ Patient estimates by phase.

___ Identify expected areas of patient density.

___ Anticipate and plan for likely patient conditions.

___ Formulate/evaluate/implement/reevaluate COA.

___ Location of supported units.

___ Location of supporting MTFs/FSTs/hospitals.

___ Location of supporting evacuation units.

___ Augmentation of personnel and vehicles coordinated supporting MTFs.

___ Class VIII supply/resupply coordinated and available.

___ Location of supported unit CCPs and AXPs.

___ Evacuation routes primary and alternate.

___ Trafficability (enemy activity, hazards, condition, obstacles).

___ Security (gunship escort, MP escort, convoy support).

___ Strip maps/overlays/maps.

___ Landing zones designated for rotary-wing evacuation aircraft (primary and alternate)

___ Ensure that platoon/company/battalion/treatment facility litter teams are available to assist ambulance crews in loading patients.

___ Available litters and types (NATO standard, folding nonrigid, SKED®, and so forth).

___ Ground ambulances available (type/number/carrying capacity).

___ Air ambulances available (type/number/carrying capacity).

___ Supporting evacuation assets available (requested/ prepositioned).

___ Nonmedical vehicles for CASEVAC (type/number/carrying capacity).

___ Available aircraft (UH-60, CH-47).

___ Medical evacuation support plan for units without evacuation resources.

___ Treatment and evacuation of EPW and detainees.

___ Chemical, biological, radiological, nuclear considerations (contaminated evacuation vehicle plan, patient decontamination site, decontamination sites, contaminated aircraft).

___ Other.

Casualty Care Considerations

___ Self-aid/buddy aid/CLS/trauma specialist.

___ Combat lifesavers trained and CLS bags stocked and issued.

___ Emergency care specialists and flight medics are rehearsed, aid bags fully stocked.

___ Trauma specialist Class VIII resupply sets fully stocked and pushed forward to company command post.

___ Locations of platoon/company/battalion/treatment facility CCPs are known.

___ Other.

Mass Casualty Plan (Resourced/Rehearsed/Coordinated)

___ Litter teams.

___ Litters (all).

___ Medical evacuation/CASEVAC vehicles (all).

___ Class VIII supplies (on hand).

___ Class VIII resupply (Prepackaged and prepositioned).

___ Company/battalion command net.

___ Brigade support medical company command net.

___ Personal effects, weapons, equipment, and property exchange.

___ Chemical, biological, radiological, and nuclear casualty/evacuation plan.

___ Location of decontamination site.

LANDING ZONE OPERATIONS

Safety

___ At all times take all the necessary safety precautions.

___ Platoon that is responsible for running the LZ provides all required safety gear. Included are—

___ Safety goggles.

___ Visual Signal, VS-17 panels/chemical lights.

___ Smoke (Appendix E).

___ Is the LZ secure.

Marking the LZ

___ Use colored smoke to identify the LZ and wind direction.

___ Use an inverted "Y" to mark the LZ.

___ During daylight, camouflage the LZ so the VS-17 panels cannot be seen until evacuation aircraft are confirmed and inbound.

___ At night, emplace chemical lights only when an AA is confirmed and is inbound.

Landing Requirements for Light Helicopters

___ Minimum real estate required is a cleared area 30 meters in diameter.

___ Area must have an approach and departure zone clear of obstructions and not over the established company area.

Appendix I
Safety on the Battlefield

The tactical environment provides ever-changing demands and unpredictable problems, often under stressful conditions. The interface of man, machine, and environment is constantly shifting. In this environment, mission accomplishment requires continuous leader involvement and flexible decision making. Not surprisingly, accidents and injuries increase during tactical operations. Safety in the tactical environment depends upon compliance with established standards. However, due to fluid conditions in the tactical environment, safe mission accomplishment relies heavily on complete integration of risk management into both planning and execution phases. Risk management assists commanders in anticipating and controlling hazards in the planning phase and in dealing with unexpected hazards as they arise in the execution phase.

SECTION I — GENERAL

COMMON ACCIDENTS

I-1. During all operations, it is critical commanders and leaders minimize risks through appropriate risk management procedures to protect the force and preserve the unit's war-fighting capability. Over half of all Army accidents during tactical operations are accredited to the following operations.

VEHICLE OPERATIONS

I-2. Most accidents in this category are caused by excessive speed for conditions (weather, traffic). Other causes are recklessness, fatigue, unfamiliarity with roads, and untrained and inexperienced drivers. A lack of equipment/vehicle handling characteristics knowledge also contributes to accidents. Strict enforcement of standards is needed for all vehicle operations. The senior occupant must be responsible for this enforcement. Only trained, licensed personnel should be assigned to operate vehicles or equipment. Ground guides are mandatory during movement in bivouac and assembly areas, when backing, and during periods of reduced visibility. Operators must be familiar with proper operation and maintenance of commercial equipment.

MATERIEL HANDLING

I-3. These accidents occur when an object is too large or too heavy to handle for those attempting to move it. As a result, someone may sustain a muscle or back injury, or may be crushed. Overconfidence in one's ability, a lack of planning, and fatigue are common factors in such accidents.

MAINTENANCE

I-4. These accidents are often caused by failure to follow procedures, using the wrong tools, and/or fatigue.

SEVERE ACCIDENTS

I-5. The following activities produce fewer accidents than those listed above; however, when they do occur, they often result in catastrophic damage or death. Commanders and leaders must pay special attention to unit missions which involve these activities or put Soldiers in situations or locations where they will come in contact with associated hazards and ensure they properly brief their personnel and mitigate risk to the fullest extent.

AMMUNITION AND EXPLOSIVES HANDLING

I-6. Horseplay, mishandling, disassembly, unauthorized use, and improper storage of ammunition and explosives account for many personnel injury accidents. It is essential to enforce accountability and security procedures for unexpended ammunition and explosives and to comply with explosive storage safety standards in AR 385–64 and DA Pam 385–64.

EXPLOSIVE SOUVENIRS

I-7. Educate Soldiers to dangers involved and the serious consequences of collecting unexploded ordnance on the battlefield or ranges. Post-tactical training shakedown inspections for this type of materiel are a must. Amnesty boxes are also useful. Platoon sergeants and squad leaders policing their Soldiers can prevent most of these accidents through proper training.

FIELD EXPEDIENTS

I-8. Be suspicious of shortcuts. Although tactical operations frequently involve employment of field expedients, risks and benefits must be carefully weighed. In many cases, field expedients are the result of a weak supply system or inadequate planning.

FIELD HEATERS AND STOVES

I-9. Operators of all types of heaters and stoves should be trained and licensed in advance. Equipment should be maintained and operated in accordance with operating instructions including use of proper fuel. Ensure combustible material is kept well away from heaters and stoves and fire fighting equipment is available for each heater and stove. Heaters or stoves with self-contained fuel supplies (small space heater) should not be refilled while the heater is on or still warm. Do not use heaters or stoves in tents or other confined spaces without use of proper ventilation, such as tent vent flaps, doors, or windows.

PETROLEUM, OILS, LUBRICANTS STORAGE AND HANDLING

I-10. The POL handlers must know and practice safety rules and procedures. Frequent inspections ensure safe storage and transfer of POL products. Proper grounding procedures must be followed. Field manual 10–67–1 describes use of protective equipment to prevent personnel exposure.

SOLDIER FATIGUE

I-11. When a Soldier's sleep time is dependent upon the tactical situation, debilitating fatigue can occur. Soldiers suffering from sleep loss experience various symptoms of fatigue including decreased coordination, narrowed attention span, and reduced performance to standard. Anticipate fatigue-related errors and take action to prevent them.

TACTICAL SLEEP PLAN

I-12. Commanders need to develop and enforce sleep plans. Control sleeping areas to prevent Soldiers from being crushed by moving vehicles in and out of the area.

WATER OPERATIONS

I-13. Plan water operations carefully, as risks of drowning and equipment loss are high during water operations. Pair strong swimmers with weak ones to protect personnel. Secure equipment and float it across rather than requiring individuals to carry their equipment. Use safety lines and personal floatation devices.

WEAPONS

I-14. Most accidents occur when cleaning or clearing individual weapons, entering or exiting vehicles, or running with loaded weapons. Provide guidance for weapons handling, loading, and clearing, and ensure it is strictly enforced. Do not load weapons that are not essential for the current mission.

WEATHER-RELATED CASUALTIES

I-15. Consider the effects of weather during planning. Unit effectiveness is quickly lost through weather-related casualties such as frostbite, heat stroke, lightning strikes, and falls. Instruct Soldiers in awareness, prevention, and first aid for weather-related injuries and when these conditions can be expected.

VEHICLE CONVOY OPERATIONS

I-16. Convoy operations can be very dangerous if not properly planned. Control of convoy speed and proper separation between vehicles is critical to reducing risk of an accident. See FM 4-01.45 for in-depth information on tactical convoy operations.

CONVOY COMMANDER RESPONSIBILITIES

I-17. Convoy accidents are most commonly caused by a leader failing to perform his or her duties as the commander of a vehicle movement as required in FM 55-30. This includes failure to:

- Control the group's movement.
- Ensure vehicles maintain proper march speed.
- Properly mark unit vehicles.

I-18. Convoy commanders are responsible for ensuring the safe movement of the convoy. To do this, they must positively control the convoy's movement by using communications equipment among the vehicles. This includes ensuring each vehicle has a properly trained, equipped, and supervised crew leading from the front in the absence of radios and/or other means of managing the march. Control of the movement includes enforcing speed limits, march intervals, and seat belt usage. It also requires the ability to stop the march if an unexpected hazard is encountered along the route.

I-19. Preparing vehicles and Soldiers for movement is a leader responsibility. Inexperienced Soldiers, personnel turbulence, and ever increasing training requirements have caused some units to become complacent in managing risks associated with vehicle movements. According to the risk management process, leaders must identify hazards associated with the mission and develop, implement, and supervise control measures to mitigate those risks. These control measures include marking the vehicles in accordance with local SOPs, briefing crews on hazardous conditions expected along the route, precombat checks of personnel and equipment, and developing preaccident emergency contingency plans.

I-20. Unit SOPs should address leader responsibilities during movements of any number of vehicles and identify means to implement common controls.

DRIVER SKILLS

I-21. Operators should be taught the specific skills needed for tactical vehicle operations, which, in addition to requirements outlined in AR 600-55, include—

- Pulling and backing trailers.
- Vehicle recovery operations.

- Loading and lashing of cargo.
- Methods of negotiating difficult terrain such as sand dunes, rice paddies, mountainous terrain, icy roads.
- Ground-guide procedures and signals.
- Methods and procedures for retrieving vehicles stuck in snow, mud, sand, or other restrictive terrain.
- Proper parking and use of the proper-sized chock blocks.

Use of Safety Equipment

I-22. When the tactical situation allows, flashers should be turned on immediately if a vehicle is disabled or impedes traffic, and every effort made to move the disabled vehicle from the roadway. Highway warning kits should be provided for each vehicle in a convoy. When a vehicle is disabled, place the warning triangle a minimum of 100 meters to the rear of the vehicle and ensure personnel remain clear of the road and rear of the vehicle.

Night Operations

I-23. Personnel who are required to operate motor vehicles while wearing NVG must be trained and tested on use and operation of such devices, and the training recorded on the individual's driver training records. Other factors include:

- Ground guides should be used when moving vehicles at night in areas where troops are present.
- Blackout driving should be prohibited on roads open to the public.

FIRE PREVENTION

I-24. The risk of fire is high in areas where a large number of Soldiers are in tents. The following guidance will reduce the risk of fires:

- Establish a fire prevention and protection plan including procedures for inspecting and recharging fire extinguishers during tactical operations.
- Appoint a fire marshal for each bivouac area and train them in their duties. Train Soldiers in fire prevention techniques as well as emergency procedures in the event of a fire.
- Establish safe distances between tents to reduce risk of multiple losses from one fire.
- Provide available fire fighting equipment (portable extinguishers, sand, water buckets, and shovels) to contain small fires. Ensure personnel are trained on their use.
- Establish procedures for sounding fire alarms.
- Ensure no-smoking areas are established and enforced.
- Establish an inspection system ensuring compliance with fire prevention standards.
- Ensure flammable materials are stored according to appropriate directives and checklists.
- Ensure vehicle fire extinguishing/suppression systems are operational and crews are proficient in using the systems.
- Provide a designated fire plan, equipment, and trained personnel for POL storage, ammunition supply points, motor pools, hospitals, and hangars.

SECTION II — ASSEMBLY AREA OPERATIONS

SAFETY OFFICER/NONCOMMISSIONED OFFICER

I-25. The safety officer and NCO must establish, in coordination with the 1SG, platoon leaders, and platoon sergeants, procedures for ground guides and incorporation of these procedures into the unit SOP. Company commanders must weigh mission requirements and personnel available while in combat operations and use risk management analysis to validate procedures for implementation into the SOP.

Operating within the modular force structure may require assistance from support units to meet necessary personnel and vehicle requirements to safely ground guide and move aircraft.

Ground Vehicles

I-26. Commanders and leaders must be aware of hazards from ground vehicles in AAs. Vehicles moving in the AA should always have a ground guide leading the vehicle until out of the AA. This prevents injury to personnel or damage to equipment. Use of a one-way path through the unit area will assist greatly in preventing accidents. Additionally units may be operating in the vicinity of armored vehicles and Soldiers must take additional precautions due to the inability of armored vehicle crews to see personnel close to their vehicles.

I-27. During initial entry into the AA, platoon ground guides lead convoy parties to their platoon locations with each vehicle dismounting one person to ground guide their vehicle. During this period of movement into the AA, there is a higher probability of accident or injury due to the amount of traffic and unfamiliarity of the area. Commanders must ensure movement into the AA is thoroughly briefed.

Aircraft

I-28. Aircraft in the AA present movement hazards and contribute to an additional hazard of blowing debris. Blowing debris is a hazard to personnel, aircraft, and equipment. The variation in environments in which a unit may operate determines the type and amount of debris. Desert environments, for instance, produce brownout conditions which are a hazard to personnel, equipment, aircraft, and crews. Preventive measures should be incorporated into the unit SOP mitigating this risk. An example of risk mitigation for aircraft operations, in and around the AA parking areas, is brownout mats. These mats reduce the brownout effect of operating in a desert environment. Commanders must ensure SOPs are implemented for each of the environments the unit may operate within. This will allow the unit to establish ground guide procedures for aircraft, whether while hovering into and out of the AA or during ground movement of the aircraft within the AA.

I-29. Ground guide procedures for flight into or out of the AA will be based on the distance between aircraft and proximity to hazards (natural or manmade). In environments where the AA allows for wide scatter of aircraft, the need for ground guides while hovering into and out of the parking area is negligible. In contrast, when the AA parking areas require close interval spacing of aircraft, the need for guides is critical to safety. Additionally, if the AA requires aircraft to park in close proximity to natural or manmade hazards, ground guides are required. The need for ground guides is more critical with operations under night or with NVG devices.

I-30. Upon initial entry into the AA, ground guides should meet incoming aircraft. The situation may dictate crews entering the AA know limited information on where they will park their aircraft. The ground guides guide the aircraft into their assigned parking spot. In desert environments, the ability of personnel to guide aircraft into the parking location is decreased. Brownout upon landing requires a thorough prebrief to the flight crew on the AA parking plan and as exact a location as possible in relation to a ground marking, such as an inverted Y, of each aircraft parking location. Certain situations may require aircraft to land directly on a parking spot. The company safety officer and NCO must be involved in the design and procedures planned for aircraft operations into and out of the AA.

Wing Walking

I-31. Ground guides will also be utilized while moving aircraft with ground handling equipment. Ground guiding aircraft while moving with a vehicle requires additional personnel to ensure rotor blades and aircraft do not come in contact with another aircraft or object. Commanders must ensure procedures designed by the safety officer/NCO are integrated into the SOP. The plan should address guides necessary for various settings (such as define how many are necessary for close proximity parking and scattered parking not in close proximity to another aircraft or hazard). Ground movement of aircraft is normally associated with aviation maintenance operations. The maintenance officer should be involved in the design of ground guide/wing walking procedures for the unit SOP. A critical component of the aircraft ground movement plan is communication. Personnel conducting these operations must conduct a team prebrief and ensure that the team leader has a device that will signal all members to immediately halt the operation (survival vest whistle).

GROUND GUIDES

I-32. Commanders and leaders must ensure ground guide operations, vehicle and aircraft, are incorporated into their company SOPs. During AA operations, ground guides perform a key role in preventing personnel injury, and vehicle and equipment damage. Ground guide operations are critical to maintaining combat readiness for the unit.

PYROTECHNICS

I-33. Aviation units normally use pyrotechnics while in an AA. Units utilize pyrotechnics along likely enemy avenues of advance within their security sector as a warning of activity along these avenues. Additionally, aircrews utilize pyrotechnics as signaling devices in downed aircrew rescue operations. Commanders, unit leaders, and the ALSE officer must train personnel in proper use of pyrotechnics and its associated hazards. These items will be addressed in the unit SOP.

I-34. Pyrotechnics require protection against moisture, dampness, and high temperature. Pyrotechnic items must be given high priority for the best available protection because of their sensitivity. Wet pyrotechnic material is hazardous to store; consequently, any boxes showing signs of dampness will be removed from a storage site and inspected. If the pyrotechnic materiel is wet, it will be destroyed. Certain kinds of materiel deteriorate with age and have expiration dates on the containers. Loose pyrotechnic tracer composition, flare composition, and similar mixtures spilled from broken containers should be carefully taken up and covered completely with SAE 10 (EO–10) engine oil and removed for appropriate disposal.

I-35. Pyrotechnics are hazard division 1.3 which includes items that burn vigorously and cannot usually be extinguished in storage situations. Explosions are confined to pressure ruptures of containers and do not produce propagating shock waves or damaging blast overpressure beyond the magazine distance specified in DA Pam 385-64, Table 5–16. Tossing about of burning container materials, propellant, or other flaming debris may cause a severe hazard of spreading fire.

I-36. The fire fighting extinguishing agent for pyrotechnic materials, Class D (combustible metals [magnesium, potassium, and so forth]) hazard, should be a dry powder agent. Pyrotechnics are a mass fire hazard due to their chemical agents. The mass fire hazard can be more hazardous based on other likely types of ammunition stored with pyrotechnics. These fires may be fought if:

I-37. Explosives are not directly involved.

- White phosphorous munitions are involved, there may be an explosion (immerse/spray with water continuously).
- Pyrotechnics and magnesium incendiaries do not use carbon dioxide, Halon extinguishers, or water on or near munitions. Allow magnesium to cool unless upon flammable materiel. In this case, use a 2-inch layer of dry sand or powder on the floor and rake the burning material onto this layer and re-smother.

WARNING

DO NOT DISCHARGE PYROTECHNICS IN THE VICINITY OF AIRCRAFT FLYING IN THE AREA.

SECTION III — PROTECTING THE SOLDIER

ENVIRONMENTAL CONSIDERATIONS

I-38. As the military looks into the next century, and even today, military units should try to avoid unnecessary environmental damage, not only in training, but also across the spectrum of operational missions. A mission's success may be determined by political or socio-economic stability, both of which are affected by environmental factors and resources. The Army must be able to identify ways to protect the natural environment while executing full range of their missions:

- Considering the environment in planning and decision making in conjunction with other essential considerations of national policy.
- Protecting the environment of home stations and training areas as a means of retaining resources for mission purposes.
- Using environmental risk assessment and management principles to integrate environmental considerations into mission performance.
- Instilling an environmental ethic in Soldiers.
- Understanding linkages between environmental protection issues and their associated impact on safety, force protection, and AHS.

I-39. Soldiers must be prepared to respond across full spectrum operations in any component (offense, defense, stability or civil support operations), sometimes within the same operation. Deployed forces must be able to conform to environmental protection requirements of the theater commander without impairing combat effectiveness. For in-depth information on environmental considerations and Soldier protection, see FM 3-100.4.

I-40. When deployed, company commanders will often deal with base camps. Base camps, while not installations, are comparable to small towns and require many of the considerations applied to installations. A mayor (often the HHC commander) assists base camp commander with control of base operations. A base camp coordination agency will provide expertise and support to the commander, largely through its subordinate base camp assistance/assessment team. Environmental expertise is resident or aligned with this team and available to support the base camp commander and designated mayor of the base camp. They provide technical recommendations and maintain appropriate standards. More information about this phenomenon is provided in the Center for Army Lessons Learned Newsletter 99-9, Integrating Military Environmental Protection.

I-41. Army Regulation 200-1 and the Army's Commander's Guide to Environmental Management specify commanders' environmental responsibilities. To carry out these responsibilities, commanders:

- Comply with an installation's environmental policies and legally applicable and appropriate federal, state, and local laws and regulations or country-specific final governing standards if outside the continental United States.
- Demonstrate a positive and proactive commitment to environmental stewardship and protection.
- Provide environmental training required by law, regulation, or command policy.
- Ensure all personnel perform their duties in compliance with environmental laws and regulations, and respond properly to emergencies.
- Promote proactive environmental measures and pollution prevention.
- Supervise compliance with environmental laws and regulations during operational, training, and administrative activities.
- Include environmental considerations in mission planning, briefings, meetings, execution, and AARs.
- Understand the requirements of Army environmental programs.
- Identify and assess environmental risks of proposed programs and activities.
- Coordinate unit activities with higher headquarters environmental elements.
- Appoint and train an environmental compliance officer (ECO) coordinator for the unit.

- Ensure SOPs contain all environmental considerations and regulatory requirements appropriate for the level of command.
- Conduct environmental self-assessment or internal environmental compliance assessments.
- Understand linkages between environmental considerations and their associated impact on safety, force protection, and AHS.

ENVIRONMENTAL DUTIES

I-42. The key to fulfilling environmental requirements successfully at unit level is the ECO. Army Regulation 200-1 directs all Army unit commanders "appoint and train ECOs at appropriate levels to ensure compliance actions take place." In company-sized units, this duty will generally translate into an extra duty. The ECO manages environmental issues at unit level and ensures environmental compliance. He also coordinates, through the respective chain of command, with the supporting installation environmental staff to clarify requirements and obtain assistance.

I-43. The ECO accomplishes environmental compliance requirements on behalf of the commander. He also coordinates with supporting installation environmental staff to clarify requirements and obtain assistance. While this position of responsibility is not a formal staff position, the ECO is critical to the commander's environmental program. The ECO:

- Advises the unit on environmental compliance during training, operations, and logistics functions.
- Serves as commander's eyes and ears for environmental matters.
- Coordinates between unit and higher/installation headquarters environmental staffs.
- Manages information concerning unit's environmental training and certification requirements.
- Performs unit environmental self-assessment inspections.
- Performs environmental risk assessments.

UNIT PLANNING

I-44. Staffs integrate environmental protection into planning for larger units. Unit leaders integrate environmental protection into unit planning for battalion- and company-level units. Soldier safety should be addressed in SOPs, OPORDS, and training plans.

STANDING OPERATING PROCEDURES

I-45. Unit leaders develop SOPs reflecting environmental protection considerations for routine tasks and activities. The SOPs provide information to Soldiers on how to accomplish routine tasks in an environmentally sound manner. The SOPs incorporate local requirements. As local requirements change, unit leaders update their SOPs. The SOPs also help define environmental protection requirements for all unit activities—facility operations, field operations, deployment, and combat. Unit leaders ensure SOPs comply with local requirements by coordinating with higher headquarters staff—usually the environmental office, the surgeon and his staff, preventive medicine personnel, and the SJA or engineer coordinator.

I-46. Unit leaders conduct environmental risk assessments when planning operations or activities. Risk assessment is a standard element of TLP. Unit leaders perform environmental risk assessments for activities not addressed in the SOP or when conditions differ significantly from those described in the SOP. A maintenance unit does not perform a risk assessment every time it performs a lubrication or service. Rather, the SOP describes the correct manner to perform these actions. Risk assessments apply to garrison operations as well as field operations.

OPERATION ORDERS

I-47. Unit leaders address environmental protection in their plans and orders including: WARNOs, OPORDs, OPLANs, concept of operation plans (CONPLANs), and FRAGOs. The higher headquarters staff develops an environmental appendix/annex to its OPORD/OPLAN/CONPLAN. Subordinate unit leaders draw environmental information from the environmental appendix to the OPORD/OPLAN/

CONPLAN, or from Annex L in a joint operations planning and execution system document. Field Manual 5-0 directs the inclusion of Appendix 2 (Environmental Considerations) to Annex F (Engineer) of the OPLAN/OPORD/CONPLAN (battalion level and above) and specifies that lower-level unit leaders/staffs include environmental information in coordinating instructions and service and support paragraphs.

Risk Management Plan

I-48. Risk management is an integral part of the military decision-making process. The risk management process provides the framework for making risk management a routine part of planning, preparing, and executing operational missions and everyday tasks. Assessing environmentally-related risks is part of the total risk management process.

I-49. Knowledge of environmental factors is a key to planning and decision making. With this knowledge, leaders quantify risks, detect problem areas, reduce risk of injury or death, reduce property damage, and ensure compliance with environmental laws and regulations. Unit leaders should conduct risk assessments before conducting any training, operations, or logistical activities.

Tactical Risk/Accident Risk

I-50. When assessing the risk of hazards in operations, the commander and leaders must look at two types of risk.

- Tactical risk is risk concerned with hazards existing because of the presence of either the enemy or an adversary, thus involving the considerations of force protection. It applies to all levels of war and across the spectrum of operations. For example, during the Gulf War, the enemy's demolition of oil fields created a significant health and environmental hazard to the surrounding countryside and those units maneuvering through the area.
- Accident risk includes all operational risk considerations other than tactical risk. It includes risk to friendly forces and risk posed to civilians by an operation, as well as the impact of operations on the environment. It can include activities associated with hazards concerning friendly personnel, civilians, equipment readiness, and environmental conditions. Examples of environmentally related accident risk are improper disposal of hazardous waste, personnel not properly trained to clean up a spill, and units maneuvering in ecologically sensitive terrain. Preventive medicine considerations also fall into this area of risk.

Note: Tactical risk and accident risk may be diametrically opposed. The commander may accept a high level of environmentally-related accident risk on reducing the overall tactical risk. For example, a commander may decide to destroy an enemy's petroleum storage area to reduce its overall tactical risk.

Training Plans

I-51. Formal planning for training culminates with publication of the training schedule. Informal planning and coordination (preexecution checks) continue until training is performed. During rehearsals, leaders ensure all safety and environmental considerations are met.

I-52. To conduct effective, meaningful training for Soldiers, leaders, and units, thorough preparation is essential. This means well prepared trainers, Soldiers, and support personnel are ready to participate, and their facilities, equipment, and materials are ready to use.

I-53. A unit executes training the same way it executes a combat mission. The chain of command is present, in charge, and responsible. During operations, leaders ensure environmental practices and preventive measures are being employed. Once Soldiers understand what is expected of them, these practices become merely another measure of unit proficiency and level of unit discipline.

FIGHTER MANAGEMENT

I-54. Fighter management describes a process essential to effective and efficient battle staff operations. At company level, commanders do not have staffs. The battle staff becomes those personnel in the unit who operate the CP. Successful continuous operations at the company level are more demanding than higher level organizations. The unit requires a TACSOP allowing rest, especially for critical personnel. At company level, the commander and 1SG will find it very demanding to try to rest. Their rest plan must be a priority for the organization to be lead effectively and their units to be successful on the battlefield.

I-55. The cycle of recurring events within a CP focuses the CP officer in charge/noncommissioned officer in charge (OIC/NCOIC) on meeting information and action requirements. Company personnel are normally required to attend battalion/company recurring meetings. This has an impact on company CP operations and schedules. These recurring events include—

- Shift changes.
- Battle update briefings to the commander.
- Battalion targeting meetings.
- Battalion reports.
- Battalion battle updates without the commander.
- Battalion commanders' collaborative sessions.
- Company CP OIC/NCOIC shift change briefing/collaborative sessions.

I-56. The company CP OIC/NCOIC must achieve battle rhythm for updating and viewing information and understand how to use it to affect operations. A well-established battle rhythm aids the commander and leaders with CP organization, information management and display, decision making, and fighting the battle. Battle rhythm demands careful planning and design. Many competing demands must be deconflicted. Even subordinate platoons and sections affect a company battle rhythm based on their needs and unit procedures.

I-57. Command discipline is required to enforce sleep cycles and create an environment where sound sleep can be achieved. Battle rhythm and associated planning is a proactive, not reactive, exercise.

SLEEP PLANS

I-58. Units must develop detailed rest plans and enforce them. Leaders must rest to maintain their effectiveness; however, some leaders attempt to get involved in every aspect of planning and execution. An integral part of the planning process is determining when senior leader presence is required. It is just as important to identify when a leader's presence is not required. The planning process should include the following supporting techniques:

- Include a sleep plan in the METT-TC analysis.
- Ensure leaders have confidence in second and third echelons of leadership and their ability to make routine decisions.
- Instill trust and confidence in the officers, junior NCOs, and specialists by effective training and SOPs.
- Consider contingencies and establish criteria for waking leaders.
- Post sleep plans in CPs, platoon, and section areas; everyone needs to know the plan.
- Synchronize sleep plans with higher headquarters and subordinate platoons/sections.

CONTINUOUS OPERATIONS

I-59. Continuous operations are combat operations continuing around the clock at a high pace, requiring Soldiers to fight without relief for extended periods. Opportunities for sleep are scattered throughout the day and night.

Sustained Operations

I-60. Sustained operations are operations conducted 24 hours a day with little or no opportunity for sleep. Sustained operations are when the same Soldiers or small units engage in combat operations with no opportunity for the unit to stand down and little time for Soldiers to sleep. Aviation units routinely conduct sustained operations to an extent even greater than the infantry due to aviation's maneuver, maneuver support and sustainment roles. To combat the negative effects of sustained operations, commanders and leaders must enforce fighter management principles on their organizations. A fighter management policy should be addressed in the unit SOP, and higher headquarters policy should correlate with company policy.

Degradation of Combat Capability

I-61. As sustained operations continue, all Soldiers begin to show effects of general fatigue and lack of sleep. Unless counteracted, unit performance of combat tasks decline. Recent studies indicate performance is degraded by 25 percent for each 24-hour period without sleep. After 96 hours, performance can be expected to be near zero. Determination to endure must be supplemented by countering the adverse effects to slow the rate of decline. It becomes more difficult to perform assigned tasks to the required standard. Leaders need to recognize signs of serious sleep deprivation in their subordinates. Studies show performance in all duty positions does not degrade the same. Performance in a duty position where there is a heavy load of mental tasks, such as aviation duties (determining, calculating, thinking, decision-making), degrades faster than the performance in a position whose tasks are mainly physical (firing, running, lifting, digging).

I-62. In addition to the degradation caused by fear, fatigue, and loss of sleep, there is a significant loss of effectiveness caused by operation in MOPP 4. When units are using full CBRN protective equipment, judgment is degraded, communications are less effective, and information flow between units is reduced.

FATIGUE IN FLYING OPERATIONS

I-63. Aviation operations are inherently dangerous. Commanders and leaders must be aware of increased dangers present in combat aviation continuous and sustained operations. Additionally, leaders must be able to recognize the symptoms of fatigue and how to deal with fatigued aircrews. Field Manual 3-04.301 provides an in-depth review of aeromedical factors associated with aviation.

OPERATIONAL TEMPO AND BATTLE RHYTHM

I-64. The aviation company should be staffed for 24-hour operations; however, it also conducts cyclical missions. Standard operating procedures establish methods of ensuring the right personnel are available for either cyclical or 24-hour operations. Regardless of methods used, practice during exercises must determine strengths and weaknesses of headquarters personnel for CP operations and training for additional personnel who may be utilized to staff the CP during continuous or sustained operations. Such knowledge allows leaders to focus on critical areas and personnel requiring additional training.

I-65. In planning, the battalion staff must consider battle rhythm requirements of subordinate companies and platoons. Depending on the situation, the battalion may schedule missions allowing company or platoon rotations to maintain their battle rhythm.

Absence of Battle Rhythm

I-66. Without procedures establishing battle rhythm, leaders and units reach a point of diminished returns. This typically occurs between 72 and 96 hours of operations. As leader fatigue sets in, information flow, planning process, execution, and sustainment suffer—often greatly. Symptoms of diminished battle rhythm include the following:

- Leader fatigue.
- Leaders not fully aware of critical decision points (DPs).
- Leaders not available at critical DPs.
- Disjointed time lines between various levels of command.

Presence of Battle Rhythm

I-67. Battle rhythm allows units and leaders to function at a sustained level of efficiency for extended periods. Effective battle rhythm permits an acceptable level of leadership at all times. It can focus leadership at critical points in the fight or during particular events. Procedures and processes facilitating efficient decision making and parallel planning are critical to achieving battle rhythm. Every component of battle rhythm contributes uniquely to sustained operations.

Training

I-68. It is difficult, if not impossible, to establish battle rhythm while simultaneously conducting operations. Preplanning makes it happen. Planning, preparing, and training before deployment lays a solid foundation for viable battle rhythm during operations. Commanders must ensure all company personnel are trained on TOC set-up and operations, and they understand the team concept required to conduct CP operations during combat operations for extended periods. Additionally, company commanders must coordinate with battalion staffs during training to deconflict staff operations and requirements which are counterproductive to the company battle rhythm.

Battle Rhythm Elements

I-69. Battle rhythm is a multifaceted concept including the following elements:

- Sleep/rest plans.
- Trained second- and third-tier leadership in CPs.
- Synchronized upward multiechelon timelines.
- Parallel planning.
- Established processes and SOPs.

Command Post Personnel Depth

I-70. Established processes and SOPs relieve many antagonistic effects of extended operations. Standard operation procedures establishing and maintaining battle rhythm by facilitating routine decisions and operations are a step in the right direction. Soldiers trained to act appropriately in the absence of leaders or orders can relieve commanders and leaders of many of the time-consuming tasks robbing them of essential rest. Examples of areas NCOs and junior officers can accomplish for the commander include—

- Battle summaries and updates during a fight.
- Intelligence updates before, during, and after a battle.
- Sustainment updates before, during, and after a battle.
- Updates to the next higher commander.
- Shift change briefings.

Challenges to Battle Rhythm

I-71. Challenges to battle rhythm include enlisted, NCO, and junior officer duties; and synchronization of planning, execution, rehearsal timelines, equipment maintenance, and sleep plans.

Noncommissioned Officer and Junior Officer Responsibilities

I-72. At the company level, all personnel are critical to the success of the operation and provide valuable contributions. It is imperative commanders make certain each Soldier understands the importance of even the most menial tasks, such as CP security and TOC setup and teardown. Given the amount of personnel in a company, 24-hour operations and force protection requirements, commanders must utilize all personnel available to successfully accomplish CP operations and continuous missions. This becomes even more critical when a company has platoons or sections staged forward in support of BCTs. The improper use of personnel produces the following results:

- Key leaders become exhausted.

- CP operations trained personnel become exhausted.
- The initiative of trained subordinates is stifled, and the incentive to train is diminished.

I-73. The following techniques ensure proper use of personnel:

- Appropriate tasks are assigned to junior NCOs and specialists.
- Effective training and SOPs instill trust in officers and confidence in junior NCOs and specialists.
- Effective command guidance to the company that it is a team effort and all personnel available contribute to the effort.

CONTINUOUS OPERATIONS AND TIMELINES SYNCHRONIZATION

I-74. Timelines for the operation at hand must allow for not only the next operation, but also extended continuous operations. Synchronized, multiechelon timelines assist units in achieving battle rhythm. If units do not address critical events at least one level up and down, disruption results. An example of an unsynchronized timeline is a battalion rehearsal that conflicts with platoon precombat inspections or other events in their internal timeline. Lower echelon units seldom recover from a poor timeline directed by a higher headquarters, as well as platoons and sections at company level. Company commanders must coordinate with their battalion staffs on development of SOPs that include planning, rehearsal, and execution timelines one level below battalion to prevent these conflicts.

STANDING OPERATING PROCEDURES UTILIZATION

I-75. SOPs must be practiced and reviewed during professional development and sergeants' time. The existence of an SOP will not resolve troop-leading challenges unless the SOP is practiced often and internalized by unit members. Checklists are critical, as many leaders will often find themselves rushed, physically fatigued, distracted, and deprived of sleep. Checklists ensure each step is considered even when leaders are exhausted.

SECTION IV — ACCIDENT/INCIDENT REPORTING

I-76. Persons involved in, or aware of, an accident must immediately report it to the commander or supervisor directly responsible for the operation, materiel, or person(s) involved. Table I-1 and Table I-2 identify reporting requirements for aviation and ground accidents/incidents (peacetime and combat). See AR 385-40, DA Pam 385-40, and AR 385-95 for in-depth instructions on Army accidents. Company commanders ensure unit SOPs cover accident/incident reporting, and it is fully addressed by the company aviation safety officer.

Table I-1. Aviation accident notification and reporting requirements and suspenses

Peacetime				*Combat **	
Accident Class	***Notification Telephonic Worksheet***	***DA Form 2397***	***Reporting AAA report***	***Notification Telephonic Worksheet***	***Reporting AAA report***
A	Immediate–to USASC (telephonic notification–no hardcopy notification required (DSN 558-2660/2539/3410 or Commercial (334) 255-2660/2539/3410	(CAI/IAI) 90 calendar days	Aircraft ground ACDTS only-30 calendar days	Same as peacetime to USASC or Safety Rep, forward	(Only when CDR determines DA Form 2397 investigation/ report not feasible) submit as soon as conditions, situation permit-Do not exceed 30 calendar days.
B	Immediate–to	(CAI/IAI) 90	Aircraft ground	Same as	(Only when CDR

Peacetime				Combat *	
Accident Class	Notification Telephonic Worksheet	DA Form 2397	Reporting AAA report	Notification Telephonic Worksheet	Reporting AAA report
	USASC (telephonic notification–no hardcopy notification required (DSN 558-2660/2539/3410 or Commercial (334) 255-2660/2539/3410	calendar days	ACDTS only 30 calendar days	peacetime to USASC or Safety Rep, forward	determines DA Form 2397 investigation/ report not feasible) submit as soon as conditions, situation permit- Do not exceed 30 calendar days.
C	Immediate–to USASC (telephonic notification–no hardcopy notification required (DSN 558-2660/2539/ 3410 or Commercial (334) 255-2660/2539/3410	N/A	30 calendar days	Same as peacetime to USASC or Safety Rep, forward	Same as peacetime
D	N/A (Unless SOF issue involved/ suspected)	N/A	10 calendar days	Same as peacetime	Same as peacetime
E	N/A (Unless SOF issue involved/ suspected)	N/A	10 calendar days	Same as peacetime	Same as peacetime
F	N/A (Unless SOF issue involved/ suspected)	N/A	10 calendar days	Same as peacetime	Same as peacetime
Submis-sion Methods	Class A-C telephonic (immediate) Class D, E, F-IF SOF	Mail	Typed or hand printed AAA reports by mail/FAX or courier/ message/ format/elec-tronic submission, Include attachments as required	Same as peacetime	Same as peacetime
Notes * Only when the senior tactical commander determines that the situation, conditions, and/or time does not permit normal peacetime investigating and reporting.					

Table I-2. Ground accidents notification and reporting requirements and suspenses (3)

Peacetime				*Combat (2)*	
Accident	***Telephonic Notification***	***AGAR***	***DA Form 285***	***Telephonic Notification***	***AGAR only*** ***By any means possible***
Class	***Worksheet***			***Worksheet***	***(Message, Electronic, FAX, Phone, Hand Carry, Mail)***
On-Duty					
A	Immediately (1)	Not required	IAI/CAI-90 days	Immediately (1)	As time permits (Not to exceed 30 days)
B	Immediately (1)	Not required	IAI/CAI-90 days	Immediately (1)	As time permits (Not to exceed 30 days)
C	Not required	Within 30 days	Not required	Not required	As time permits (Not to exceed 30 days)
D	Not required	Within 30 days	Not required	Not required	As time permits (Not to exceed 30 days)
Off-Duty					
A	Immediately (1)	Within 30 days	Not required	Immediately (1)	As time permits (Not to exceed 30 days)
B	Immediately (1)	Within 30 days	Not required	Immediately (1)	As time permits (Not to exceed 30 days)
C	Not required	Within 30 days	Not required	Not required	As time permits (Not to exceed 30 days)
D	Not required	Within 30 days	Not required	Not required	As time permits (Not to exceed 30 days)

Notes:
(1) USASC must be notified IMMEDIATELY by phone at DSN 558-2660/2539-3410 or Commercial (334) 255-2660/2539/3410 or notify USASC Safety Rep forward (during combat).
(2) ONLY when the senior tactical commander determines that the situation, conditions, and/or time does not permit normal peacetime investigating and reporting.
(3) Army civilian injury only accidents should be reported on appropriate Department of Labor Forms according to this regulation.

Glossary

SECTION I — ACRONYMS AND ABBREVIATIONS

A2C2	Army airspace command and control
AA	air ambulance/active Army
AAA	abbreviated aviation accidents
AAR	after-action review
AATF	air assault task force
ABCA	American, British, Canadian, and Australian
ACDTS	accidents
ACR	armor cavalry regiment
AD	air defense
AE	aeromedical evacuation
AECT	aeromedical evacuation control team
AELT	aeromedical evacuation liaison team
AF	Air Force
AGAR	abbreviated ground accident report
AGPU	aviation ground power units
AHS	Army Health System
ALCT	airlift control team
ALSE	aviation life support equipment
amb	ambulance
AMEDD	Army Medical Department
AO	area of operations
AOC	air operations center
AOE	Army of Excellence
APOD	aerial port(s) of debarkation
AR	Army regulation
ASI	additional skill identifier
ARSOF	Army special operations forces
ARNG	Army National Guard
ASCC	Army service component command
ASF	aeromedical staging facility
ASMC	area support medical company
ASOS	Army support to other Services
ASTS	aeromedical staging squadron
ATM	advanced trauma management
ATS	air traffic service

attn	attention
aug	augmentation
avail	available
AVIM	aviation intermediate maintenance
AVN	aviation
AVUM	aviation unit maintenance
AXP	ambulance exchange point
BAE	brigade aviation element
BAS	battalion aid station
BCT	brigade combat team
bde	brigade
BH	behavioral health
BIFV	Bradley infantry fighting vehicle
bn	battalion
BOA	basis of allocation
BSA	brigade support area
BSB	brigade support battalion
BSMC	brigade support medical company
BSS	brigade surgeon section
BTU	British thermal unit
C2	command and control
CA	civil affairs
CAB	combat aviation brigade
CAI	centralized accident investigation
CASEVAC	casualty evacuation
CASF	contingency aeromedical staging facility
CBRN	chemical, biological, radiological, and nuclear
CBRNE	chemical, biological, radiological, nuclear, and high yield explosives
CCATT	critical care air transport team
CCIR	commander's critical information requirements
CCP	casualty collecting point
CE	communications-electronic
CJA	Command Judge Advocate
CJTF	combined joint task force
CLS	combat lifesaver
CLT	casualty liaision team
CM	campaign module
CMCC	corps movement control center
CMD	command
CMO	civil-military operations

CO	company
COA	course(s) of action
COCOM	combatant command
COMSEC	communications security
CONPLAN	concept of operation plan
CONUS	continental United States
COSC	combat and operational stress control
COSR	combat and operational stress reaction
CP	command post
CRTS	casualty receiving and treatment ships
CSH	combat support hospital
CSAR	combat search and rescue
CTC	Combat Training Center
CZ	combat zone
DA	Department of the Army
DCO	defense coordination officer
DCSCOMPT	Deputy Chief of Staff, Comptroller, US Army
DCSIM	Deputy Chief of Staff for Information Managment, US Army
DCSLOG	Deputy Chief of Staff for Logistics, US Army
DCSPER	Deputy Chief of Staff for Personnel, US Army
DCSSPO	Deputy Chief of Staff for Support Operation, US Army
DD	Department of Defense
DHS	Department of Homeland Security
div	division
DMOC	division medical operations center
DNBI	disease and nonbattle injuries
DOD	Department of Defense
DODD	Department of Defense Directive
DP	decision point
DS	direct support
DSN	defense switched network
DSS	division surgeon section
EAB	echelons above brigade
EAC	echelons above corps
EAD	echelons above division
ECO	environmental compliance officer
EEE	early entry element(s)
EEH	early entry hospital
EEM	early entry module
EMEDS	expeditionary medical support

EMT	emergency medical treatment
EPS	essential personnel services
EPW	enemy prisoner(s) of war
1SG	first sergeant
F	Fahrenheit
FARP	forward arming and refueling point
FAX	facsimile
FEMA	Federal Emergency Management Agency
FHP	force health protection
FM	field manual/frequency modulated
FMC	field medical card
FMI	field manual interim
FRAGO	fragmentary order
FRSS	forward resuscitation
FSB	forward support battalion
FSMC	forward support medical company
FSMT	forward support MEDEVAC team
FST	forward surgical team
G3	Assistant Chief of Staff (Operations and Plans)
GA	ground ambubance
GAMC	ground ambulance medical company
GCC	geographic combatant commander
GRREG	graves registration
GS	general support
GSAB	general support aviation battalion
GWS	Geneva Convention for the Amelioration of the Condition of the Wounded and Sick in Armed Forces in Field
HBCT	heavy brigade combat team
HD	homeland defense
HHD	headquarters and headquarters detachment
HMMWV	high mobility multipurpose wheeled vehicle
HN	host nation
hosp	hospital
HQ	headquarters
HR	human resources
HS	homeland security
HSS	health service support
IAI	installation-level accident investigation
IED	improvised explosive device
IG	Inspector General
IM	information management

IPB	intelligence preparation of the battlefield
IPMC	intratheater patient movement center
ISB	intermediate staging base
IT	information technology
JFC	joint force commander
JHSS	joint health service support
JMD	joint manning document
JMTB	Joint Military Transportation Board
JOA	joint operations area
JP	joint publications
JPMRC	joint patient movement requirements center
JTF	joint task force
LHA	amphibious helicopter assault carrier
LHD	amphibious helicopter assault carrier dock
LIN	line item number
LOC	lines of communication
LOTS	logistics over-the-shore
LPD	amphibious transport dock
LPH	amphibious assault helicopter carrier–aviation
LSD	dock landing ship
LST	tank landing ship
LZ	landing zone
MACA	military assistance to civil authorities
MASCAL	mass casualty
MASF	mobile aeromedical staging facility
MBA	main battle area
MC3T	medical command, control, and communication telemedicine
MC4	medical communications for combat casualty care
MCB	movement control battalion
MCO	major combat operation
MCP	main command post
med	medical
MEDBDE	medical brigade
MEDCOM	medical command
MEDEVAC	medical evacuation
MEDLOG	medical logistics
MEF	Marine expeditionary force
MES	medical equipment sets
METT-TC	mission, enemy, terrain and weather, troops and support available, time available, and civil considerations

MEV	medical evacuation vehicle
MH	mental health
MHS	military health system
MIA	missing in action
MIJI	meaconing, intrusion, jamming, and interference
MMB	multifunctional medical battalion
MOC	medical operations cell
MOPP	mission-oriented protective posture
MOS	military occupational specialty
MP	military police
MRE	meals, ready-to-eat
MRI	medical reengineering initiative
MRO	medical regulating office
MSB	main support battalion
MSC	Military Sealift Command
MTF	medical treatment facility
MTOE	modification table of organization and equipment
MWD	military working dogs
MWR	morale, welfare, and recreation
N/A	not applicable
NATO	North Atlantic Treaty Organization
NAVAIDS	navigational aids
NBC	nuclear, biological, and chemical
NCO	noncommisioned officer
NOE	nap-of-the-earth
NP	neuropsychiatric
NSHS	National Strategy for Homeland Security
NVG	night vision goggles
OCP	operational command post
OEG	operational exposure guidance
OEH	occupational and environmental health
OMF	originating medical facility
OPCON	operational control
OPLAN	operation plan
OPORD	operation order
OPS	operations
OPSEC	operations security
OPTEMPO	operational tempo
OR	operating room
OVE	on vehicle equipment

PASR	personnel accountability and strength reporting
PAD	patient administrator
PCC	precombat checks
PCI	precombat inspections
PDS	personnel daily summary
PERS	personnel
PIM	personnel information management
PIR	priority intelligence requipment
PMCS	preventive maintenance checks and services
PMI	patient movement items
PMM	preventive medicine measures
PMRC	patient movement requestment center
POI	point of injury
POL	petroleum, oils, and lubricants
PR	personnel recovery
prep	preparation
PRM	personnel readiness management
PT	platoon
PSYOP	psychological operations
PZ	pickup zone
PVNTMED	preventive medicine
QSTAG	Quadripartite Standardization Agreement
R5	reception, replacement, return to duty, rest and recuperation, redeployment
R&R	rest and recuperation
RC	reserve component
rep	representative
RPG	rocket propelled grenade
RSTA	reconnaissance and target acquisition
RSOI	reception, staging, onward movement, and integration
RTD	return to duty
S&RO	stability and reconstruction operations
S1	Adjutant (US Army)
S2	Intelligence Officer (US Army)
S3	Operations and Training Officer (US Army)
S4	Supply Officer (US Army)
S6	Information Management (US Army)
SBCT	Stryker Bridage Combat Team
SEC	section
SF	special forces
SIMLM	single integrated medical logistics manager

SINCGARS	Single Channel Ground/Airborne Radio System
SJA	Staff Judge Advocate
SOA	special operations aviation
SOF	special operations forces
SOP	standing operating procedures
SPT	support
SRC	standard requirements code
STANAG	standardization agreement
SURG	surgical
SUST	sustainment
TAA	tactical assembly area
TACC	tanker airlift control center
TAC CP	tactical command post
TAC SOP	tactical standing operating order
TAES	Theater Aeromedical Evacuation System
TC	training circular
tm	team
TMIP	Theater Medical Information Program
tmt	treatment
tng	training
TO	theater operations
TOC	tactical operations center
TOE	tables of organization and equipment
TPFDL	time-phased force deployment list
TPMC	theater patient movement center
TPMRC	theater patient movement requirements center
TSOC	Theater Special Operations Command
TSOP	tactical standing operating procedures
UO	urban operations
US	United States
USAF	United States Air Force
USAMEDDC&S	United States Army Medical Department Center and School
USASC	United States Army Safety Center
USN	United States Navy
USNS	United States Navy Ship
USNORTHCOM	United States Northern Command
USTRANSCOM	United States Transportation Command
vet	veterinary
WARNO	warning order
WIA	wounded in action

WMD	weapons of mass destruction
XO	executive officer
ZULU	universal time

References

SOURCES USED

These are the sources quoted or paraphrased in this publication.

NATO STANAGs

2040. *Stretchers, Bearing Brackets, and Attachment Supports*, 23 September 1982 (Latest Amendment, 6 October 2004)

2087. *Medical Employment of Air Transport in the Forward Area*, 29 June 1983 (Latest Amendment, 10 September 1999)

2128. *Medical and Dental Supply Procedures*, 15 November 1982 (Latest Amendment, 21 November 1991)

2132. *Documentation Relative to Medical Evacuation, Treatment, and Cause of Death of Patients*, 7 August 1974 (Latest Amendment, 15 September 1986)

2454. *Road Movements and Movement Control*, 27 January 2005

2931. *Orders for the Camouflage of the Red Cross and Red Crescent on Land in Tactical Operations*, 18 October 1984 (Latest Amendment, 3 April 1998)

3204. *Aeromedical Evacuation*, 26 September 1973 (Latest Amendment, 10 September 2001)

ABCA QSTAGs

436. *Minimum Labeling Requirements for Medical Materiel* (Amendment 1), 8 October 1980

470. *Documentation Relative to Medical Evacuation, Treatment, and Cause of Death of Patients*, 23 February 1979 (Latest Amendment, 14 August 1989)

PRESIDENTIAL DIRECTIVE

Homeland Security Presidential Directive-5, *Management of Domestic Incidents*, February 28, 2003

DODD

DoD Directive 3025.15, *Military Assistance to Civil Authorities*, February 18, 1997

DoD Directive 5200.31E, *DoD Military Working Dog (MWD) Program*, March 29, 2006

JOINT AND MULTISERVICE PUBLICATIONS

Joint Pub 1-02, *DOD Dictionary of Military and Associated Terms*, 23 March 1994
Joint Pub 4-02, *Doctrine for Health Service Support in Joint Operations*, 26 April 1995
Joint Pub 5-0, *Joint Operation Planning*, 26 December 2006
Joint Pub 5-00.1, *Joint Doctrine for Campaign Planning*, 25 January 2002
Joint Pub 5-00.2, *Joint Task Force (JTF) Planning Guidance and Procedures*, 13 January 1999

ARMY PUBLICATIONS

AR 40-5, *Preventive Medicine*, 22 July 2005

AR 71-32, Force Development and Documentation-Consolidated Policies, 3 March 1997

AR 200-1, *Environmental Protection and Enhancement*, 21 February 1997

AR 385-40, *Accident Reporting and Records*, 1 November 1994

AR 385-64, *U.S. Army Explosives Safety Program*, 1 February 2000

AR 385-95, *Army Aviation Accident Prevention*, 12 December 1999

AR 600-55, *The Army Driver and Operator Standardization Program (Selection, Training, Testing, and Licensing)*, 31 December 1993

DA Pam 385-40, *Army Accident Investigating Reporting*, 1 November 1994

DA Pam 385-40, *Ammunition and Explosives Safety Standards*, 15 December 1999

FM 1-202, *Environmental Flight*, 23 February 1983

FM 1-564, *Shipboard Operations*, 29 June 1007

FM 3-0, *Operations*, 14 June 2001

FM 3-04.301, *Aeromedical Training for Flight Personnel*, 29 September 2000

FM 3-05, *Army Special Operations Forces*, 20 September 2006

FM 3-06.11, *Combined Arms Operations in Urban Terrain*, 28 February 2002

FM 3-07, *Stability Operations and Support Operations*, 20 February 2003

FM 3-11.5, *Multiservice Tactics, Techniques, and Procedures for Chemical, Biological, Radiological, and Nuclear Decontamination*, 4 April 2006

FM 3-34.2, *Combined-Arms Breaching Operations (Including Changes 1—3)*, 31 August 2000

FM 3-97.6, *Mountain Operations*, 28 November 2000

FM 3-100.4, *Environmental Considerations in Military Operations (Including Change 1)*, 15 June 2000

FM 4-01.45, *Multi-Service Tactics, Techniques, and Procedures for Convoy Operations*, 24 March 2005

FM 4-02, *Force Health Protection in a Global Environment*, 13 February 2003

FM 4-02.1, *Combat Health Logistics*, 28 September 2001

FM 4-02.4, *Medical Platoon Leader's Handbook—Tactics, Techniques, and Procedures (Including Change 1)*, 24 August 2001

FM 4-02.6, *The Medical Company—Tactics, Techniques, and Procedures (Including Change 1)*, 4 September 2004

FM 4-02.7, *Health Service Support in a Nuclear, Biological, and Chemical Environment—Tactics, Techniques, and Procedures*, 10 January 2002

FM 4-02.12, *Health Service Support in Corps and Echelons above Corps*, 2 February 2004

FM 4-02.17, *Preventive Medicine Services*, 28 August 2000

FM 4-02.18, *Veterinary Services—Tactics, Techniques, and Procedures*, 30 December 2004

FM 4-02.21, *Division and Brigade Surgeon's Handbook (Digitized)—Tactics, Techniques, and Procedures*, 15 November 2000

FM 4-02.24, *Area Support Medical Battalion—Tactics, Techniques, and Procedures*, 28 August 2000

FM 4-02.25, *Employment of Forward Surgical Teams—Tactics, Techniques, and Procedures*, 28 March 2003

FM 4-25.12, *Unit Field Sanitation Team*, 25 January 2002

FM 4-02.43, *Force Health Protection Support for Army Special Operations Forces*, 28 November 2006

FM 5-0, *Army Planning and Orders Production*, 20 January 2005

FM 8-42, *Combat Health Support in Stability Operations and Support Operations*, 27 October 1997

FM 8-55, *Planning for Health Service Support*, 9 September 1994

FM 8-250, *Preventive Medicine Specialist (Including Change 1)*, 27 January 1986
FM 10-67.1, *Concepts and Equipment of Petroleum Operations*, 2 April 1998
FM 21-10, *Field Hygiene and Sanitation*, 21 June 2000
FM 31-70, *Basic Cold Weather Manual*, 12 April 1968
FM 55-30, *Army Motor Transport Units and Operations (Including Change 1)*, 15 September 1999
FM 90-3, *Desert Operations*, 24 August 1993
FM 90-13, *River Crossing Operations*, 26 January 1998
ST 4-02.121, *Multifunctional Medical Battalion*, October 2005
TC 1-400, *Brigade Aviation Element Handbook*, 27 April 2006
TB MED 507, *Heat Stress Control and Heat Casualty Management*, 7 March 2003
CALL Newsletter 99-9, *Integrating Military Environmental Protection*, 1999
FMI 4-90.1, *Heavy Brigade Combat Team Logistics*, 15 March 2005
FMI 5-0.1, *The Operations Process*, 31 March 2006

DOCUMENTS NEEDED

These documents must be available to the intended users of the publication.

DA Form 285 Series, *U.S. Army Accident Report*
DA Form 1156, *Casualty Feeder Card*
DA Form 2397 Series, *Technical Report of U.S. Army Aircraft Accidents*
DA Form 2404, *Equipment Inspection and Maintenance Worksheet*
DD Form 601, *Patient Evacuation Manifest*
DD Form 1380, *U.S. Field Medical Card*

Index

References are to paragraph numbers unless otherwise stated.

S

T

Made in United States
Orlando, FL
04 November 2026

90203575R10251